PREFACE

The impressive increases in crop productivity achieved over the last several decades have resulted from genetic improvements in crop cultivars and from advances in agricultural technology and management practices. Genetic engineering in agriculture, which in a broad sense refers to any practice that leads to the development of improved cultivars, began with the discovery of recombinant DNA technology. Although initially applicable to only bacteria and fungi, recombinant DNA techniques have been extended to plants with ease. As a result, isolation, amplification, and *in vitro* manipulations of plant genes have now become routine.

During the 1980s, the research achievement having probably the greatest potential significance for plant breeding was the ability to insert DNA into plant cells and to recover an intact plant that carried a new characteristic encoded by that DNA. Using this novel technology, plants have now been produced that carry genes conferring insect and virus resistance, increased tolerance to herbicides, and the deposition of novel storage proteins within their seeds. This small selection of characters will soon be supplemented by more complex multigene traits, which will significantly improve the performance of crops. In the longer term, this ability to modify a particular trait will provide new options for plant breeders and should improve both the range and quality of plant products available to industrial and domestic consumers.

Initially, plants like tobacco, potato, and cucumber were manipulated with tools of genetic engineering. Now these techniques have been extended to cereals such as wheat, rice, and maize. Although there are still problems with cereals, the technology is developing so rapidly that their solution does not seem far off. In fact, the field of plant biotechnology is expanding at an enormous rate, making it extremely difficult to cope with the recent developments. Thus, a related danger in writing a book of this kind is that it may rapidly become outdated. Fortunately, the recent exciting advances in plant genetic engineering do not require scientists with a formal background in botany, thus making larger participation of scientists of different disciplines possible. Once a method is developed for the isolation and manipulation of plant DNA, plant genes, like those of any other organism, can be isolated and manipulated *in vitro*.

Manipulation of resistance to herbicides was among the first traits to which new genetic approaches were applied. In some cases, a specific target for herbicide action had been identified through physiological and biochemical studies. In other cases, genetic studies had shown that resistance to herbicide was a dominant trait exhibiting the simple Mendelian inheritance pattern of a mutation in a single nuclear gene. Dominance makes genetic selection of herbicide-resistance mutants or transformants easier. The potential utility of herbicide-resistant genes as dominant selectable genetic markers for research in plants, in a manner analogous to that for antibiotic resistance genes in bacteria, has provided an incentive for research. In addition to these technical considerations, the agronomic importance of herbicides has been a major driving force behind the development of herbicide-resistant plants.

Transgenic biopesticides are created by transfering a gene that expresses a naturally available toxin (biotoxin) into the cells. This is achieved by genetically splicing the gene into bacteria that commonly colonize the plant. Crop geneticists estimate that 5 to 15 years after they introduce a new form of genetic resistance into a strain, the resistance collapses in the face of a newly evolved form of disease or pest. Scientists are therefore attempting to produce plants with multiple biotoxic genes, with the hope that a particular insect may not acquire resistance to all these genes. Scientists have also developed plants that are not prone to attack by viruses. In this case, a plant contains a virus gene that produces protein (CP) to protect it from disease in much the way polio vaccinations protect people. In addition, sufficient progress has been also made in modifying the storage proteins of seeds in plants.

Among the concerns associated with the release of transgenic plants into the environment is the possibility that genes transformed into a crop plant by recombinant DNA methods could spread into its wild relatives. If these genes confer some selective advantage on the wild relatives, then they could contribute to the development of one or more troublesome weedy forms. For instance, genes for disease, drought, and herbicide resistance could all produce problems by enhancing the fitness of the wild relatives. Thus, there is need to generate data on such parameters for realistic assessment of the risk of introgression of engineered genes into wild crop relatives. The success of genetic engineering for crop improvement will also depend on public confidence that there is sufficient oversight to minimize the possibility that transgenic plants might cause adverse environmental effects.

THE AUTHORS

Rup Lal, Ph.D., recently joined the University of Delhi as Associate Professor in the Department of Zoology. Earlier he was a lecturer for nearly 12 years in the Department of Zoology, Sri Venkateswara College, New Delhi. Born on September 27, 1957, in Kanoh, Himachal Pradesh, India, he obtained his B.Sc. degree from D.A.V. College, Jullundur, in 1973. He obtained his M.Sc. degree from Kurukshetra University and his Ph.D. degree in 1980 from the University of Delhi. Since then he has been actively engaged in research and teaching.

Dr. Lal has been the recipient of many research grants from the government of India. Currently, he is the Principal Investigator of a project, "Indegenization and Genetic Manipulations of Rifamycin Production in *Amycolatopsis mediterranei* (*Nocardia mediterranei*)," funded by the Department of Biotechnology, Government of India.

Dr. Lal was awarded the Alexander Von Humboldt Fellowship from July 1988 to August 1990 to work with Prof. R. Eichenlaub in the Department of Genetechnology/Microbiology, University of Bielefeld, Germany. During his visit he obtained his basic training in molecular biology and microbiology. He has authored 3 books and over 50 research papers in the areas of microbiology, biotechnology, and molecular biology.

Sukanya Lal, Ph.D., is a Lecturer in the Department of Zoology, Kirori Mal College, University of Delhi. She was born on October 28, 1957, in Delhi, India. She graduated from the Government Degree College, Hamirpur, in 1975 and received her M.Sc. and Ph.D. degrees in Zoology from University of Delhi, India. She was a Visiting Scientist at the University of Bielefeld, Germany, from November 1988 to August 1990. During her stay in Germany, she, along with her team, succeeded in the construction of a plasmid cloning vector for *Amycolatopsis mediterranei*. She has authored 2 books and over 15 research papers and review articles in the areas of microbiology, biotechnology, and molecular biology.

ACKNOWLEDGMENTS

We would like to express our gratitude to Drs. P. S. Dhanraj and V. V. S. W. Rao for their help during the preparation of the book. We are especially indebted to Prof. D. M. Saxena for his continuous encouragment during the project. We must also express our thanks to Ms. Monisha Khanna and Mr. Vachaspati Mishra, who read every chapter carefully. Thanks are also due to Raji and Vinod for typing the manuscript.

TABLE OF CONTENTS

Chapter 1
Gene Transfer Methods for Crop Improvement
I. Introduction 1
II. Plant Viruses as Biological Vectors 1
III. DNA Transfer via *Agrobacterium tumefaciens* 2
    A. Selection Markers 5
    B. *Agrobacterium* Cocultivation 9
IV. Chemically Mediated DNA Uptake 12
V. Electroporation 14
    A. Mechanism of Electroporation 15
    B. Factors Influencing Electroporation 15
        1. Electrical Parameters 16
        2. Size of Plasmid 19
        3. Plasmid DNA Concentration and Carrier DNA 21
        4. Electroporation Media and Incubation Temperature 22
        5. Effect of Physiology of Protoplasts and Incubation Period 23
        6. Influence of Cell Cycle and Irradiation on Transformation 23
VI. Transient and Stable Transformation 24
VII. Microinjection 27
VIII. Microprojectile Bombardment Method 28
IX. Electric Discharge Particle Acceleration Method 34
X. Conclusions 35
References 36

Chapter 2
Genetic Engineering of Insect Control Agents and Production of Insect-Resistant Plants
I. Introduction 49
II. Proteinase Inhibitors 49
III. *Bacillus thuringiensis* and Its Endotoxin 55
    A. Characterization of Bt Endotoxin 55
    B. Plasmid Coded Genes 58
    C. Chromosomal Coded Genes 59
    D. Characterization and Cloning of Bt Endotoxin Genes 60
    E. Regulation of Bt Endotoxin Genes 61
    F. Transgenic Plants Expressing the δ-Endotoxin Gene 64

IV. Genetically Engineered Insect-Resistant Plants and
    Ecological Considerations...71
V. Conclusions..73
References..76

Chapter 3
Engineering Herbicide Resistance into Plants
I.   Introduction..85
II. Strategies for the Development of Herbicide-Resistant
     Plants..87
     A. Altering the Level and Sensitivity of the Target
         of the Herbicide...88
         1. Mutant *psb*A Gene......................................88
         2. Acetolactate Synthase (ALS) Genes.......................91
            a. Microbial ALS Mutant Genes...........................93
            b. Plant Mutant ALS Genes...............................95
         3. EPSPS and Mutant *Aro*A Genes...........................97
III. Alternative Strategies for Achieving Herbicide
     Resistance 98
     A. Gene Amplification and Overexpression......................98
     B. Herbicide-Detoxifying Genes...............................100
         1. Herbicide-Detoxifying Enzymes from
            Plants..101
         2. Herbicide-Detoxifying Genes from
            Bacteria..102
IV. Conclusions...104
References...107

Chapter 4
Engineered Resistance against Plant Virus Diseases
I.   Introduction...117
II. Strategies for Engineering Genes for Plant Protection..........117
     A. Coat Protein Expression and Resistance in
         Transgenic Plants...118
     B. Use of Satellite RNAs.....................................121
     C. Use of Antisense RNA 123
III. Construction and Design of Chimeric Genes Encoding
     Coat Proteins and Their Introduction into Plants..............124
     A. Isolation of the CP Gene: Limitations.....................124
     B. Selection of Appropriate Promoter 125
     C. Importance of the 3' End Sequence 125
     D. Selection of Transformed Plants...........................126
IV. Detection of CP-Mediated Protection and Its
     Characteristics 126

	A.	Development of Disease Symptoms on CP(−) and CP(+) Plants 127
	B.	Factors Affecting Viral Resistance in Transgenic Plants....128
		1. Viral Inoculum Concentrations....128
		2. Nature of the Host Plant....128
		3. Growth Conditions and Plant Resistance....129
		4. Specificity of the CP-Mediated Protection....129
V.	Examples of Coat Protein-Mediated Resistance....131	
	A.	Potato Viruses X and Y....132
	B.	Potato Virus S....134
	C.	Alfalfa Mosaic Virus (AlMV)....136
	D.	Cucumber Mosaic Virus (CMV)....139
	E.	Tobacco Streak Virus (TSV)....139
	F.	Soybean Mosaic Virus (SMV)....140
	G.	Tobacco Rattle Virus (TRV)....142
	H.	Tobacco Mosaic Virus (TMV)....143
VI.	Mechanisms of CP-Mediated Resistance....147	
	A.	Role of the CP Gene in Cross-Protection....147
	B.	Role of Nonstructural Proteins in Cross-Protection....151
	C.	Viral Genome Replication and Dissemination of the Virus through the Plant....153
VII.	Field Trials....159	
VIII.	Conclusions....161	
References....163		

Chapter 5
Improvement of the Nutritional Quality of Plants by Manipulation of Seed Storage Genes

I.	Introduction....171	
II.	Genetic Manipulations of Seed Storage Proteins....172	
	A.	Legume Storage Proteins....174
		1. Glycinin and Conglycinin....174
		a. General Characteristics....174
		b. Organization and Structure....175
		2. Legumin....177
		3. Vicilin and Convicilin....180
		4. Phytohemagglutinin and Phaseolin....184
		a. Synthesis and Glycosylation....184
		b. Intracellular Transport and Processing....185
		c. Transport and Biological Activities....186
		d. Phytohemagglutinin and Phaseolin Genes....187

		5.	Arcelin...192
	B.	Cereal Storage Proteins....................................192
		1.	Storage Proteins of Wheat and Barley 193
		2.	Rice Storage Proteins 196
		3.	Maize Storage Proteins 197
			a. Maize Prolamines 197
			b. Molecular Organization and Structure................................197
			c. Transgenic Plants........................202
III.	Production of Small Peptides of 2S Albumin 206
IV.	Conclusions and Future Prospectus 207
References..211

Chapter 6
Impact of Genetically Modified Crops in Agriculture and Ecological Risk

I.	Introduction...219
II.	Transgenic Plants and Their Commercial Value.................220
	A. Herbicide Resistance 222
	B. Insect Resistance 222
	C. Disease Resistance 223
	D. Field Performance of Transgenic Plants 223
III.	Ecological and Environmental Impacts and Legislation 224
References..229

Index 231

Chapter 1

GENE TRANSFER METHODS FOR CROP IMPROVEMENT

I. INTRODUCTION

Plant breeders and genetic engineers share the common goal of plant improvement. While plant breeders traditionally use selective breeding for varietal enhancement, genetic engineers continue to develop techniques for the isolation and insertion of genes for desirable traits. Genes which are unavailable to a particular plant species due to sexual incompatibility may be obtained from other organisms (plants, microorganisms, or animals) and transferred into plants.

Several steps using both molecular and cellular biology techniques are involved in producing a genetically engineered plant. A trait must be chosen and the gene(s) encoding the trait identified and isolated from a donor organism. The functional gene must include regulatory regions in order to be correctly expressed in the plant. The isolated gene may be transferred directly or inserted into vectors. For successful transformation the gene sequences introduced into the plant cell must be inserted into the plant genome, expressed, and maintained throughout subsequent cell divisions. Finally, the transformed plant cells must be regenerated into whole plants. Various gene transfer methods which introduce DNA sequences into plant cells will be discussed in this chapter.

II. PLANT VIRUSES AS BIOLOGICAL VECTORS

Certain naturally occurring plant DNA viruses can be used to introduce DNA into normal, healthy plants. Viral vectors and their application in gene transformation in plants have already been described in detail.[1] Among all viral vectors, cauliflower mosaic virus (CaMV) is becoming very popular. CaMV, a caulimovirus, contains double-stranded DNA which lends itself easily to the manipulations involved in recombinant DNA technology. The ability of CaMV to infect plants and then move systematically through the host avoids the need for cell culture. Plants can be inoculated by rubbing the leaves with engineered CaMV. One region of the CaMV genome (gene II) has been shown to be dispensable for viral replication. When this gene is replaced with a bacterial gene the resulting engineered CaMV is able to systematically infect and express the bacterial gene in inoculated plants.[1]

Brisson et al.[1a] replaced the nonessential open reading frame II of CaMV with the dihydrofolate reductase gene and used the construct to transform turnip plants. However, the vector had a restricted host range, limited space for inserting DNA, and it was necessary to eliminate 5' and 3' noncoding sequences. In order to circumvent these defects, pKR612B1 was constructed, containing the NPTII (neomycin phosphotransferuse) gene under the control of the CaMV gene VI promoter.[2] This construct was similar to, but more versatile than, pABD1[3] and was used to transform protoplasts of *Brassica campestris* var. *rapa* by PEG-induced plasmid uptake. However, transformation was achieved only when the hybrid gene was supplied to protoplasts with wild-type viral DNA.

Viral vectors, however, have very little potential for production of transgenic cereals although they can be easily used for amplification of genes and gene products. The discovery that RNA virus genomes can be reverse-transcribed to yield cDNA clones that again are infective opened the possibility of using genetic engineering not only with the small group of DNA viruses, but also with the larger group of RNA viruses.[4-7] According to available evidence, these viruses do not integrate into the host genome and they are excluded from meristems and thus from transmission to sexual offspring. These facts determine the advantages and disadvantages of viruses as vectors for possible genetic engineering in plants. Although it would be very difficult to use viral vectors for integrative transformation, they have invaluable potential for gene amplification and systemic spread within individual plants.[8,9]

Because viral vectors are limited to the host species which they naturally infect, and may not be sexually transmitted, a more general mechanism of DNA incorporation into plant cells is needed. At present, *Agrobacterium tumefaciens* is the most common vehicle used for the transformation of plants.

III. DNA TRANSFER VIA *AGROBACTERIUM TUMEFACIENS*

Agrobacterium tumefaciens is a soil bacterium that naturally infects many dicotyledonous and gymnospermous plants predisposed by some form of wounding. *Agrobacterium* infection causes tumorous plant growth, commonly called crown galls, by introducing DNA into the plant cells at the wound site. The tumor-forming ability of *A. tumefaciens* depends on the presence of a large plasmid called the tumor-inducing (Ti) plasmid. Molecular analysis has revealed a small region of the Ti plasmid, referred to as the T-DNA, which is transferred into the plant cell and covalently integrated into a plant chromosome.[10] A region of the Ti plasmid outside the T-DNA, referred to as the virulence region, carries genes (*vir* genes) that are involved in tumor induction. Expression of the *vir* genes in the bacterium may be required for conditioning of the plant cells during infection and for subsequent transfer of the T-DNA. T-DNA itself contains genes referred to as oncogenes, which are responsible for tumor induction. The oncogenes code for the production of indoleacetic acid and zeatin riboxide, natural plant hormones. Overproduction of these hormones results in tumorous growth of plant cells. The T-DNA also encodes several genes responsible for the synthesis of compounds called opines which are metabolic substrates for the bacteria.[11] By transferring the opine genes to the plant cell genome, bacteria are able to subvert the plant's own metabolism in order to produce substances (opines) which are not used by the plant but on which the bacteria can proliferate.

The removal of sequences in the T-DNA required for tumor promotion does not interfere with transfer of the T-DNA to the plant genome, and their replacement with chimeric genes constitutes the basis for *Agrobacterium*-mediated transformation of plants. Although the nature of T-DNA transfer is not completely understood, molecular characterization of the T-DNA sequence reveals flanking direct repeats at the boundary regions of the T-DNA (25 nucleotide imperfect border sequences). Genes that are to be introduced into plant cells must be inserted between the borders or adjacent to at least one border of the T-DNA. By cloning foreign DNA into the T-DNA of the Ti plasmid, plant molecular biologists are able to exploit the natural ability of *Agrobacterium* to transfer new DNA into the plant genome. The area of *Agrobacterium*-mediated gene transfer to plants has seen exciting progress, and several aspects of gene transformation have been described from time to time.[12-20]

As mentioned earlier, plant tissue transformed by wild-type *Agrobacterium* is identified by its characteristic tumorous growth, which proliferates rapidly on hormone-free media due to an altered internal hormonal balance. However, plant genetic manipulation with the Ti plasmid is of limited value if the resulting transformed plant cells are tumorous and incapable of regenerating into normal fertile plants. Therefore it has been necessary to eliminate the oncogenic properties of the T-DNA which inhibit the normal differentiation of plant cells. So-called "disarmed" Ti plasmids have had the oncogenes deleted from the T-DNA, yet retain the border regions. These disarmed plasmids not only maintain the ability to transfer the T-DNA to the plant genome but also allow for the regeneration of the healthy plants.

A major advance in the development of useful genetic engineering vectors was the discovery that the T-DNA and the *vir* region could be separated onto two different plasmids, without loss of the DNA transfer capability of the T-DNA.[21] This achievement has allowed development of binary T-DNA vectors for the transfer of foreign genes into the plant. The binary method employs two plasmids: a Ti plasmid containing the *vir* genes with oncogenes or without the oncogenes ("disarmed")[22] and a separate genetically engineered T-DNA plasmid. This binary T-DNA plasmid is constructed *in vitro* using a wide-host-range plasmid which can replicate in both *Escherichia coli* and *Agrobacterium* cell systems.[23] Also contained on the binary T-DNA border are sequences required for successful transfer of DNA from the *Agrobacterium* cell into the plant genome. The desired foreign DNA (gene) is cloned into the engineered binary T-DNA plasmid containing border sequences. Binary vectors typically have a broad host range origin of replication and a marker gene (generally drug resistance marker) for selection and maintenance in both *Agrobacterium* and *Escherichia coli* in addition to a T-DNA segment containing a plant selectable marker and one or more restriction sites for cloning DNA insert (Figures 1 A and B).[24] The use of these vectors has eliminated

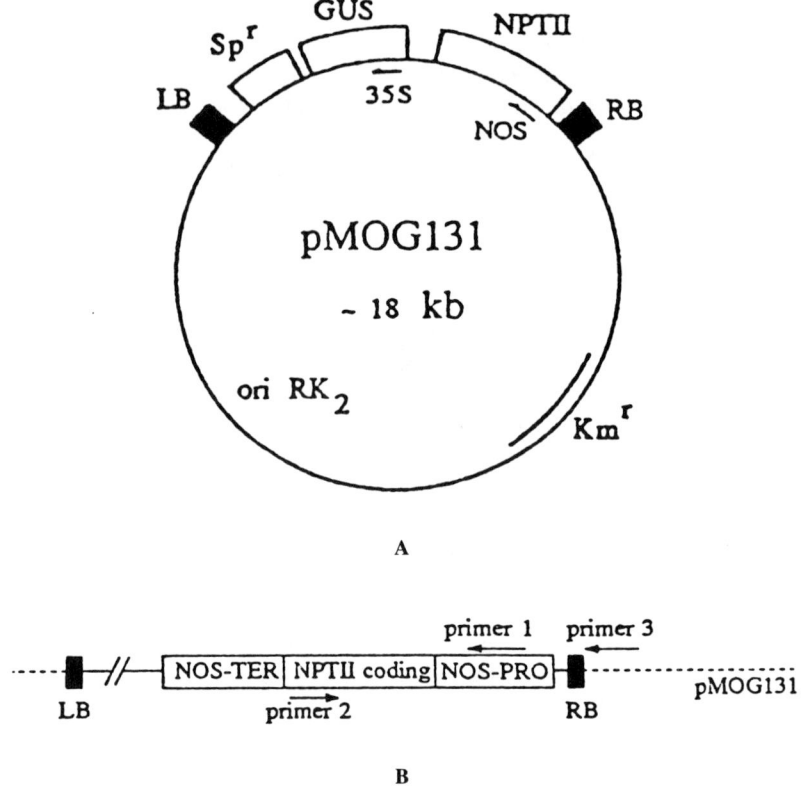

FIGURE 1. (A) Structure of binary vector. pMOG131 was constructed from pBI121 by inserting a 5.7-kb *Eco*RI fragment carrying the spectinomycin resistance (Sp_r) gene from transposon Tn1831. (B) Segment of the linear map of plasmid pMOG131, in which the position of the primers 1, 2, and 3 are indicated. (From Schrammeijer, B., Sijmons, P. C. van den, Elzen, J. M., and Hoekema, A., *Plant Cell Rep.*, 9, 55, 1990. With permission.)

the requirement for homologous recombination into the Ti plasmid T-DNA region, thus simplifying the task of introducing and analyzing recombinant T-DNA in *Agrobacterium*. One major drawback is that cloning into these vectors is more difficult than for standard *E. coli* cloning vehicles. Binary vectors either have a low copy number in *E. coli*, with very few restriction sites in the T-DNA for inserting genes of interest or no insertional inactivating marker for detection of cloned DNA. All these problems have been dealt with simultaneously in a few recent vectors, and certain vectors do contain a ColE1 origin of replication for increased copy in *E. coli*[22,25] or a *lacZ'* gene from pUC 19 incorporated into the T-DNA to supply a number of unique restriction sites for cloning genes of interest as well as insertional inactivation markers.[23]

A. SELECTION MARKERS

In an attempt to create more compact vectors, smaller derivatives of the pRK290 broad host range vector have been utilized that are unstable in many host bacteria including *E. coli* and *Agrobacterium*. A small segment of DNA containing the pRiHR1 origin of replication has also been utilized to construct a shuttle cloning vector that is stable in *A. tumefaciens* without selection.[25] Recently, McBride and Summerfelt[26] constructed a set of binary vectors that combined the previous ColE1 origin of replication, the *lacZ'* gene segment incorporated into T-DNA, and small size with new improvements such as addition of the pRiHR1 origin of replication for high stability in *A. tumefaciens*, a gentamycin resistance gene for selection in *E. coli* and disarmed *A. tumefaciens* strains, and a choice between two types of chimeric NPTII (neomycin phosphotransferase) genes as plant selectable markers (Figure 2).[26]

In general, plants regenerated from tissue transformed by disarmed Ti plasmids containing the opine synthesis genes can be screened and transformants identified by the detection of opine production in the tissue. A more effective method for identification of transformed plant cells is achieved by the addition, within the T-DNA, of a plant selectable marker gene (such as a gene conferring resistance to an antibiotic, e.g., kanamycin). In this case, the transformed cells grow on medium containing the selective agent while nontransformed cells do not. The major difficulty of using neomycin/kanamycin, as selection marker is that these agents also influence the regeneration of transformed cells and cause some damage to transgenic plants.

Attempts are also being made to use the *bar* gene as a selection marker in plants.[27-29] The *bar* gene codes for enzyme phosphinothricin acetyltransferase (PAT), which inactivates the herbicide phosphinothricin by acetylating it.[30] Phosphinothricin is a glutamate analogue that inhibits glutamate synthetase. The inhibition results in accumulation of NH_4^+ which is toxic for plant cells. The idea of using *bar* as a marker gene emerges from the fact that transgenic plants expressing a chimeric *bar* gene are resistant to high doses of phosphinothricin.[27,28] De Block et al.[29] have also used *bar* and *neo* genes as selectable markers.

The *E. coli* chloramphenicol acetyl transferase (CAT) gene is also widely used as a reporter gene for plant systems.[31] This gene has proved useful because plant cells have no background CAT activity and the assays developed for CAT are straight-forward and unambiguous.[32] The most commonly used assay for CAT involves incubating the enzyme with ^{14}C-labeled chloramphenicol and acetyl coenzyme A. The acetylated reaction products are separated from unacetylated chloramphenicol by thin layer chromatography.[33] Quantitative data can be obtained by liquid scintillation counting after scraping spots off the thin layer plates. But this procedure is time consuming. Another assay uses radiolabeled acetyl coenzyme A as a substrate.[34] Radiolabeled, acetylated chloramphenicol derivatives are separated from radiolabeled acetyl coenzyme A by extraction with ethyl acetate. These derivatives are then quantitated

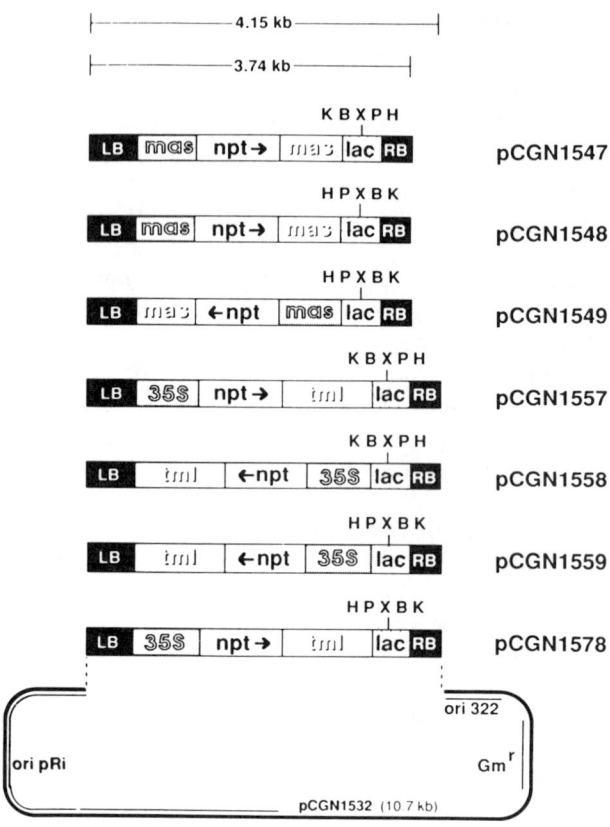

FIGURE 2. Scheme of the backbone and T-DNA components for the binary vector series. LB, left T-DNA border; RB, right T-DNA border; mas, mannopine synthase gene (dark stippled box = 5′ region and white box = 3′ region); 35S, CaMV 35S promoter; tml, tml 3′ region; npt, neomycin phosphotransferase gene (arrow denotes orientation of coding region); lac, lacZ′ locus (encodes the lac alpha peptide). The unique restriction sites of the polylinker region (from pUC18) are listed: K, KpnI; B, BamHI; X, XbaI; P, PstI; H, HindIII. Other restriction sites present in the polylinker region (EcoRI, SacI, SmaI, SalI and SphI) are not shown since they are not unique. (From McBride, K. E. and Summerfelt, K. R., *Plant Mol. Biol.*, 14, 269, 1990. With permission.)

directly by liquid scintillation. This procedure is simpler and quicker than the first assay but still requires several manipulations and handling of radioactive materials. A rapid qualitative immunoassay was developed by Burns and Erickson,[35] demonstrating some of the advantages of immunochemical methods for CAT detection. Recently Gendloff et al.[36] demonstrated the development of ELISA for CAT activity and its use for quantitation of CAT in transgenic tobacco plants (Figure 3).[36] Using this assay, they have observed a correlation between the level of CAT gene expression and CAT gene copy number in plants analyzed.

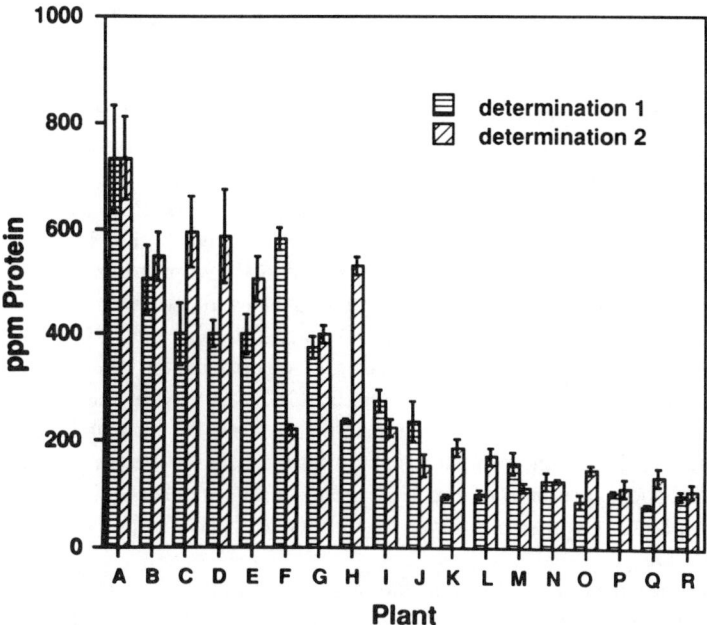

FIGURE 3. CAT expression of 18 transgenic *N. tabacum* cv. *xanthi* as determined by ELISA. Expressed as ppm protein. Bars indicate ± 1 standard deviation of 3 determinations made on each of 2 days. CAT was never detected in untransformed plants or plants transformed without the CAT gene. (From Gendloff, E. H., Bowen, B., and Bucholz, W. G., *Plant Mol. Biol.*, 14, 575, 1990. With permission.)

It is worthwhile to note that CAT is also unsuitable as a reporter gene in *Brassica napus* and *B. juncea,* since plant tissues contain high levels of endogenous CAT activity.[37] Although CAT has been used most extensively as a reporter gene in transient systems, NPTII is preferred. This gene has also proved useful in cereals. Maximum NPTII activity was detected 4 days after PEG-induced plasmid uptake into protoplasts of *Triticum monococcum,* with gene expression still detectable after 10 days.[38] The majority of the DNA introduced into protoplasts remained extrachromosomal, with the plasmid DNA changing from supercoiled to the open circular linear forms within the plant cells. Other studies involving the NPTII gene fused to four different promoters confirmed that PEG was effective in delivering plasmids to protoplasts of barley, maize, rice, and rye.[39]

However, there is little information which compares the levels of sensitivity of CAT, β-glucuronidase (GUS), and NPTII as reporter genes. In experiments using nearly identical constructs with the CaMV 35S promoter and CAT, GUS, and NPTII reporter genes introduced into tobacco SR1 protoplasts by $Ca(NO_3)_2$-PEG treatment, CAT gene expression was the easiest to detect, followed by NPTII and GUS.[40] GUS has been used to study promoter function. With this system, Marcotte et al.[41] reported the normal regulation of a wheat promoter, inducible by abscissic acid, in a transient expression assay using rice protoplasts. This study provided direct evidence for phytohormone activity to regulate the initiation of gene transcription.

Recently, there has been increased interest in protoplasts of woody species, with GUS expression in those of *Alnus incana*[42] and *Picea glauca*,[43] and CAT expression in those of *P. glauca*.[43,44] Gene expression was influenced by the DNA concentration, the amplitude and duration of the electric pulse, and, in the case of *Alnus* protoplasts, by the presence of PEG in the electroporation medium. Linearized plasmid constructs gave higher levels of CAT activity than circular plasmids in electroporated *Picea* protoplasts,[44] but gene expression was reduced with PEG-mediated DNA delivery to the same protoplasts.[43] However, herring sperm DNA as carrier increased GUS expression by 50% with PEG uptake.[43] Interestingly, electroporated *Picea* protoplasts released a small molecule that mimicked GUS activity in the fluorescence assay, masking GUS gene expression in this system.[44] Using a chimeric gene construction, in which corn sucrose synthase-1 gene (*sh*) promotor was used to control the expression of the GUS gene, Yang and Russell[45] transformed tobacco plants. Expression of *sh-gus* activity in these plants was found to be cell specific.

The glucuronidase *gus* gene has many advantages.[46-49] Its gene product can be localized *in vivo* in tissue sections using histochemical techniques. The *gus* reporter gene is also useful in situations where an antibiotic like kanamycin is toxic to normal shoots and regeneration of plants is not possible. For instance, the *gus* gene when used as a reporter gene provided reliable information of transformants[49] as compared to the kanamycin gene (the selection on which showed both transgenic and nontransgenic cells).

Reporter genes such as CAT, NPTII, and *gus* have certain disadvantages. Thus attempts have been made to introduce reporter genes, the activity of which can be easily detected in transgenic plants. One such reporter gene is the luciferase gene (*luc*) which has been used to visualize gene expression in the whole plant[50] and plant cells.[51] Firefly luciferase produced by the *luc* gene catalyzes the ATP-dependent oxidation of luciferin in a reaction that produces light.[48] The *luc* gene encodes a single, catalytically active polypeptide of 62 kDa.[52] The cDNA for *luc* has been cloned from fireflies[53] and has been expressed in a variety of systems, including bacterial, plant, and animal cells.[53,54] Luciferase activity is also a suitable "reporter" to monitor transcriptional regulation of chimeric genes in plant cells. Such experiments also demonstrated that transgenic plant cells are able to assemble correctly complex heterodimeric bacterial enzymes. The *in vitro* luciferase assay is also sensitive, has virtually no background, and is very rapid and inexpensive, but usually requires a luminometer to detect.[50] New luciferase has been, therefore, cloned from click beetle encoded enzymes that produce light of different colors[55] which, with appropriate instrumentation, could allow the simultaneous detection of several reporter genes. Recently Schneider et al.[56] used the *luc* gene (Figure 4A) and found that the distribution of luciferase activity was closely correlated with the tissue-specific pattern of luciferase on RNA (Figure 4B and 4C).

B. AGROBACTERIUM COCULTIVATION

Agrobacterium cocultivation techniques are currently being developed for a wide range of dicotyledonous and gymnosperm species. Cocultivation is coincubation of plant cells with bacteria. Following a determined length of cocultivation, plant cells are washed free of the bacteria and cultured in media containing bacteriostatic antibiotics. Of prime importance in cocultivation is the ability to regenerate plants from the cultured material. There are two systems commonly used for *Agrobacterium* cocultivation: regenerable protoplasts and explant material. Each protoplast is in a state of wound repair making it susceptible to *Agrobacterium* infection. A large number of protoplasts can be treated in each cocultivation, making detection of low frequency transformation events feasible. In addition, this material is uniformly exposed to the selective agent, resulting in fewer escapes (growing nontransformed cells) and single transformation events are more easily identified. Regenerable excised plant material (explant) for co-cultivation with *Agrobacterium* is especially helpful in situations where regeneration of plants through protoplasts is not possible. The following is a typical example of transforming plants with *Agrobacterium* cocultivation, in which selection, confirmation, and enzyme assay for a marker gene was provided by Fang and Grumet.[57] Explants are inoculated with *Agrobacterium tumefaciens* bearing a Ti plasmid with a marker gene. After cocultivation for 3 to 4 days, explants are transferred to regeneration medium with antibiotic to select the transformed tissues. Southern blot analysis and leaf callus assay are used further to confirm the transgenic plants.[57] The effectiveness of this system has been demonstrated in several species and has two primary advantages:[28,48,57] it can be applied to species for which no regenerable protoplast culture techniques exist, and leaf discs or other explant material will regenerate plants more rapidly than protoplast systems.

In addition, suspension culture is also used for inoculation with *Agrobacterium tumefaciens*-bearing Ti plasmids. This is proving successful in woody plants.[57-60] These methods are primarily applicable to dicots because they show the pronounced wound response needed for efficient transformation.[8] However, dicots that have not been transformed probably do not show the appropriate wound response.

There have been several difficulties in transforming cereals or monocots with the *Agrobacterium* cocultivation method. Most attempts have not been published because they are negative. Transformation of cereals is not difficult because they are monocots. In fact monocots with a wound response are as easy to transform as a dicot with a wound response, and monocots without a wound response are as difficult to transform as are dicots.[9] Wounding of differentiated cereal tissue does not lead to the wound response-induced dedifferentiation in wound-adjacent cells. Therefore, no competent cells are available. Instead, wounding leads to the death of the wound-adjacent cells. Even though *Agrobacterium* efficiently transfers T-DNA into cereal cells, integration of this T-DNA can not lead to transgenic clones because the receptor cells die.

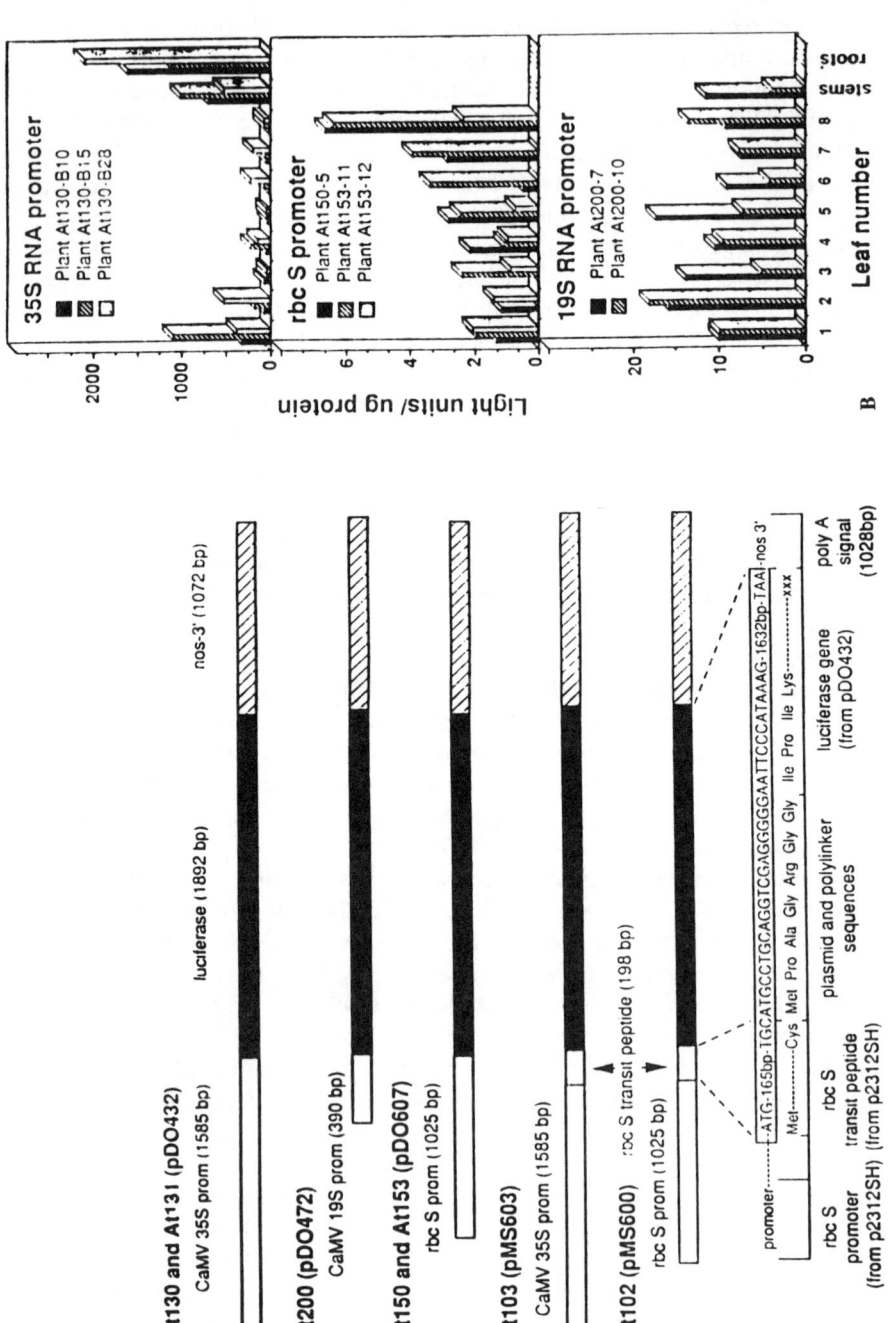

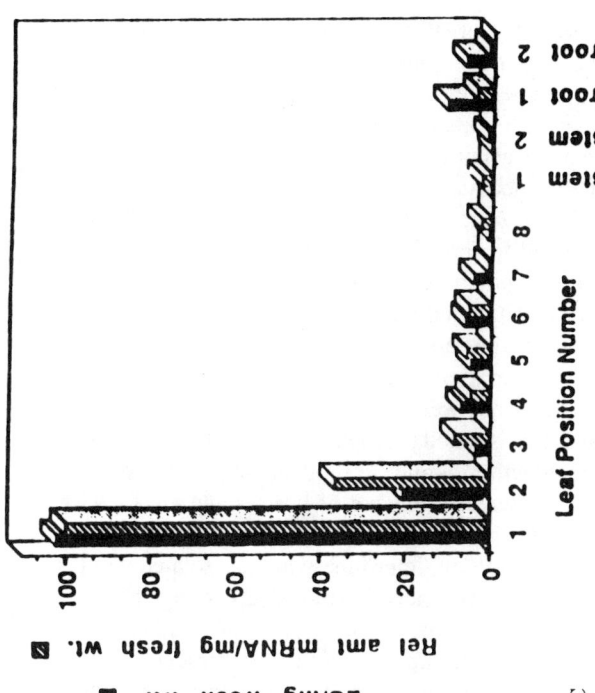

FIGURE 4. (A) Constructs of luciferase gene expression vector. The names given to the *E. coli* plasmids, all pUC19 derivatives, are indicated in parentheses. The constructs were inserted as plasmid cointegrates or DNA fragment inserts into the *Agrobacterium* binary vector, Bin19. The binary vectors were transformed from *E. coli* to *A. tumefaciens*, and the resulting *A. tumefaciens* lines were given an "At" designation. (B) Distribution of luciferase activity in transgenic plants bearing *luc* driven by different promoters. Luciferase activity in extracts was assessed in *in vitro* assays and expressed on a per protein basis. Leaf number refers to the position of the leaf on the plant, starting with leaf 1 as the first unfolded leaf at the top of the plant. (C) Correlation between the organ-specific distribution of luciferase activity and *luc* mRNA. Luciferase activity in extracts from a transgenic plant bearing *luc* driven by the CaMV 35S promoter (At 130-B10 line) was assessed in *in vitro* assays and expressed on a fresh weight basis. *Luc*-specific mRNA was determined by extracting RNA from the leaves, stems, and roots as indicated, subjecting RNA from the same amount of fresh weight tissue to electrophoresis on formaldehyde gels, probing blots to which the RNA was transferred with a ^{32}P-labeled luciferase probe (the insert from pD0432), and quantifying the amount of RNA by densitometer scans of the resulting autoradiograms. The amount of *luc* mRNA present in a single band at 1.9 kb was expressed on the basis of the fresh weight of the sample from which the RNA was derived. (From Schneider, M., Ow, D. W., and Howell, S. H., *Plant Mol. Biol.*, 14, 935, 1990. With permission.)

Therefore, other methods of transformation are being developed. Several alternatives involve direct transformation accomplished by introducing large amounts of DNA into the plant cell. Although the transferred DNA may be expressed for only a short period of time following the DNA transfer process (transient expression), stable expression occurs when DNA is integrated into the plant nuclear or plasmid genomes and expression continued in long-term callus-culture-regenerated plants. Methods of direct transformation such as chemically mediated DNA uptake, electroporation, and microinjection are currently being fully exploited for producing transgenic plants.

IV. CHEMICALLY MEDIATED DNA UPTAKE

The chemicals polyethylene glycol (PEG) and polyvinyl alcohol (PVA), together with Ca^{2+} and high pH, induce protoplasts to take up naked DNA. This technique has resulted in stable integration of foreign DNA in regenerated *Nicotiana tabacum* plants. The integrated gene was shown to be inherited in Mendelian fashion in subsequent generations.[18] The transformation frequencies obtained using chemically mediated DNA uptake have been low when compared to those following *Agrobacterium* cocultivation. The technique of direct DNA uptake is most useful for graminaceous monocotyledons or other species that are not susceptible to or difficult to transform with *Agrobacterium*. Protoplasts of *Lolium multiforum* (Italian rye grass) and *Triticum monococcum* (diploid wheat) have been efficiently transformed using chemically mediated DNA uptake.[61-65] However, regeneration of plants from protoplasts of monocotyledons remains difficult and not well documented.

The interaction of viruses with isolated protoplasts demonstrates that poly-L-ornithine (PLO) stimulated infection, while subsequent work confirmed that polyethylene glycol (PEG) also increased uptake of viruses and viral nucleic acids into protoplasts. Thus, during the late 1970s, it was appropriate to combine the techniques then available to determine whether isolated Ti plasmid could transform plant protoplasts. Indeed, studies involving the interaction of supercoiled Ti plasmid from an octopine strain of *A. tumefaciens* with cell suspension protoplasts of *Petunia hybrida* in the presence of PLO resulted in protoplast-derived colonies that expressed the crown gall characteristics of growth on hormone-free medium and the synthesis of octopine, encoded by T-DNA genes.[66,67]

Work in several laboratories confirmed PEG-mediated transformation of protoplasts from leaves of a streptomycin-resistant line (SR1) of *N. tabacum* by isolated Ti plasmids.[68,69] Interestingly, the T-DNA structure and organization in petunia and tobacco protoplasts transformed with Ti plasmid was different from T-DNA in cells transformed by intact agrobacteria. Analysis of transformed-petunia tissues showed the presence of truncated T-DNA,[70-72] while in tobacco the T-DNA was often fragmented.[69] Overlapping copies of T-DNA of varying size, as well as T-DNA fragments, were also detected. T-regions in the Ti plasmid were used as the preferred sequences for integration into the plant genome. Transformation could be achieved using a T-region clone lacking the left border of the TL-DNA and the right border of the TR-DNA. In addition, the use of Ti plasmid carrying a mutated *vir* region confirmed that the *vir* region was not essential for transformation by direct DNA uptake. Such studies[69,73] also provided evidence for cotransformation. The transformation frequency in all these experiments with the Ti plasmids was of the order of 1 in 10^4 to 10^5 cells.[66,68] In an evaluation of chemical methods to deliver pTiACH5 to suspension cell protoplasts of petunia, Draper[70] obtained transformation frequencies of 2.9, 6.7, 2.2, and 2.5×10^{-5} after treatment of protoplast-plasmid mixtures at high pH with PLO, PEG, high salts, and Ca^{2+}, respectively.

In an attempt to protect the Ti plasmid from degradation during plasmid uptake, Hasezawa et al.[74] used *Agrobacterium* spheroplasts treated with PEG or PVA to transform *Vinca rosea*. Transformation mainly occurred due to endocytosis and to a less extent by direct fusion with the plasma membrane of recipient protoplasts. The transformation frequency was higher than with naked Ti plasmids (7.6×10^{-4}). Similar studies with tobacco protoplasts[75,76] and *E. coli* spheroplasts carrying cloned CaMV DNA[77] also confirmed the results of the above-mentioned experiments.

Deshayes et al.[78] encapsulated a plasmid carrying the aminoglycoside phosphotransferase type II gene (APHII; neomycin phosphotransferase, NPTII) from Tn5 conferring resistance to kanamycin in plant cells, into negatively charged liposomes and fused plasmid-containing vesicles with tobacco protoplasts. This resulted in successful transformation, and the restriction pattern of DNA inserted into transgenic plants indicated a tandem integration of plasmid sequences, implying homologous recombination between the sequences during transformation. Maximum association of plasmid with protoplasts occurred when the protoplasts were interacted with naked DNA or with plasmid-containing liposomes, prepared by the reversed-phase evaporation method, and 15% w/v PEG 6000, or with *E. coli* spheroplasts containing chloramphenicol-amplified plasmid in the presence of 25% w/v PEG 6000.

Chemically mediated transformation has also made it possible to transform protoplasts with vectors other than that of *Agrobacterium*. The difficulty with the Ti plasmid is the isolation of its DNA, which has a low copy number, compared to *E. coli* vectors. It was also realized that T-DNA sequences were unnecessary for stable transformation and expression of foreign DNA in plant cells.[2] Thus a selectable hybrid gene (comprising the protein-coding region of the Tn5, NPTII gene under the control of the CaMV gene VI promoter) was introduced into *N. tabacum* SR1 protoplasts as part of an *E. coli* plasmid (pABD1) by treating protoplast-plasmid mixtures with PEG 6000. The foreign gene was stably integrated into the plant genome and was retained in several generations. The efficiency of gene transfer was, however, later increased more than 1000-fold with a linearized pABD1 by inactivating the restriction barriers by heat treatment of recipient protoplasts (5 min at 45°C) prior to addition of the plasmid DNA followed by PEG treatment. Hain et al.[76]

incubated protoplasts with pLGVneo 2103 in calcium phosphate coprecipitate, followed by fusion of the protoplasts with PVA and exposure to high pH. Kanamycin-resistant transformants were obtained in 0.01% of the protoplast-derived colonies, which was a significant improvement in transformation. Efforts have also been made to improve the vectors used in direct DNA uptake. Balazas et al.[79] constructed pKR612, which had two unique Sma 1 and BamHI restriction sites for gene insertion into the vector pKR612N1. Turnip and tobacco protoplasts were transformed to kanamycin resistance with pKR612N1 using the PEG procedure of Krens et al.[68]

Peng et al.[80] found PEG-mediated cotransformation of protoplasts with *neo* and *gus* genes on separate plasmids coupled with selection on kanamycin as an effective way of transferring foreign genes into the rice genome. Several experiments have demonstrated that cotransformation of protoplasts is possible with *E. coli*-based vectors by mixing plasmids during the uptake procedure, or by constructing a vector carrying two genes. Transformation of tobacco protoplasts with pABD1 carrying the NPTII gene and pMS1 carrying the *zein* gene, resulted in 88% of the kanamycin-resistant cell colonies with sequences from the *zein* genomic clone,[81] whereas with pABD1 and pGV0422 carrying the napoline synthase (*nos*) gene, 47% of kanamycin-resistant colonies contained the *nos* gene. Czernilofsky et al.[82,83] also studied integration of two separate plasmids, one containing the NPTII gene, the other containing the maize transposable element Ac. Other workers also used the DNA uptake procedure essentially as described to show the transformation of *Petunia hybrida* leaf protoplasts.[3] The grain legume *Vigna aconitifolia* was transformed directly and kanamycin-resistant calli were regenerated from heat-shocked protoplasts treated with PEG and pLGVneo2103, giving rise to transgenic plants.[84,85] The plant cultivar was important in maximizing the transformation frequency, with the highest value (58% of resistant colonies) being obtained in cultivar 560.

V. ELECTROPORATION

Electroporation, developed by animal researchers for the introduction of DNA into mammalian cell lines, is the application of high voltage electrical pulses to a solution containing cells and DNA. The commonly used term "electroporation" was introduced by Neumann et al.[86] The term emphasizes the postulated formation of pores in the membrane and has no bearing on the integrity/viability of the exposed membrane/cell. DNA enters the cells through reversible pores which are created in the membrane of the cells by the action of short electrical pulses. This technique has recently been applied to plants that were either recalcitrant to traditional transformation methods or had no standard transformation method. Because of success in the plant system this technique is described here in detail. Many species have been transformed either transiently or stably, and several features about gene regulation and expression have been elucidated in plants, using electroporation.[87,88]

A. MECHANISM OF ELECTROPORATION

The mechanism for induction of the high-permeability state of membranes/cells is not fully understood. The reversible permeabilization of the plasma membranes of whole plant cells to charged molecules, which are otherwise unable to enter the cell, can be achieved by application of electrical fields. Such molecules can also be introduced into cells after permeabilization with organic solvents and other chemical agents, although these may reduce viability.[89] However, the plant cell wall is a barrier to DNA uptake by intact cells.[90-92] The level of transformation is about 20 to 50 times lower than that obtained by electroporated intact cells. Several workers have attempted to explain the phenomenon, and different theories have been proposed: the electromechanical,[93] electroinduced,[86,94] and the viscoelastic.[95,96] According to electromechanical models, electroporation increases membrane permeabilization. The increase in permeability to DNA depends on the thickness of membrane, voltage across the membrane, and hydrostatic pressures. It also proposes the occurrence of a critical potential, below which no membrane permeabilization can occur. This model predicts the existence of a critical hydrostatic pressure, at which the intrinsic membrane potential is sufficiently high to induce mechanical breakdown of the membrane. The experimental evidence in support of this phenomenon comes from observation in human erythrocytes subjected to high pressures.[97,98]

Neumann et al.[86] proposed the concept of electroinduced pores in the membrane. According to this model, the primary action of the electric fields is on the charges and dipole configurations associated with overall dipole moment (Figure 5).[99] This is also known as the concept of electroinduced pores in the membrane. As a result, thinning/compression of the membrane, and an increase in the defects of the membrane structure and formation of holes/pores occurs. In this model the boundaries of the pores are believed to consist of lipid blocks tilted at 90°. Continued exposure to the external electric field increases the number of tilted lipids. Eventually, irreversible membrane destruction occurs with the increase in the pore size.

The third model is known as the viscoelastic model. This model incorporates the motion of the two monolayers leading to fluctuations in membrane thickness. Continuous exposure to an external electric field brings about molecular rearrangements in the membranes, which results in the formation of discontinuous pores. Mechanical breakdown of the membrane may occur if the membranes are further exposed to an electric field.

B. FACTORS INFLUENCING ELECTROPORATION

Electroporation has proved to be a very exciting and successful method for introducing DNA into cells which are either difficult or recalcitrant with traditional transformation methods. After its initial success in dicotyledonous plants, this technique has now been extended to several cereals and legumes. This technique has been standardized in different species to introduce DNA

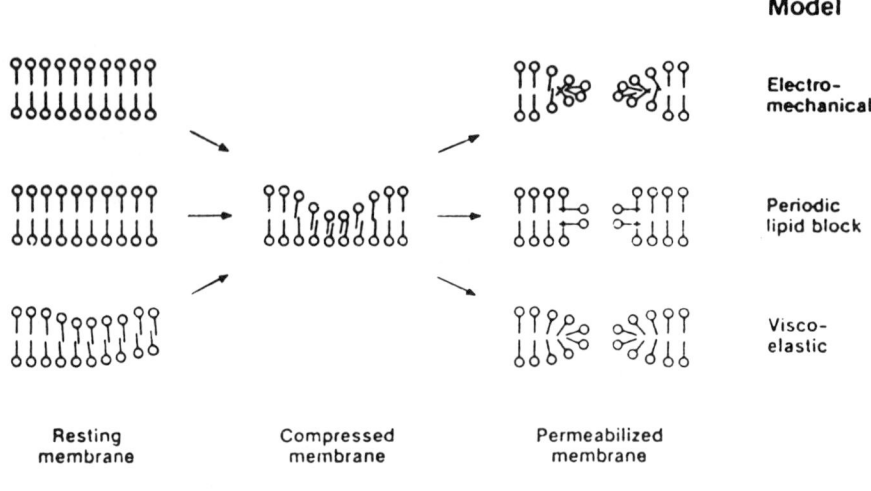

FIGURE 5. Schematic presentation in molecular terms of the three major models for electropermeabilization of bilayer membranes. The holes/pores in the membrane are according to the electromechanical model formed in response to extensive membrane compression and thinning. The viscoelastic model assumes the formation of pores with continuous borders, while according to the periodic lipid-block the lipids are rearranged into pores with discrete boundaries. (From Joersbo, M. and Brunstedt, J., *Physiol. Plant.*, 80, 1, 1990. With permission.)

into a large proportion of protoplasts while maintaining reasonably sufficient cell viability.

1. Electrical Parameters

Whereas strong direct electric fields reduce cell viability, weak direct current/electric fields, applied over extended periods of several days, have been shown to enhance plant regeneration from callus,[100,101] this response being attributed to an increase in the polar transport of auxin.[102] Electroporation, involving exposure of freshly isolated protoplasts to micro- or millisecond high voltage electric pulses, has already been shown to enhance division and plant regeneration in several species.[103,104] Electromagnetic fields also stimulate division and DNA synthesis in cultured animal cells,[105,106] and for both animal and plant systems the application of an electric field has been correlated with an increased uptake of ions and nucleic acids from the surrounding medium.[86,107-112]

Three basically different pulse types have successfully been used for transferring genes to plant protoplasts: exponentially decaying, rectangular, and alternating current pulses. Exponentially decaying pulses are delivered by a device consisting of a bank of capacitors (usually in the range of 10 to 1000 μF) and a DC power supply. After charging the selected capacitor, usually to 100 to 1000 V, it is discharged over a chamber/cuvette holding the protoplasts. Pulse duration varies with the capacitance of the medium in the electroporation chamber. Pulse duration capacitor discharge machines are commercially available: Bio-Rad, BRL, Hoefer, BTX, Promega, and ISCO.[113]

Rectangular (square) pulse machines include electronic components for the modification of electric pulse to square waves. In these machines the pulse duration can usually be selected in the range of 10 to 1000 μs and the voltage from 1000 to 2000 V. Within limits, duration can essentially be independent of medium conductivity, etc. Machines producing square pulses can be purchased from Kruss, Zimmermann, BTX, and Baekon.[113]

Electroporation by alternating current from the mains (220 V) controlled by blow-out of the small fuse (50 to 400 mA) has been reported.[114] Duration of the pulse is 0.5 to 10 ms. Although this type of equipment is not commercially available, it can easily be assembled from an electro cord, a small fuse, and a suitable electroporation chamber.[113]

Electroporation of sugar beet protoplasts by alternating current and exponentially decaying pulses resulted in 3- to 4-fold higher transient expression compared to rectangular pulses (Figure 6A, B).[114] In sugar beet protoplasts, 4-fold higher transient expression was obtained by alternating current pulses than by square pulses. Joersbo and Brunstedt[114] also found in electroporated tobacco and sugar beet protoplasts that exponentially decaying pulses resulted in 30 to 280% higher transient expression than did the rectangular pulses.

In order to induce the uptake of DNA through permeabilization of the plasma membrane, the electric field strength must exceed a certain minimum value (critical field strength) to induce the uptake of DNA (Figure 7).[99] Extremely low levels of transient expression were found in tobacco mesophyll protoplasts when electroporated at field strength up to 200 V/cm (exponentially decaying pulses with a total duration of 200 ms).[115,116] Increasing the field strength to 300 V/cm resulted in 20- to 30-fold higher CAT activity. Guerche et al.[115] found that the transient expression increased approximately linearly with the number of pulses up to ten consecutive pulses. More pulses reduced the CAT activity in the tobacco mesophyll protoplasts, probably due to reduced viability. For barley (*Hordeum vulgare* L.), optimal electrical conditions were five pulses of 100 to 400 μs at 120 V/mm (Figure 8).[117] Microspore samples subjected to electroporation were able to develop microcalli or proembryos and subsequently plants. Similar results were obtained by Ou-Lee et al.,[118] who pulsed wheat protoplasts with a very large number of pulses (up to 80 pulses). Similarly, very low transient expression was observed at 200 V/mm irrespective of the pulse length.[118] Doubling of the field strength resulted in an 8-fold increase in transient expression. This suggests that above a certain critical field strength, an increased pulse duration can to some extent compensate for a decreased field strength.

Riggs and Bates[119] found the maximum number of kanamycin-resistant tobacco colonies (2.2×10^{-4}) after electroporation by discharge of a 10-nF capacitor charged to 2.0 kV (4.0 kV/cm). This electric pulse reduced the

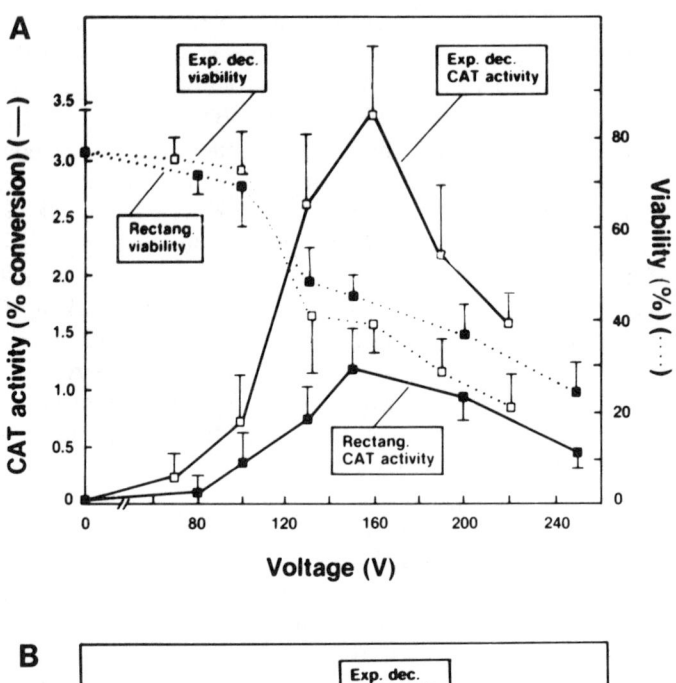
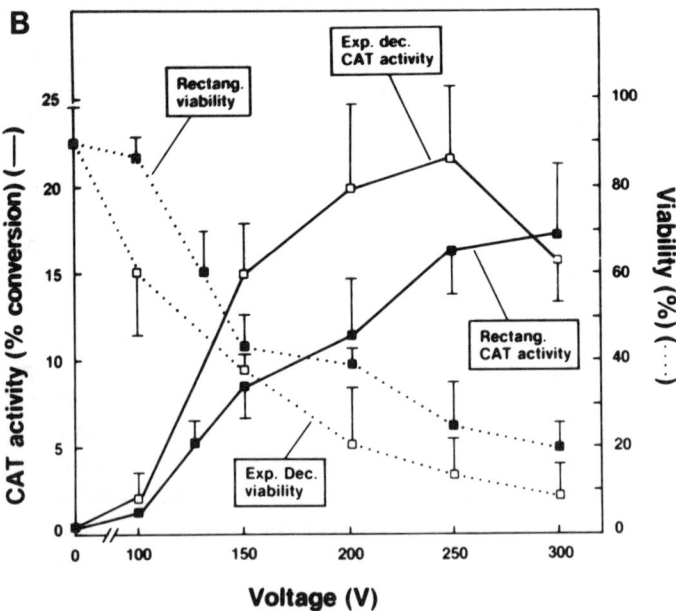

FIGURE 6. (A) The transient expression, measured as CAT activity, and the viability of sugar beet protoplasts (M1), electroporated at increasing voltages by rectangular (100 μs) or exponentially decaying (800 μF) pulses. (B) The transient expresssion, measured as CAT activity, and the viability of tobacco protoplasts, electroporated at increasing voltages by rectangular (100 μs) or exponentially decaying (800 μF) pulses. (From Joersbo, M. and Brunstedt, J., *Plant Cell Rep.*, 8, 701, 1990. With permission.)

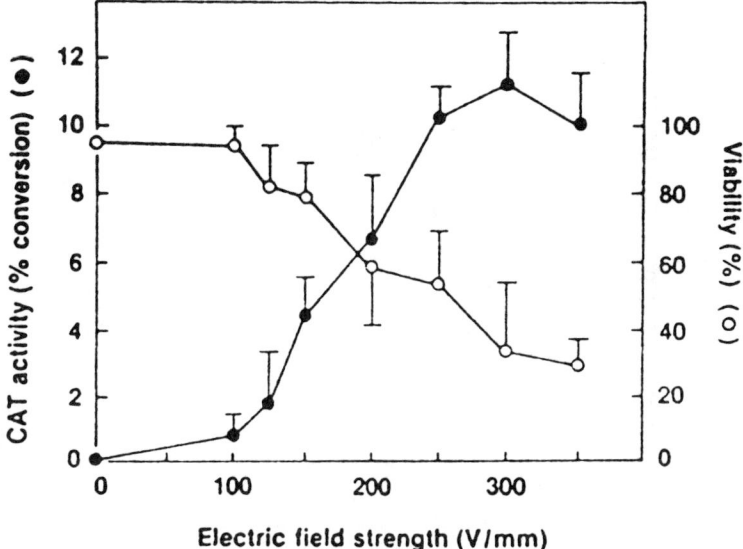

FIGURE 7. Transient expression of a CAT gene and viability of electroporated (sequare pulses) tobacco protoplasts. At this pulse duration (100 μs) the critical field strength is 80 V/mm. (From Joersbo, M. and Brunstedt, J., *Physiol. Plant.*, 80, 1990. With permission.)

viability of the protoplasts to 36%. More gentle electric conditions (5 nF capacitor) resulted in a transformation frequency of 0.4×10^{-4} with a reduction in viability to 51%. Discharging a 15-nF capacitor over the protoplasts gave both fewer transformants (0.6×10^{-4}) and fewer surviving protoplasts (22%). Similarly, Guerche et al.[115] found maximal transient expression after electroporation of tobacco protoplasts by 8 pulses (300 V/cm, 16 nF capacitor), while the maximum number of stable transformants was obtained by five of these pulses. The plating efficiency was reduced by 35% at 5 pulses and by 53% at 7 pulses as compared to non-electroporated control samples.

2. Size of Plasmid

There is not much information available on the size limit of the DNA constructs above which direct stable transfer to protoplast becomes inefficient.[120,121] Several authors[66,68,70] have used chemical methods to introduce Ti plasmids into protoplasts, while Langridge et al.[122] obtained transgenic somatic embryos of carrot after electroporation with about 194 kb pTiC58. However, while the presence of T-DNA (about 20 kb) was demonstrated in the transformed tissues, these studies gave no clear indication whether large plasmids, other than the "naturally transferable" plasmids of *Agrobacterium*, might be delivered to protplasts by such methods.

As already discussed, the recent trend has been to improve transformation efficiency by reducing the size of the vector DNA. Most groups have confined

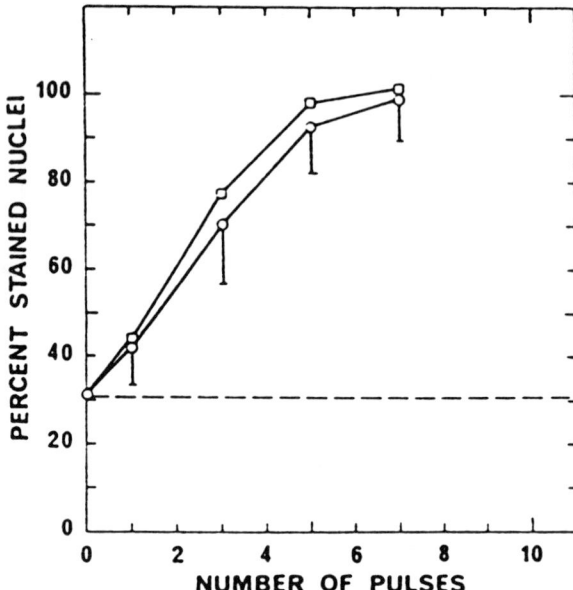

FIGURE 8. The effect of the number of pulses on the percentage of microspores with nuclei stained with propidium iodide. The field strength applied was 120 V/mm and pulse duration 950 μs. PI was added just before (□) or 2 h after (○) electroporation. Bars represent SD for at least two independent experiments. (From Joersbo, M., Jorgensen, R. B., and Olesen, P., *Plant Cell Tissue and Organ Culture*, 23, 125, 1990. With permission.)

their attention to plasmids of less than 20 kb in size when transforming protoplasts by direct DNA uptake, the 16-kb pCT1T3[123] being one of the largest plasmids to have been used in such studies. By comparison, using *Agrobacterium*-mediated delivery, somewhat larger fragments of foreign DNA may be introduced into plant cells by employing suitable cosmid shuttle vectors. For example, a vector, pOCA18, was used to deliver 20-kb fragments of *Arabidopsis* DNA into tobacco.[124] Some (10/16) transformants produced in these experiments contained full-length fragments of *Arabidopsis* DNA, while others (6/16) possessed partial fragments. It would be of interest to determine whether such recombinant cosmids could be introduced into protoplasts using direct methods. Obviously, to be of general use, chemical methods or electroporation would have to deliver the cosmids efficiently.

The question of the upper size limit for stable DNA transfer and integration by chemical or electrical methods is particularly topical in light of recent developments in DNA cloning and electrophoresis methodology. Burke et al.[125] have developed a cloning system based on the construction of yeast artificial chromosomes (YACs), which can accept foreign DNA inserts of more than 100 kb at a suitable restriction site; the foreign DNA-containing YACs can be maintained in *Saccharomyces cerevisiae* after PEG-induced transformation of spheroplasts. The YAC vectors themselves can be maintained as circular molecules in *E. coli,* which permits the isolation of milligram quantities for cloning procedures. YACs containing 100-kb inserts can be distinguished and separated from native yeast chromosomes by pulsed field gel electrophoresis.[126] A preliminary investigation of the use of the YAC system in constructing libraries of large (greater than 100 kb) fragments of *Arabidopsis* and carrot DNA has been described.[127] If such libraries can be made truly representative of the entire genome of plants such as *Arabidopsis,* this YAC methodology should be a stimulus in attempts to isolate genes by chromosome walking. In addition, it will be of interest to determine whether such plant DNA-containing YACs can be introduced into protoplasts by chemical methods or electroporation and maintained without deletion or rearrangement. Although at present somewhat speculative, success in this technical challenge would greatly facilitate the development of strategies for isolating genes by complementation of genetic mutant, especially mutant species with relatively small genomes such as *Arabidopsis*.

3. Plasmid DNA Concentration and Carrier DNA

The frequency of stable transformation of protoplasts has been shown to be influenced by the concentration and type of exogenous DNA. By type, we mean whether the plasmid is linear or supercoiled, or whether carrier DNA (such as herring or salmon sperm DNA) is supplied with the plasmid. Shillito et al.[108] for example, in their studies on the transformation of tobacco mesophyll protoplasts by a combination of PEG and electroporation, observed a strong effect of both plasmid and carrier DNA on the transformation. Although it was reported that transformants of tobacco could be obtained with as low a concentration of plasmid DNA as 0.001 µg/ml, the highest frequencies of transformation were obtained by using 10 to 100 µg/ml DNA. At 10 µg/ml plasmid DNA, the addition of 50 µg/ml carrier DNA resulted in a more than 10-fold increase in transformation frequencies. Linearized plasmid DNA has also been reported to give higher transformation frequencies than does supercoiled.[108,128,129] Increasing the concentration of DNA in the electroporation medium also increases the transformation frequency. A linear correlation between the transient expression in electroporated carrot and *Alnus* protoplasts and the plasmid DNA concentration up to 40 µg/ml has been reported.[42,109] Similarly, Tautorus et al.[130] found very low transient expression in electroporated black spruce and jack pine protoplasts below 50 µg/ml, and above this DNA concentration the transient expression in black spruce increased rapidly (the CAT activity being 30-fold higher with 150 µg/ml than with 50 µg/ml DNA). On the contrary, in sugar beet protoplasts, significant CAT activity after electroporation was observed only at 10 µg/ml.[90]

The role of carrier DNA in enhancing transformation frequencies is not known with certainty due to the presence of DNase activity in the plasma membrane and cytosol of some plant species, e.g., maize.[43,131,132] Addition of carrier DNA has been used in some studies in order to protect the plasmid

DNA from degradation. Tautorus et al.[130] found a 27% stimulation of transient expression in jack pine protoplasts, when using 50 μg/ml sonicated calf thymus DNA. However, at 150 μg/ml carrier DNA it did not have any effect on transient expression. It is possible that even higher concentrations of carrier DNA may reduce the uptake of plasmid DNA due to the presence of a limited number of sites available for penetration of DNA into the electropermeabilized protoplasts. The carrier DNA may protect the plasmid from degradation by acting as a substrate for endogenous nucleases. However, it has been observed that the carrier can not always be replaced by excess plasmid. The carrier may recombine with the marker gene before and carrier sequences may be expressed more efficiently than the marker gene. In addition, the carrier may integrate into the genome alone and create integration sites for the marker, perhaps by disrupting chromatin secondary structure. The carrier may disrupt genomic DNA replicon or repair mechanisms creating opportunity for marker gene integration. Closed circular supercoiled plasmid DNA has been used in most studies for transformation by electroporation, but linearized plasmids have also been used. In jack pine, linearized plasmids resulted in 2.5-fold higher expression than supercoiled plasmids, while in black spruce supercoiled plasmids gave 35% higher transient expression than linearized plasmids.[130]

4. Electroporation Media and Incubation Temperature

Interrelated with the electrical parameters and plasmid size is also the composition of the electroporation media. Several workers have used different media for electroporation. The pulse strength decreases with increasing conductivity of the medium when exponentially decaying pulses are used. However, the duration is independent of the electroporation medium for square pulses. Fromm et al.[109] used a high conductivity medium, phosphate-buffered saline (PBS) supplemented with 0.2 M mannitol, while Seguin and Lalonde[42] used a medium with low conductivity (13% mannitol). However, machine breakdown may occur if the conductivity is too high. Addition of Ca^{2+} to the electroporation medium is advantageous due to their membrane-stabilizing properties. Fromm et al.[109] found that 4 to 5 mM Ca^{2+} significantly enhanced transient expression, this being strongly reduced by the absence of Ca^{2+} and by addition of 10 mM Ca^{2+}. Equal conductivity of the tested media was obtained by addition of KCl.

In some studies the protoplasts are incubated on an ice-bath some minutes before and immediately after electroporation[109] in order to enhance DNA uptake by prolonging the duration of the high permeability state of the membranes. However, the diffusion rate is highly dependent on temperature and DNA uptake is believed to take place predominantly during the electric exposure.

It has been found that electroporated protoplasts of sugar beet retain their permeability longer when maintained at 4°C and when at 25°C or 35°C,[90] and some workers routinely electroporated at the lower temperature to increase DNA uptake.[109,122] A short incubation of protoplasts at high temperatures

Chapter 1

before electroporation is included in some protocols. However, in protoplasts of carrot[133] and *Alnus incana*[42] no detectable effect of a heat treatment at 45°C for 5 min found transient expression.

5. Effect of Physiology of Protoplasts and Incubation Period

The ability to be transiently transformed varies enormously with species and also among varieties/genotypes within one species. Variability among genotypes of black spruce was 300-fold with respect to transiently expressed CAT activity, and transient expression varied 2- to 3-fold among different genotypes of jack pine.[130]

Also with respect to stable transformation, large differences in the competence of protoplasts are found. Using different cultivars of moth bean, Koehler et al.[85] obtained transformation frequencies from zero to 1.2×10^{-5}. Similar results were obtained with tobacco by Tyagi et al.[134] and with maize by Huang and Dennis,[135] emphasizing the necessity of screening a range of genotypes in order to increase the probability of obtaining some transformants.

Due to very low regeneration of most monocotyledonous protoplasts, intact cells and other cell types have been used for electroporation. However, cell wall pectin acts as a barrier for DNA uptake through electroporation. Lindsey and Jones[92] also demonstrated that the removal of pectin from the plant cell wall increases the amount of DNA which can be introduced by electroporation. There are two factors which appear to play a significant role in electroporation in plants. The first is that the pectin itself may inhibit DNA transfer, either as a physiological barrier or chemical one.[136] Baron-Epel et al.[137] have indeed implicated pectin as the major determinant of wall porosity in soybean cells. Second, the size of the trans-wall channels may be enlarged by treatment with pectinase.

6. Influence of Cell Cycle and Irradiation on Transformation

The cell cycle of recipient protoplasts influences transformation. Meyer et al.[138] synchronized tobacco leaf protoplasts by culturing them 2 to 5 days in the presence of aphidicolin, a mycotoxin that inhibits DNA polymerase, and 2,6- dichlorobenzonitrile, an inhibitor of cell wall synthesis. When transformed with a plasmid derived from pLGVneo 11, up to 3% of the surviving cells were transformed at the S or M phase. This frequency was twice that with unsynchronized protoplasts. In similar studies involving aphidicolin, G418-resistant colonies were obtained with M phase protoplasts at frequencies two to eight times those obtained with protoplasts at other stages of the cell cycle, indicating that the absence of a nuclear membrane in mitotic cells favors delivery to the nucleus of exogenous DNA introduced into the cytoplasm.[139-141] Irradiation of protoplasts prior to treatment with plasmid and PEG has also resulted in a 3- to 6-fold enhancement of hygromycin- or kanamycin-resistant colonies without affecting colony production and plant regeneration.[142] The higher integration rates of foreign DNA may result from increased recombination in irradiated cells.

VI. TRANSIENT AND STABLE TRANSFORMATION

The standardization of electroporation conditions has resulted in the production of transgenic calli or plants in several species of plants. Fromm et al.[109] reported for the first time the expression of genes transferred to plant protoplasts by electroporation. They electroporated carrot, tobacco, and maize protoplasts in the presence of plasmid DNA encoding the bacterial *cat* gene and measured the amount of expression of the introduced gene by assaying *cat* activity 48 h after electroporation (transient expression). Since then transient expression in plant protoplasts after electroporation has been reported in rice (*Oryza sativa*), wheat (*Triticum monococcum*), and sorghum (*Sorghum bicolor*);[118] soya bean (*Glycine max*), petunia (*Petunia hybrida*), napier grass (*Pennisetum purpureum*), *Panicum maximum*, and sugar cane (*Saccharum officinarum*);[118,143] sugar beet (*Beta vugaris*);[90] moth bean (*Vigna aconitifolia*);[84] white spruce (*Picea glauca*);[44] black spruce (*Picea mariana*) and jack pine (*Pinus banksiana*);[130] breadwheat (*Triticum aestivum*);[144] tomato (*Lycopersicon esculentum*);[145] wild potato (*Solanum brevidens*); and cultivated *Solanum tuberosum*.[146]

Transient expression assays have been used to compare the regulation of plant promoters coupled to coding regions of a number of reporter genes including CAT,[11,109] GUS,[147,148] NPTII,[147] and luciferase.[50] Promoter sequences fused with such genes can be analyzed within days or even hours after electroporation by determining expression of the reporter gene using an enzyme assay. The short time frame required for transient expression after electroporation is a major advantage compared to regeneration of stably transformed plants.

Rapid procedures, such as those for CAT, GUS,[46,47] and NPTII[147] and luciferase[50] useful for detection of transformants, are also available. Fromm et al.,[109] using electroporation to induce DNA uptake, reported that gene transfer efficiency into protoplasts increased with DNA concentration and was influenced by the amplitude and duration of the electrical pulse and composition of the electroporation medium. Comparison of two plasmids (pNOSCAT and pCaMVCAT) with the *nos* and CaMV 35S promoters, respectively, showed that CAT expression was higher with the CaMV promoter in carrot and maize protoplasts. Both constructs were expressed in tobacco protoplasts. Other studies used carrot protoplasts to optimize the electroporation conditions before detecting CAT expression in protoplasts of *Glycine max*, *Petunia monococcum*, *Pennisetum purpureum*, *Panicum maximum*, *Saccharum officinarum*, and a double-cross, trispecific hybrid between *Pennisetum purpureum*, *P. americanum*, and *P. squamulatum*.[143] In an extension of the work with *Panicum maximum*, gene expression was evaluated in protoplasts from two cell lines after delivery of pMON563 by PEG and electroporation. CAT expression was higher after electroporation, but PEG-treated protoplasts exhibited optimum plating efficiencies.[120]

CAT has also been used as a reporter gene in constructs carrying promoters isolated from a developmentally regulated *Zea mays* seed storage protein gene and from the mannopine synthase gene of an octapine Ti plasmid. CAT activity was correlated with DNA concentration, and gene transfer was stimulated by treatment of protoplast-plasmid mixtures with PEG prior to electroporation and by optimization of the voltage used during electroporation. In carrot and petunia protoplasts, transformation efficiency was independent of $CaCl_2$ in the system; however, transformation efficiency was stimulated by carrier DNA.[121] In studies with sugar beet, Lindsey and Jones[90] reported that mesophyll protoplasts were more susceptible to damage by electroporation than were cell suspension protoplasts. CAT activity was also demonstrated after electroporation of intact suspension cells with plasmid, gene expression being increased by pectinase treatment of the cells prior to plasmid uptake. CAT expression has also been used to demonstrate gene expression in plant cell systems.[149] Through electroporation, transgenic plants have been produced in tobacco,[118,119] in *Brassica napus*,[116] in orchard grass (*Dactylis glomerata*),[150] in moth bean (*Vigna aconitifolia*),[84,85] in lettuce (*Lactuca sativa*),[151] in rice (*Oryza sativa*),[152,153] in *Lycopersicon peruvianum*,[9] and in soybean (*Glycine max*).[154-156]

A major constraint in the production of transgenic plants in most of the cereals is the inability to induce plant regeneration from protoplasts treated with DNA. The experimental approach used most extensively involves DNA uptake into protoplasts by treatment with polyethylene glycol or electroporation. These techniques have been used to study transient gene expression[39,109,143] and have also resulted in the production of stably transformed tissues of gramineous species, including *Lolium multiflorum*,[64] *Oryza sativa*,[123,157,158] *Panicum maximum*,[118] *Saccharum* spp.,[159] *Triticum monococcum*,[62,118] and *Zea mays*.[11,160,161] Stable transformation appears to be obtained by most if not all methods leading to transient expression. By some largely unknown mechanisms, the extrachromosomal exogenous DNA can be incorporated into the genome of plant cells, leading to stable transformation in which the introduced gene will be inherited as a single dominant trait.

Although several dicotyledonous plants have been transformed stably through electroporation, this discussion primarily focuses on the stable transformation of monocotyledons through electroporation. Some of the earlier transient gene expression studies, especially those with cereal protoplasts, optimized the DNA uptake conditions for subsequent attempts to produce stable transformants. Although high levels of natural resistance to antibiotics such as kanamycin, shown by cells of some monocotyledons, were believed to preclude the use of kanamycin for selection of transformed cells,[161] this has been shown not to be the case. In studies published simultaneously by Potrykus et al.[63,64] and Lorz et al.[61,62] kanamycin-resistant tissues of *Lolium multiflorum* and *Triticum monococcum* were obtained after PEG-induced uptake of pABD1 and pBL1103-4 carrying the NPTII gene with CaMV and *nos* promoters, respectively.

In an extension of their transient gene expression studies, Fromm et al.[111] obtained kanamycin-resistant tissues when protoplast-derived colonies were exposed to 100 μg/ml of kanamycin weeks after electroporation of pCaMVneo into maize cell suspension protoplasts. Likewise, Ou-Lee et al.[118] assessed the use of vectors conferring resistance to hygromycin, kanamycin, or methotrexate with cell suspension protoplasts of *Triticum, Panicum, Pennisetum,* and *Saccharum,* and obtained resistant clones at 1.0×10^{-5} to 2.0×10^{-6}. Southern blot hybridization confirmed stable integration and expression of DNA in *Triticum monococcum* protoplasts using the methotrexate vector, with 1 to 10 copies integrated per haploid genome. Other studies with sugarcane cell suspension protoplasts[159] produced kanamycin-resistant tissues at 8 in 10^7 after PEG-induced uptake of *Sma*I linearized pABD1 (the plasmid used extensively in earlier transformation studies with tobacco protoplasts).[18]

Much attention has been devoted to the production of transgenic plants in maize and rice, following development of protoplast-to-plant systems. Rhodes et al.[160] electroporated protoplasts from embryogenic cell suspensions of *Zea mays* (Black Mexican Sweet) and selected cell growth of protoplasts on filter over feeder layers of maize suspension cells. Unfortunately, the transgenic plants were sterile, but the production of fertile nontransformed plants from protoplasts of elite lines of maize[84,161,162] should result in fertile transgenic plants. Sugarcane protoplasts were found to be unstable when electroporated either in the presence or absence of PEG.[108] In rice, Uchimiya et al.[123] selected protoplast-derived colonies on medium with 100 μg/ml kanamycin upon PEG-induced plasmid uptake. Such experiments formed the basis for the production of transgenic rice plants by the same group.[152] In the latter experiments, transformed cells were selected by their tolerance to G418 after electroporation-mediated plasmid delivery. In a similar series of experiments with *Oryza sativa* cv. Taipei 309, Yang et al.[158] optimized the conditions for transformation of cell suspension protoplasts by pCaMVneo. Electroporation of heatshock protoplasts gave transformation frequencies of 19.9×10^{-5} and was more efficient than PEG or PEG combined with electroporation. Southern blot analysis showed some clones to have a single insert of the NPTII gene; others had a series of bands in addition to that of the NPTII gene. These optimum electroporation conditions resulted in transgenic plants when applied to protoplasts of an embryogenic cell suspension.[163,164] Of six regenerated plants selected for analysis, all contained the NPTII gene, but only two expressed NPTII activity. One of these NPTII-positive plants was fertile and produced seed. Hygromycin B at 20 μg/ml, used 5 weeks after electroporation of *Oryza sativa* cell suspension protoplasts with pGL2, has also been used to select transformed colonies from which fertile transgenic plants have been regenerated.[153] Seeds from transformed colonies also expressed GUS activity in their embryos and endosperm, confirming co-transformation for an unlinked nonselectable gene with pGL2.

VII. MICROINJECTION

Microinjection uses microcapillaries and microscopic devices to deliver DNA into defined cells in such a way that the infected cell survives and can proliferate. This technique has provided transgenic clones from protoplasts[165] and transgenic chimeras from microspore-derived proembryos in oilseed rape.[166,167] With this technique injection into the nucleus through cytoplast is possible, and cells can be cultured indivdually in some species to produce callus or plants. In this way selection of transformants by drug resistance or marker genes may be avoided. For one experiment[168,169] nuclei or cytoplasm from leaf-derived tobacco protoplasts were injected, and cells were grown individually in microliter drops of culture medium to generate calli for DNA analysis. Seven percent of injected DNA was in the nucleus. Efficiency was reduced when transferred DNA integrated as fragmented incomplete copies, presumably due to handling during the injection process.

In another approach, Neuhaus and Schweiger[170] separated nuclei and cytoplasm from rapeseed protoplasts before injection. Nuclei were then grown individually or rejoined with cytoplasm by electrofusion. Most of the nuclei that were injected (80%) survived to produce callus, whereas growth of fused cells was only 22%. Of all viable cells, 60 to 70% exhibited positive reactions to NPTII antibodies after injection.

Procedures to deliver functional DNA across intact callus cell walls to the nucleus through microinjection have also been reported.[171] One advancement over holding cells with a pipette was to place single cells in agarose for easier manipulation. Up to 98% of cells survived DNA injection into the nucleus and 22% of those grew in the presence of the drug kanamycin. All resistant clones carried marker DNA.

The potential of microinjection as a method to obtain stable transgenic plants was first demonstrated in rapeseed.[167] Injection across intact cells into the cytoplasm of microspore-derived embryoids (4- to 12-cell stage) was performed with an equal mixture of supercoiled and linear DNA. Fifty embryoids could be injected per hour, after which they were detected in 28 to 50% of regenerants with approximately 70% being chimeric for the transferred gene. Not much progress has been made in the application of this method. Low regeneration rates and competency of cells for DNA uptake may be prime reasons for lack of progress in other species.[8] Rigorous studies to address these problems are urgently needed if microinjection is to serve as a general and reliable tool for plant transformation.

This method may prove more useful because the DNA can not only be injected into the nucleus of the other cells but also, the potential exists for injecting DNA directly into pollen grains, microspores, cells in suspension culture, embryos, or ovules. In addition, with this method the transfer of cell organelles and the manipulation of isolated chromosomes has been proposed.[168,169]

Microinjection definitely delivers DNA into plant cells, but this technique has some disadvantages. The major disadvantage is that only one cell receives DNA per injection and handling requires more skill and instrumentation. Thus, delivery of the DNA to protoplasts or intact cells via microinjection is labor intensive. The procedure also requires special capillary needles, pumps, micromanipulators, and other equipment.

Additional difficulties of this technique lie in the physical coordination of both an instrument to inject the DNA as well as a means to immobilize the cell to be injected. Immobilization of protoplasts using methods such as agarose embedding,[172] agar embedding,[173] polylysin-treated glass surfaces,[174] and suction holding pipettes[168,169] have been investigated. Once injection has been achieved, the injected cell must be properly cultured to ensure its continued growth and development. Intranuclear injections of tobacco protoplasts, using the holding pipette and hanging drop system, have resulted in the integration of foreign DNA in 14% of the samples.[168,169] However, with this method the operator has the ability to maneuver the cell and thus more accurately target the nucleus. This added flexibility may prove critical for the transfer of this technology to plant material other than protoplasts.

VIII. MICROPROJECTILE BOMBARDMENT METHOD

In this procedure, micron-size tungsten or gold particles are accelerated in a gun barrel to velocities sufficient for nonlethal penetration of cell walls and membranes.[174-178] The particle gun was initially used to monitor transfer of chimeric DNA and viral RNA molecules into intact onion cells.[174,175] Tungsten acts as a carrier of nucleic acids because it is nontoxic to cells, and dense enough for rapid penetration of target material. For each transfer, 50 μg tungsten is needed, which is accelerated up to 430 m/s in a partial vacuum. Approximately one half of the bombarded tissue remains viable when only 1 to 5 projectiles penetrate a cell, but mortality rates reach 80% with 21 or more particles.[175] Successful transfer of RNA and DNA was detected by the presence of viral inclusion bodies or CAT enzyme activity. These first experiments demonstrated that the biolistics approach could successfully deliver viable nucleic acids to intact tissue, but refinements in concentration and distribution of tungsten were needed to reduce cell trauma.

Friable embryogenic and nonembryogenic cultured maize cells were utilized as model systems to optimize various parameters.[176,177] The highest transformation efficiency (2×10^{-3}) occurred when 1.25 to 2.5 μl of a 2 μg/mg DNA-tungsten preparation was used. Larger amounts of DNA led to severe clumping of projectiles and lower transformation rates. Levels of CAT gene expression ranged from 17 to 36% with relative activity from single bombardments of up to 70%. Gene expression for CAT was decreased, however, to 2 to 8% in embryogenic cells. Distribution and activity of transformed material were also monitored with the GUS gene, whose level of expression was variable both in single cells and aggregates. These important studies identified several parameters that must be tested empirically for different cell types.

One of the first applications of the biolistics technique involved molecular studies of structural (*A1, Bz1, Bz2*) and regulatory genes (*B, C1, P1,* and *R1*) for production of anthocyanin pigment in maize.[178,179] In one study, cloned *A1* or *Bz1* genes were transferred to the aleurone layer of maize kernels by bombardment amd complemented genetic mutants for anthocyanin production in a tissue-specific manner.[178] Equally important was the fact that the transferred genes were regulated in a manner similar to their endogenous counterparts. Cloned *B* genes were also delivered to intact aleurone and embryo tissue,[179] where their products coordinated the expression of the *A1* and *Bz1* structural genes. Mutants at the *Bz2* locus in the anthocyanin pathway were identified and cloned when they contained endogenous transposons.[180] A wild type allele without transposon insertions was required, to examine structure and regulation of the *Bz2* gene. To accomplish this the polymerase chain reaction was also used to clone a portion of reconstructed *Bz2* gene. This method made it possible to assess activity of reconstituted genes derived from transposon-induced mutants.

The availability of cloned genes that are specifically expressed in pollen is an important step toward understanding the molecular basis of pollen development and pollen-specific gene expression.[181-183] One approach to characterize cis-acting DNA sequences that are required for pollen-specific gene expression depends upon the ability to regenerate fertile transgenic plants containing modified versions of pollen-specific genes.[184] A complementary approach is the transient expression of genes introduced into plant protoplasts.[14,185] However, genes introduced into plant protoplasts do not always respond to the factors that regulate their expression in whole plants.[186] This fact, together with the difficulty in preparing large numbers of protoplasts from pollen, precludes transient expression studies of pollen-specific genes. The development of the microprojectile bombardment method[175] for the delivery and expression of gene in intact plant cells has overcome some of these limitations.[176-178] This method has received considerable attention and has been used for delivering nucleic acids to intact plant cells,[175] negating the requirement to isolate protoplasts for transformation. As mentioned earlier, tungsten particles (4 μm in diameter) were used to carry 35S CAT or RNA into epidermal tissue of onion.[175] Extracts from epidermal tissue bombarded with microprojectiles showed high levels of CAT activity. In subsequent work,[187] these transient gene expression studies were extended by monitoring GUS expression studies in the rice and wheat cells, followed by detailed studies of CAT gene expression in cells of rice, wheat, and soybean. In studies with soybean, Christou et al.[155] used immature embryos as targets for DNA- coated gold particles. Protoplasts prepared from these embryos and cultured under selective conditions for NPTII gene expression produced kanamycin-resistant calli at a frequency of 1 in 10^5. Subsequently, DNA-coated gold particles were introduced into the meristems of immature soybean seeds.[188] Approximately 2% of the shoots derived from the meristems by organogenesis were chimeric for expression of the introduced gene. R0 and R1 plants expressed both NPTII and GUS genes; one R0 plant expressed both NPTII and GUS activity after transformation with a plasmid carrying both these genes.

Kanamycin-resistant tobacco plants have also been obtained after plant regeneration from cells of leaves and suspension cultures treated with tungesten microprojectiles.[176,177] Other studies, involving an assessment of factors influencing the efficiency of DNA delivery to suspension cells of *Zea mays* by GUS expression, showed that several parameters were important in maximizing transformation, including the nature of the support for the recipient cells in the bombardment apparatus, microprojectile number and velocity, and the concentration of $CaCl_2$ used to absorb DNA to the microprojectiles.[176,177] These authors suggested that particle bombardment may be useful for the stable transformation of monocotyledonous species.

Several independent laboratories have investigated the microprojectile bombardment process and have produced stable transgenic plants. Stable transformation of tobacco plants and suspension cultures[176,177] and of soybean cells and plants using a different system to accelerate gold particles[188] have been reported. The inheritance pattern in soybeans does not appear to be Mendelian[188] and progeny segregation data were not presented for tobacco.[176,177] Twell et al.[182] described the ability of intact pollen to support transient gene expression directed by 5' flanking DNA of the pollen-specific gene LAT52 from tomato using the particle bombardment method. The tissue specificity of gene regulation in this system was investigated by comparison of the activity of the LAT52 and the CaMV 35S promoters in the bombardment leaves and pollen (Figure 9). Transgenic tobacco plants and progeny carrying the coding sequence for NPTII and GUS were recovered following microprojectile bombardment of tobacco leaves.[189] Among 160 putative transgenic plants, 76 expressed NPTII and 50% expressed GUS (Table 1).[189] These genes were found to be transmitted to the next generation (Table 2). Although one transformant demonstrated a Mendelian ratio for seed germination, DNA analysis of the progeny indicated rearrangement as a result of DNA integration.

In plants, organelle transformation has been reported using microprojectile methods. Transformation of three mutants of the chloroplast *apt*B gene of *Chlamydomonas reinhardtii* with tungsten particles coated with cloned chloroplast DNA carrying the wild-type gene restored the photosynthetic capacity of the mutants.[190] Bombardment also helped to provide molecular evidence that different patterns of anthocyanin pigmentation were due to variation in promoter, but not structural coding sequences in the *R* gene family. Intact maize cells penetrated with cloned *R* genes showed promise as convenient visible indicators without the use of added substrates or enzyme assays.

Chapter 1

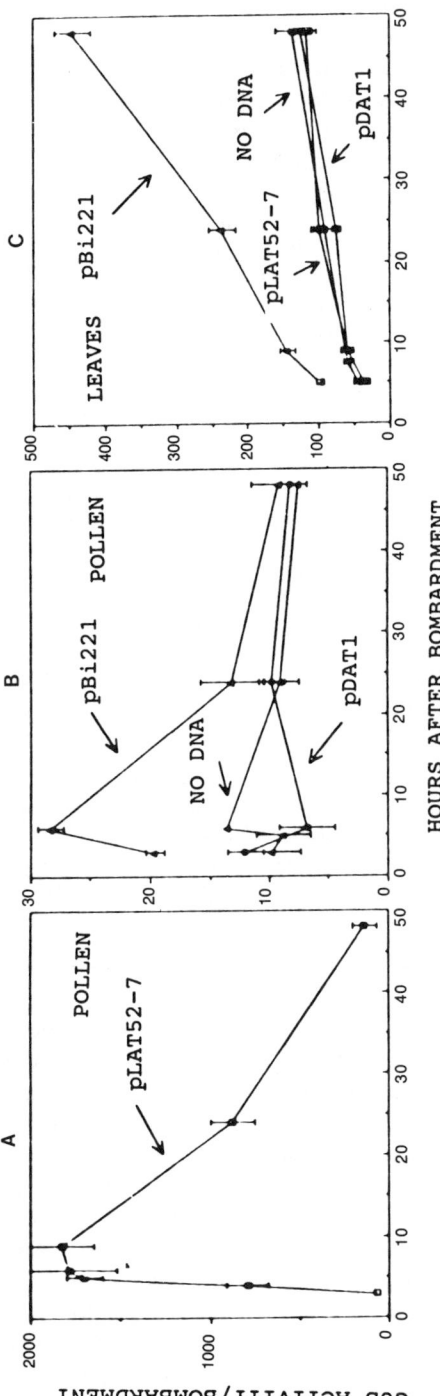

FIGURE 9. Transient expression of GUS activity in pollen and leaves. Approximately 10⁶ mature tobacco grains or individual leaves were bombarded with tungsten microprojectiles coated with plasmid DNA. Total GUS activity (arbitrary fluorescence units) in pollen and leaf extracts is shown per bombardment. The combined results of two separate experiments are shown in which duplicate bombardments were performed for each time point. (A) Bombardment of pollen with pLAT52-7. Bombardment of pollen (B) or leaves (C) with pDAT1, pBI221, or microprojectiles not coated with DNA (NO DNA). (From Twell, D., Klein, T. M., Fromm, M. E., and McCormick, S., *Plant Physiol.*, 91, 1270, 1989. With permission of the American Society of Plant Physiologists.)

TABLE 1
Expression of Neomycin Phosphotransferase (NPTII) and β-Glucuronidase (GUS) Enzymes and/or DNA Analysis by Southern Blot for NPTII or GUS in Plants Regenerated from Tobacco Leaves after Microprojectile Bombardment and Selection against Kanamycin

Expt.	DNA	Cultivar	Colony	GUS[a]	NPTII[a]	Southern
PG2054	pMON9909	KY17	21	2/2	6/7	—
			23	—	6/7	—
			24	13/21	11/21	2/2
			26	8/16	15/17	0/1
			28	2/2	5/13	—
			29	2/10	17/17	1/1
PG2064	pMON9909	KY17	5	11/45	19/26	3/3
			5C2	1/13	13/13	1/1
			21	14/18	10/12	3/3
PG2077	pMON9749	Xanthi	10	0/1	4/5	—
			20	2/3	3/6	1/1
			21	3/3	6/6	—
			28	5/5	5/5	1/1
PG2077	pMON9909	Xanthi	22	1/1	0/2	—
			23	1/1	1/1	1/1
			26	0/1	0/1	—

[a] Regenerated plants were considered positive for NPTII if net CPM > 500 and 2 × negative plant control and positive GUS if 10 × the negative plant control. Each assay contained nontransformed plant controls of either cv. 'Ky 17' or 'Xanthi'. Both assays were normalized for amount of protein in the assay. CPM for NPTII of negative plant controls varied between 0 and 200 CPM while GUS varied between 0 and 10 fluorescent units.

From Tomes, D. T. et al., *Plant Mol. Biol.*, 14, 261, 1990. With permission.

The first stable transformation and regeneration in maize was achieved with the particle gun,[191] using suspensions from friable embryogenic A188 × B73 cultures as targets. Transformed DNA consisted of a supercoiled plasmid with a 35S promoter and the BAR gene as the selectable marker DNA was precipitated onto tungsten particles in an equimolar ratio, and distribution of projectiles was improved by placement of a 100-μm mesh screen between the stopping plate and cells during bombardment. Southern hybridization indicated that about 50% of the selected callus lines arose from independently transformed lines. Surprisingly, selection efficiency increased 2.5-fold with cryopreserved cells, although there was greater variation in the number of transformants per experiment. In two other studies, 96% of the plants regenerated from callus reached maturity, but little viable pollen was

TABLE 2
Progeny Analysis of Seed Germination on 200 µg/ml of Kanamycin from Transgenic Tobacco Plants via Microprojectile Bombardment with pMON9909 or pMON9749

				Seedling number				
Expt.	Colony	Parent	Cross	Green	White	Ratio	χ^2	P^a
PG2054	29	MS0108	Male	2	76	3:1	109.2	0.001
		MS0108	Female	9	2	3:1	0.2	0.750
		MS0108	Self	23	3	15:1	1.3	0.250
PG2077	28	MS0163	Male	40	4	15:1	0.8	0.500
		MS0163	Self	66	4	63:1	4.9	0.050
PG2077	28	MS0165	Self	131	2	63:1	0.2	0.750
PG2077	28	MS0183	Self	293	52	3:1	18.1	0.000
PG2077	28	MS0196	Self	125	20	3:1	9.7	0.000
PG2077	35	MS0144	Self	1	401	Ctrl	Ctrl	Ctrl

[a] $P = \chi^2$ probability with d.f.

From Tomes, D. T. et al., *Plant Mol. Biol.*, 14, 261, 1990. With permission.

produced.[191] Progeny seed was obtained only when regenerated plants were pollinated by embryo rescue. Sixteen percent of regenerated plants reached maturity in separate experiments, and one quarter of those produced an average of less than one seed per original transformed plant. Approximately 50 to 60% of those transformants maintained the BAR gene in the next generation, but presence of the marker gene was not associated with phenotypic abnormalities observed in some plants.

The utility of the biolistic approach for stable transformation in maize was again shown in other experiments.[110] As in the study described earlier, friable embryogenic cultures with similar A188 and B73 parentage served as target cells. Thirty-three bombardments with three different vectors yielded eight calli expressing *luc*, *gus*, or maize *als* genes, but only one callus produced transgenic fertile mature plants positive for both *luc* and *gus* gene activity. During these experiments, no selection was applied during plant regeneration. However, greater numbers of transgenic plants were obtained when a drug was added during the regeneration phase. Fertilization of primary transformants with untransformed pollen and embryo rescue was necessary to obtain viable progeny. In the following generation 44% of the progeny expressed *luc* and *als* genes and all plants were pollen fertile. The particle gun was also used to produce transgenic cotton plants.[192] About 30 stable drug-resistant clones were selected after each bombardment, which represented 0.7% of transient *gus*-expressing cells. Data for the transformed mature plants were not presented.

The combination of a dominant plasmid-specific marker, homologous recombination, and the particle gun was exploited for stable transformation of chloroplasts in tobacco.[193] The vector DNA harbored a mutated 16S rDNA gene that conferred resistance to the nonlethal drug spectinomycin. Bombarded leaves were induced to grow calli under selection, which resulted in a transformation rate of only one resistant clone per 50 leaf bombardments. A majority of resistant lines selected initially suggested that drug-sensitive wild type rDNA was replaced by homologous recombination with transforming tolerant rDNA. Inheritance of drug resistance in genetic crosses was under maternal control as expected.

Recently Vasil et al.[194] have observed stable transformation in callus by direct delivery of DNA into plated suspension cultures of wheat (*Triticum aestivum*) using high velocity microprojectile bombardment. They introduced three different reporter or selectable marker genes jointly present on single or on separate plasmids (*gus,* NPTII, and 5-enol-pyruylshikimate phosphate). However, transgenic plants could not be raised from these calli.

IX. ELECTRIC DISCHARGE PARTICLE ACCELERATION METHOD

This method of gene transfer employs a high voltage discharge delivered to a small water droplet which quickly vaporizes and releases energy to propel DNA-coated gold spheres into target cells. Experiments to obtain transgenic soybean plants were conducted by particle acceleration into excised meristems, which produced shoots by organogensis.[188] Grafting to normal seedlings was required for continued growth of some shoots, but no selection scheme for primary transformants was employed. Most transgenic plants exhibited roots, stems or leaves that were chimeric for GUS expression. From 1845 treated meristems, approximately 2% of regenerated shoots expressed GUS activity. In different experiments nearly 4% of shoots originating from transgenic meristems were scored positive for NPTII genes, but only 0.26% of the regenerated adult plants continued to display the drug-resistant phenotype. Of the surviving plants, 10% maintained NPTII activity in the next generation.

Inheritance of GUS and NPTII transferred into the soybean by electric discharge was further examined in a separate study.[156] Most primary transformants, as expected, were chimeric for GUS. Progeny tests of one chimeric individual indicated non-Mendelian segregation of the marker gene. However, some nonchimeric transformants did inherit GUS and NPTII as single dominant genes. Tight linkage between GUS and NPTII was observed in most plants, even when introduced on independent plasmids. Christou et al.[195] described a simple and effective procedure using electric discharge particle acceleration to introduce DNA-coated gold particles into scuteller tissue of rice with subsequent recovery of transgenic plants from both *indica* and *japonica* varieties.

X. CONCLUSIONS

Numerous laboratories have transformed genes in a wide collection of plants, including not only those considered easy experimental models such as *N. tabacum* and *Petunia*, but also those considered recalcitrant, such as rice, maize, and wheat.[196-224] Now one can choose between several established methods starting from *Agrobacterium* cocultivation to electroporation. One could even use microinjection or the biolistic approach. The *Agrobacterium*-mediated gene transformation method of co-cultivation is still applicable on a larger scale and a list of transgenic plants produced through this method is increasing at a faster rate. Molecular analysis of events leading to transfer of T-DNA into the host plant cells is close to completion. However, there are still difficulties in transforming monocotyledonous plants through this method. Recently, Potrykus[9] has discussed in detail the difficulties encountered while using this method. It was also pointed out that the key problem in *Agrobacterium*-mediated transformation is neither with *Agrobacterium* nor with the host, but rather lies with the availability and accessibility of cells competent for integrative transformation and regeneration.

Direct DNA uptake is used routinely for introducing genetic information into plants, with the merits and limitations of this transformation approach being discussed.[128,129] Chemical procedures and electroporation are used extensively for species in which a protoplasts-to-plant system is available, while injection and biolistic approaches are being assessed for several gene expression studies using protoplasts as recipient cells. The commonly accepted term "electroporation" refers to induction of pores by a high-voltage electric pulse, but the existence of such pores has not yet been demonstrated directly. Gene delivery and expression are influenced by a number of parameters, including plasmid configuration, carrier DNA, the concentration and molecular weight of PEG used, and, in the case of electroporation, the voltage, pulse duration, and buffer solution. The nature of the protoplasts themselves also influences transformation.[129] Although each of these parameters must be optimized for the protoplast-DNA combination under investigation, this approach will continue to be applicable for the transformation of a range of crop species, especially monocotyledons.

However, even though some features of the mechanism of electroporation remain to be established, the extensive amount of indirect and phenomenological evidence for formation of localized structures of high permeability provides a fairly good picture of the mechanism and allows some predictability of the electroporation experiments. The advantages of employing electroporation for direct gene transfer appear to be inexpensive instrumentation, good reproducibility, and avoidance of addition of toxic chemicals.

The major drawbacks of electroporation are the requirement of efficient protoplast isolation and regeneration systems. The virtual absence of reports on transformation of intact plant cells by electroporation tends to indicate that the permeability of the cell wall towards DNA is not increased by electroporation. Apparently, other methods producing holes or cracks in the cell wall must be used for transformation of intact plant cells.

Microprojection and particle acceleration are clearly two "successful" approaches for physical gene transfer. Other techniques have also been investigated that include the "pollen tube pathway," needle injection of tillers,[218] sonication,[114] bombardment using an airgun device,[208] passive inhibition of DNA,[40] and lasers.[217] These alternatives for transformation have demonstrated either transient or stable gene expression, but their reproducibility or potential advantage over microinjection/particle acceleration has yet to be proven.

Stable gene transfer with the biolistics, electric discharge, or microinjection methods has been obtained with reliability in maize, soybean, cotton, and rapeseed. While results are encouraging, there are persistent problems of uneven particle distribution or cell death by bombardment that contribute to low transformation efficiency. Long periods of cell culture may also elicit unwanted phenotypic abnormalities or low pollen viability. Transformed cultured cells or intact meristems can give rise to chimeric plants that may distort the segregation patterns of introduced genes. A related issue for future research is competency of somatic or zygotic embryogenic cells during DNA uptake and integration. These and other challenges must be addressed before particle acceleration or microinjection are accepted as reliable tools for plant biologists. Recent improvements in microprojectile distribution using pressurized helium gas may reduce cell mortality after bombardment and thus increase transformation rates up to 15-fold. Whether this advancement is enough to offset the above-mentioned problems remains to be seen.

REFERENCES

1. **Lal, R. and Lal, S.,** *Crop Improvement Utilizing Biotechnology,* CRC Press, Boca Raton, FL, 1990.
1a. **Brisson, N., Paszkowski, J., Penswick, J. R., Gronenborn, J., Potrykus, I., and Hohn, T.,** Expression of a bacterial gene in plants using a viral vector, *Nature,* 310, 511, 1984.
2. **Paszkowski, J., Pisan, B., Shillito, R. D., Hohn, T., Hohn, B., and Potrykus, I.,** Genetic transformation of *Brassica campestris* var. *rapa* protoplasts with an engineered cauliflower mosaic virus genome, *Plant Mol. Biol.,* 6, 303, 1986.
3. **Tagu, D., Bergousnioux, C., Cretin, C., Perennes, C., and Gadal, P.,** Direct gene transfer in *Petunia hybrida* electroporated protoplasts: evidence for co-transformation with a phosphoenolpyruvate carboxylase cDNA from *Sorghum* leaf, *Protoplasma,* 146, 101, 1988.
4. **Ahlquist, P., French, R., and Bujarski, J. J.,** Molecular studies of Brome mosaic virus using infectious transcripts from cloned cDNA, *Adv. Virus Res.,* 32, 214, 1987.
5. **Ahlquist, P. and Pacha, R. F.,** Gene amplifciation and expression by RNA viruses and potential for further application to plant gene transfer, *Physiol. Plant,* 79, 163, 1990.

6. **Hull, R. and Davies, J. W.**, Genetic engineering with plant viruses and their potential as vectors, *Adv. Virus Res.,* 28, 1, 1983.
7. **French, R., Janda, M., and Alquist, P.**, Bacterial gene inserted in an engineered RNA virus: efficient expression in monocotyledonous plant cells, *Science,* 231, 1294, 1986.
8. **Potrykus, I.**, Gene transfer to plants: assessment and perspectives, *Physiol. Plant,* 79, 125, 1990.
9. **Potrykus, I.**, Gene transfer to plants. Assessment of published approaches and results, *Annu. Rev. Plant Physiol., Plant Mol. Biol.,* 42, 205, 1991.
10. **Chilton, M. D., Drummond, M. H., Merlo, D. J., Sciaky, D., Montoya, A. L., and Gordon, M. P. E. W.**, Stable incorporation of plasmid DNA into higher plant cells: the molecular basis of crown gall tumorigenesis, *Cell,* 11, 263, 1977.
11. **Nester, E. W., Gordon, M. P., Amasino, R. M., and Yanofsky, M. F.**, Crown gall: a molecular and physiological analysis, *Annu. Rev. Plant Physiol.,* 35, 387, 1984.
12. **Davey, M. R., Rech, E. L., and Mulligan, B. J.**, Direct DNA transfer to plant cells, *Plant Mol. Biol.,* 13, 273, 1989.
13. **Fraley, R. T., Rogers, S. G., and Horsch, R. B.**, Genetic transformation in higher plants, *Crit. Rev. Plant Sci.,* 4, 1, 1986.
14. **Fromm, M. and Walbot, V.**, Transient expression of DNA in plant cells, in *Plant Gene Research: Plant DNA Infectious Agents,* Hohn, T. and Schell, J., Eds., Springer, Vienna, 1987, 304.
15. **Gasser, C. S. and Fraley, R. T.**, Genetically engineering plants for crop improvement, *Science,* 244, 1293, 1989.
16. **Klee, H., Horsch, R., and Rogers, S.**, *Agrobacterium*-mediated plant transformation and its further applications in plant biology, *Annu. Rev. Plant Physiol.,* 38, 467, 1987.
17. **Klein, T. M., Roth, B. A., and Fromm, M. E.**, Regulation of anthocyanin biosynthetic genes introduced into intact maize tissues by microprojectiles, *Proc. Natl. Acad. Sci. U.S.A.,* 86, 6681, 1989.
18. **Paszkowski, J., Shillito, R. D., Saul, M., Mandak, V., Hohn, T., Hohn, B., and Potrykus, I.**, Direct gene transfer to plants. *EMBO J.,* 3, 2717, 1984.
19. **Paszkowski, J., Saul, M. W., and Potrykus, I.**, Plant gene vectors and genetic transformation: DNA-mediated direct gene transfer to plants, in *Cell Culture and Somatic Cell Genetics of Plants,* Vol. 6, *Molecular Biology of Plant Nuclear Genes,* Schell, J. and Vasil, I. K., Eds., San Diego, Academic Press, 1989, 52.
20. **Weising, K., Schell, I., and Kahl, G.**, Foreign genes in plants: transfer, structure, expression, and applications, *Annu. Rev. Genet.,* 22, 421, 1988.
21. **Hoekema, A., Hirsch, P. R., Hooykaas, P. J. J., and Schilperoort, R. A.**, A binary plant vector strategy based on separation of vir-and T-region of the *Agrobacterium tumefaciens* Ti plasmid, *Nature,* 303, 179, 1983.
22. **An, G., Watson, B. D., Stachel, S., Gordon, M. P., and Nester, E. W.**, New cloning vehicles for transformation of higher plants, *EMBO J.,* 4, 277, 1985.
23. **Bevan, M.**, Binary *Agrobacterium* vectors for plant transformation, *Nucleic Acids Res.,* 12, 8711, 1984.
24. **Schrammeijer, B., Sijmons, P. C., van den Elzen, J. M., and Hoekema, A.**, Meristem transformation of surface via *Agrobacterium, Plant Cell Rep.,* 9, 551, 1990.
25. **Jouanin, L., Vilane, F., d'Enfert, C., and Casse-Delbart, F.**, Localization and restriction maps of the replication origin regions of the plasmids of *Agrobacterium rhizogene* strain A_4, *Mol. Gen. Genet.,* 201, 370, 1985.
26. **McBride, K. E. and Summerfelt, K. R.**, Improved binary vectors for *Agrobacterium*-mediated transformation, *Plant Mol. Biol.,* 14, 269, 1990.
27. **De Block, M., Botterman, J., Vandewiele, M., Dockx, J., Thoen, C., Gossele, V., Movva, N. R., Thompson, C., Van Montagu, M., and Leemans, J.**, Engineering herbicide resistance in plants by expression of detoxifying enzyme, *EMBO J.,* 6, 2513, 1987.

28. **De Block, M.**, Genotype-independent leaf disc transformation of potato (*Solanum tuberosum*) using *Agrobacterium tumefaciens*, *Theor. Appl. Genet.*, 76, 767, 1988.
29. **De Block, M., De Brouwer, D., and Tenning, P.**, Transformation of *Brassica napus* and *Braassica oleracea* using *Agrobacterium tumefaciens* and the expression of the *bar* and *neo* genes in the transgenic plants, *Plant Physiol.*, 91, 694, 1989.
30. **Murakami, T., Anzai, H., Imai, S., Satoh, A., Nagaoka, K., and Thompson, C. J.**, The bialaphos biosynthetic genes of *Streptomyces hydroscopicus*: molecular cloning and characterization of a gene cluster, *Mol. Gen. Genet.*, 205, 42, 1986.
31. **Muller, A. J., Mendel, R. R., Schiemann, J., Simoens, C., and Inze, D.**, High meiotic stability of a foreign gene introduced into tobacco by *Agrobacterium*-mediated transformation, *Mol. Gen. Genet.*, 207, 171, 1987.
32. **Gorman, C. M., Moffat, L. F., and Howard, B. H.**, Recombinant genomes express chloramphenicol acetyltransferase in mammalian cells, *Mol. Cell Biol.*, 2, 1044, 1982.
33. **Shaw, W. V.**, Chloramphenicol acetyltransferase from chloramphenicol resistant bacteria, *Methods Enzymol.*, 53, 737, 1985.
34. **Sleigh, M. J.**, A nonchromatographic assay for expression of chloramphenicol acetyltransferase gene in eukaryotic cells, *Anal. Biochem.*, 156, 251, 1986.
35. **Burns, D. K. and Erickson, R. L.**, An immunological assay for chloramphenicol acetyltransferase, *Anal. Biochem.*, 162, 399, 1987.
36. **Gendloff, E. H., Bowen, B., and Bucholz, W. G.**, Quantitation of chloramphenicol acetyl transferase in transgenic tobacco plants by ELISA and correlation with gene copy number, *Plant Mol. Biol.*, 14, 575, 1990.
37. **Charest, P. J., Iyer, V. N., and Miki, V. L.**, Factors affecting the use of chloramphenicol acetyltransferase as a marker for *Brassica* genetic transformation, *Plant Cell Rep.*, 7, 297, 1989.
38. **Werr, W. and Lorz, H.**, Transient gene expression in a Gramineae cell line. A rapid procedure for studying plant promoters, *Mol. Gen. Genet.*, 202, 471, 1986.
39. **Junker, B., Zimmy, J., Luhers, R., and Lorz, H.**, Transient expression of chimaeric genes in dividing and non-dividing cereal protoplasts after PEG-induced DNA uptake, *Plant Cell Rep.*, 6, 329, 1987.
40. **Topfer, R., Groneenborn, B., Schafer, S., Schell, J., and Steinbiss, H. H.**, Expression of engineered wheat dwarf virus in seed-derived embryos, *Physiol. Plant*, 79, 158, 1990.
41. **Marcotte, W. R., Bayley, C. C., and Qatrano, R. S.**, Regulation of a wheat promoter by abscisic acid in rice protoplasts, *Nature*, 335, 454, 1988.
42. **Seguin, A. and Lalonde, M.**, Gene transfer by electroporation in Betulaceae protoplasts: *Alnus incana*, *Plant Cell Rep.*, 7, 367, 1988.
43. **Wilson, S. M., Thorpe, T. A., and Moloney, M. M.**, PEG-mediated expression of GUS and CAT genes in protoplasts from embryogenic suspension cultures of *Picea glauca*, *Plant Cell Rep.*, 7, 704, 1989.
44. **Bekkaoui, F., Pilon, M., Lanie, E., Raju, D. S. S., Crosby, W. L., and Dunstan, D. I.**, Transient gene expression in electroporated *Picea glauca* protoplasts, *Plant Cell Rep.*, 7, 481, 1988.
45. **Yang, N. S. and Russell, D.**, Maize sucrose synthase-1 promoter directs phloem cell specific expression of GUS gene in transgenic tobacco plants, *Proc. Natl. Acad. Sci. U.S.A.*, 87, 4144, 1990.
46. **Jefferson, R. A.**, Assaying chimeric genes in plants: the GUS gene fusion system, *Plant Mol. Biol. Rep.*, 5, 387, 1987.
47. **Jefferson, R. A., Kavanagh, T. A., and Bevan, M. W.**, GUS fusions: β-glucurinidase as a sensitive and versatile gene fusion marker in higher plants, *EMBO J.*, 6, 3901, 1987.
48. **Baribault, T. J., Skene, K. G. M., and Scott, N. S.**, Genetic transformation of grapevine cells, *Plant Cell Rep.*, 8, 137, 1989.
49. **Baribault, T. J., Skene, K. G. M., Cain, P. A., and Scott, N. S.**, Transgenic grapevines. Regeneration of shoots expressing β-glucuronidase, *J. Exp. Bot.*, 41, 1045, 1990.

50. Ow, D. W., Wood, K. V., DeLuca, M., DeWet, J. R., Helisinki, D. R., and Howell, S. H., Transient and stable expression of the firefly luciferase gene plant cells and transgenic plants, *Science*, 234, 856, 1986.
51. Gallie, D. R., Lucas, W. J., and Walbot, V., Visualizing mRNA expression in plant protoplasts: factors influencing efficient mRNA uptake and translation, *Plant Cell*, 1, 301, 1989.
52. DeWet, J. R., Wood, K. V., DeLuca, M., Helisinki, D. R., and Subramani, S., Firefly luciferase gene: structure and expression in mammalian cells, *Mol. Cell Biol.*, 7, 725, 1987.
53. DeWet, J. R., Wood, K. V., Helisinki, D. R., and DeLuca, M., Cloning of firefly luciferase cDNA and the expression of active luciferase in *Escherichia coli*, *Proc. Natl. Acad. Sci. U.S.A.*, 82, 7870, 1985.
54. Koncx, C., Olsson, O., Langridge, W. H. R., Schell, J., and Szalay, A. A., Expression and assembly of functional bacterial luciferase in plants, *Proc. Natl. Acad. Sci. U.S.A.*, 84, 131, 1987.
55. Wood, K. V., Lam, Y. A., Seliger, H. H., and McElroy, W. D., Complementary DNA coding lick beetle luciferases can elicit bioluminescence of different colors, *Science*, 244, 700, 1989.
56. Schneider, M., Ow, D. W., and Howell, S. H., The *in vitro* pattern of firefly luciferase expression in transgenic plants, *Plant Mol. Biol.*, 14, 935, 1990.
57. Fang, G. and Grumet, R., *Agrobacterium tumefaciens* mediated transformation and regeneration of muskmelon plants, *Plant Cell Rep.*, 9, 160, 1990.
58. Pythouds, F., Sinkar, V. P., Nester, E. W., and Gordon, N. P., Increased virulence of *Agrobacterium rhizogenes* conferred by the *vir* region of pTiBo542: application to genetic engineering of poplar, *Bio/Technology*, 5, 1323, 1987.
59. McGranahan, G. H., Leslie, C. A., Urastu, S. L., Martin, L. A., and Dandekar, A. M., *Agrobacterium*-mediated transformation of walnut embryos and regeneration of transgenic plants, *Bio/Technology*, 6, 800, 1988.
60. James, D. J., Passey, A. J., Barbara, D. J., and Bevan, M., Genetic transformation of apple (*Malus pumila* Mill.) using a disarmed Ti-binary vector, *Plant Cell Rep.*, 7, 658, 1989.
61. Lorz, H., Baker, B., and Schell, J., Gene transfer to cereal cells mediated by protoplast transformation, *Mol. Gen. Rep.*, 6, 165, 1985.
62. Lorz, H., Baker, B., and Schell, J., Gene transfer to cereal cells mediated by protoplast transformation, *Mol. Gen. Genet.*, 199, 178, 1985.
63. Potrykus, I., Paszkowski, J., Saul, M. W., Petruska, J., and Shillito, R. D., Molecular and general genetics of a hybrid foreign gene introduced into tobacco by direct gene transfer, *Mol. Gen. Genet.*, 199, 169, 1985.
64. Potrykus, I., Saul, M. W., Petrusak, J., Paszkowski, J., and Shillito, R. D., Direct gene transfer to cells of a graminaceous monocot, *Mol. Gen. Genet.*, 199, 183, 1985.
65. Potrykus, I., Shillito, R. S., Saul, M. W., and Paszkowski, J., Direct gene transfer: state of the art and future potential, *Plant Mol. Biol. Rep.*, 3, 117, 1985.
66. Davey, M. R., Cocking, E. C., Freeman, J., Pearce, N., and Tudor, I., Transformation of *Petunia* protoplasts by isolated *Agrobacterium* plasmids, *Plant Sci. Lett.*, 18, 307, 1980.
67. Davey, M. R. and Kumar, A., Higher plant protoplasts, retrospect and prospect, *Int. Rev. Cytol. Suppl.*, 16, 219, 1983.
68. Krens, F. A., Molendijk, L., Wullems, G. J., and Schilperoort, R. A., *In vitro* transformation of plant protoplasts with Ti-plasmid DNA, *Nature*, 296, 72, 1982.
69. Krens, F. A., Mans, R. M. W., van Slogteren, T. M. S., Hoge, J. H. C., Wullems, G. J., and Schilperoort, R. A., Structure and expression of DNA transferred to tobacco via transformation of protoplasts with Ti-plasmid DNA: co-transfer of T-DNA and non T-DNA sequences, *Plant Mol. Biol.*, 5, 223, 1985.

70. **Draper, J.**, Transformation of Plant Protoplasts by *Agrobacterium* and Isolated Ti Plasmid, Ph.D. thesis, University of Nottingham, Nottingham, 1982.
71. **Draper, J., Davey, M. R., Freeman, J. P., and Cocking, E. C.**, Isolation of plasmid DNA from *Agrobacterium* by isopycnic density centrifugation in vertical rotors, *Experientia*, 38, 101, 1982.
72. **Draper, J., Davey, M. R., Freeman, J. P., Cocking, E. C., and Cox, B. J.**, Ti plasmid homologous sequences present in tissues from *Agrobacterium* plasmid-transformed *Petunia* protoplasts, *Plant Cell Physiol.*, 23, 451, 1982.
73. **Peerbolte, R., Krens, F. A., Mans, R. M. W., Floor, M., Hoge, J. H. C., Wullems, G. J., and Schilperoort, R. A.**, Transformation of plant protoplasts with DNA: co-transformation of non-selected calf thymus carrier DNA and meiotic segregation of transforming DNA sequences, *Plant Mol. Biol.*, 5, 235, 1985.
74. **Hasezawa, S., Nagata, T., and Syono, K.**, Transformation of *Vinca* protoplasts mediated by *Agrobacterium* spheroplasts, *Mol. Gen. Genet.*, 182, 206, 1981.
75. **Hain, R., Steinbiss, H. H., and Schell, J.**, Fusion of *Agrobacterium* and *E. coli* spheroplasts with *Nicotiana tabacum* protoplasts-direct gene transfer from microorganism to higher plant, *Plant Cell Rep.*, 3, 60, 1984.
76. **Hain, R., Stabel, P., Czernilofsky, A. P., Steinbiss, H. H., Herrera-Estrella, L., and Schell, J.**, Uptake, integration, expression and genetic transmission of a selectable chimaeric gene by plant protoplasts, *Mol. Gen. Genet.*, 199, 161, 1985.
77. **Tanaka, N., Ikegami, M., Hohn, T., Matsui, C., and Watanabe, I.**, *E. coli* spheroplast-mediated transfer of cauliflower mosaic virus DNA into plant protoplasts, *Mol. Gen. Genet.*, 195, 378, 1984.
78. **Deshayes, A., Harrera-Estrella, L., and Caboche, M.**, Liposome-mediated transformation of tobacco mesophyll protoplasts by an *Escherichia coli* plasmid, *EMBO J.*, 4, 2731, 1985.
79. **Balazas E., Bouzoubaa, S., Guilley, H., Jonard, G., Paskowski, J., and Richards, K.**, Chimaeric vector construction for higher plant transformation, *Gene*, 40, 343, 1987.
80. **Peng, J., Luzmik, L. A., Lee, L., and Holdges, T. K.**, Co-transformation of *indica* rice protoplasts with *gus*A and *neo* genes, *Plant Cell Rep.*, 9, 168, 1990.
81. **Schocher, R. J., Shillito, R. D., Saul, M. W., Paszkowski, J., and Potrykus, I.**, Co-transformation of unlinked foreign genes into plants by direct gene transfer, *Bio/Technology*, 4, 1093, 1986.
82. **Czernilofsky, A. P., Hain, R., Baker, B., and Wirtz, U.**, Studies of the structure and functional organization of foreign DNA integrated into the genome of *Nicotiana tabacum*, *DNA*, 5, 473, 1986.
83. **Czernilofsky, A. P., Hain, R., Harrera-Estrella, L., Lorz, H., Goyvaerts, E., Baker, B. J., and Schell, J.**, Fate of selectable marker DNA integrated into the genome of *Nicotiana tabacum*, *DNA*, 5, 101, 1986.
84. **Koehler, F., Golz, C., Eapen, S., Kohn, H., and Schieder, O.**, Stable transformation of moth bean *Vigna aconitifolia* via direct gene transfer, *Plant Cell Rep.*, 6, 313, 1987.
85. **Koehler, F., Golz, C., Eapen, S., and Schieder, O.**, Influence of plant cultivar and plasmid DNA on transformation rates on tobacco and moth bean, *Plant Science*, 53, 87, 1987.
86. **Neumann, E., Shaefer-Ridder, M., Wang, Y., and Hofschneider, P. H.**, Gene transfer into mouse lyoma cells by electroporation in high electric fields, *EMBO J.*, 1, 841, 1982.
87. **Perani, L., Radke, S., Wilke-Douglas, M., and Bossert, M.**, Gene transfer methods for crop improvement: introduction of foreign DNA into plants, *Physiol. Plant*, 68, 566, 1986.
88. **Zachrisson, A. and Bornman, C. H.**, Electromanipulation of plant protoplasts, *Physiol. Plant.*, 67, 507, 1986.
89. **Felix, H. R.**, Permeabilized cells, *Anal. Biochem.*, 120, 211, 1982.
90. **Lindsey, K. and Jones, M. G. K.**, Transient gene expression in electroporated protoplasts and intact cells of sugar beet, *Plant Mol. Biol.*, 10, 43, 1987.

91. **Lindsey, K. and Jones, M. G. K.**, Stable transformation of sugar beet protoplasts by electroporation, *Plant Cell Rep.*, 8, 71, 1989.
92. **Lindsey, K. and Jones, M. G. K.**, Electroporation of cells, *Physiol. Plant.*, 79, 168, 1990.
93. **Coster, H. G. L. and Zimmermann, U.**, The mechanism of electrical breakdown in the membranes of *Valonia utricularis*, *J. Membrane Biol.*, 22, 73, 1975.
94. **Sugar, I. P. and Neumann, E.**, Stochastic model for electric field-induced membrane pores, *Biophys. Chem.*, 19, 211, 1984.
95. **Dimitrov, D. S.**, Electric field-induced breakdown of lipid bilayers and cell membranes: a thin viscoelastic film model. *J. Membrane Biol.*, 78, 53, 1984.
96. **Ditrov, D. S. and Zhelev, D. V.**, Stability of multilayered thin liquid films and applications to membrane systems, *J. Coll. Interface Sci.*, 99, 324, 1983.
97. **Zimmermann, U. and Benz, R.**, Dependence of the electrical breakdown on the charging time in *Valonia utricularis*, *J. Membrane Biol.*, 53, 33, 1980.
98. **Zimmermann, U., Piwat, G., Pequenux, A., and Gilles, R.**, Electromechanical properties of human erythrocyte membranes: the pressure dependence of potassium permeability, *J. Membrane Biol.*, 54, 103, 1980.
99. **Joersbo, M. and Brunstedt, J.**, Electroporation: mechanism and transient expression, stable transformation and biological effects in plant protoplasts, *Physiol. Plant.*, 80, 1990.
100. **Rathore, K. S. and Goldsworthy, A.**, Electrical control of shoot regeneration in plant tissue cultures, *Bio/Technology*, 3, 1107, 1985.
101. **Gill, R., Mishra, K. P., and Rao, P. S.**, Stimulation of shoot regeneration of *Vigna aconitifolia* by electrical control, *Ann. Bot.*, 60, 399, 1987.
102. **Goldsworthy, A. and Rathore, K. S.**, The electrical control of growth in plant tissue cultures: the polar transport of auxin., *J. Exp. Bot.*, 36, 1134, 1985.
103. **Rech, E. L., Ohchatt, S. J. Chand, P. K., Davey, M. R., Mulligan, J. B., and Power, J. B.**, Electroporation increases DNA synthesis in cultured plant protoplasts, *Bio/Technology*, 6, 1091, 1988.
104. **Rech, E. L., Ohchatt, S. J., Chand, P. K., Power, J. B., and Davey, M. R.**, Electroenhancement of division of plant protoplast-derived cells, *Protoplasma*, 141, 169, 1987.
105. **Takahashi, K., Kaneko, I., Date, M., and Fukuda, E.**, The effects of pulsing electromagnetic fields on DNA synthesis in mammalian cells in culture, *Experimentia*, 42, 185, 1986.
106. **Byus, C. V., Pieper, S. E., and Adey, W. R.**, The effects of low-energy 60 Hz environmental electromagnetic fields upon the growth-related enzyme ornithine decarboxylase, *Carcinogenesis*, 8, 1385, 1987.
107. **Kinosita, K. and Tsong, T. Y.**, Voltage-induced pore formation and hemolysis of human erythrocytes, *Biochim. Biophys. Acta.*, 471, 227, 1977.
108. **Shillito, R. D., Saul, M. W., Paszkowski, J., Muller, M., and Potrykus, I.**, High efficiency direct gene transfer to plants, *Bio/Technology*, 3, 1099, 1985.
109. **Fromm, M. E., Taylor, L. P., and Walbot, V.**, Expression of genes transferred into monocot and dicot plant cells by electroporation, *Proc. Natl. Acad. Sci. U.S.A.*, 82, 5824, 1985.
110. **Fromm, M. E., Morrish, F., Armstrong, C., Williams, R., Thomas, J., and Klein, T. M.**, Inheritance and expression of chimeric genes in the progeny of transgenic maize plants, *Bio/Technology*, 8, 833, 1990.
111. **Fromm, M. E., Taylor, L. P., and Walbot, V.**, Stable transformation of maize after gene transfer by electroporation, *Nature*, 319, 791, 1986.
112. **Hibi, T., Kano, S., Syguira, M., Kazami, T., and Kimura, S.**, High efficiency electotransfection of tobacco mesophyll protoplasts by tobacco mosaic virus RNA, *J. Gen. Virol.*, 67, 2037, 1986.
113. **Potter, H.**, Electroporation in biology: Methods, applications and instrumentation, *Anal. Biochem.*, 174, 361, 1988.

114. **Joersbo, M. and Brunstedt, J.**, Direct gene transfer to plant protoplasts by electroporation by alternating, rectangular and exponentially decaying pulses, *Plant Cell Rep.*, 8, 701, 1990.
115. **Guerche, P., Bellini, C., Le Moullec, J.-M., and Gaboche, M.**, Use of a transient expression assay for the optimization of direct gene transfer into tobacco mesophyll protoplasts by electroporation, *Biochemie*, 69, 621, 1987.
116. **Guerche, P., Charbonnier, M., Jouanin, L., Tourneur, C., Paszkowski, J., and Pelletier, G.**, Direct gene transfer by electroporation in *Brassica napus*, *Plant Sci.*, 52, 111, 1987.
117. **Joersbo, M., Jorgensen, R. B., and Olesen, P.**, Transient electropermeablization of barley (*Hordeum vulgare* L.) microspores to propidium iodide, *Plant Cell Tissue Organ Cult.*, 23, 125, 1990.
118. **Ou-Lee, T.-M., Turgeon, R., and Wu, R.**, Expression of a foreign gene linked to either a plant-virus or a Drosophila promoter after electroporation of protoplasts of rice, wheat, and sorghum, *Proc. Natl. Acad. Sci. U.S.A.*, 83, 6815, 1986.
119. **Riggs, C. D. and Bates, G. W.**, Stable transformation of tobacco by electroporation, *Proc. Natl. Acad. Sci. U.S.A.*, 83, 5602, 1986.
120. **Vasil, V., Hauptmann, R. M., Morrish, F. M., and Vasil, I. K.**, Comparative analysis of free DNA delivery and expression into protoplasts of *Panicum maximum* Jacq. (Guinea grass) by electroporation and polyethelene glycol, *Plant Cell Rep.*, 7, 499, 1988.
121. **Ballas, N. and Loyter, N.**, Transient expression of the plasmid pCaMVCAT in plant protoplasts following transformation with polyethyleneglycol, *Exp. Cell Res.*, 170, 228, 1987.
122. **Langridge, W. H. R., Li, B. J., and Szalay, A. A.**, Electric field mediated stable transformation of carrot protoplasts with naked DNA, *Plant Cell Rep.*, 4, 355, 1985.
123. **Uchimiya, H., Fushimi, T., Hashimoto, H., Harada, H., and Syono, K.**, Expression of a foreign gene in callus derived from DNA-treated protoplasts of rice (*Oryza sativa* L.), *Mol. Gen. Genet.*, 206, 204, 1986.
124. **Olszoski, N. E., Martin, F. B., and Ausbel, F. M.**, Specialized binary vector for plant transformation: expression of the *Arabidopsis thaliana* AHAS gene in *Nicotiana tabacum*, *Nucleic Acids Res.*, 16, 10765, 1988.
125. **Burke, D. T., Carle, G. F., and Olson, M. V.**, Cloning of large segments of exogenous DNA into yeast by means of artificial chromosomes, *Science*, 236, 802, 1987.
126. **Schwartz, D. C. and Cantor, C. R.**, Separation of yeast chromosome-sized DNAs by pulsed field gradient gel electrophoresis, *Cell*, 37, 67, 1984.
127. **Guzman, P. and Ecker, J. R.**, Development of large DNA methods for plants. Molecular cloning of large segments of *Arabidopsis* and carrot DNA into yeast, *Nucleic Acids Res.*, 16, 11091, 1988.
128. **Negrutiu, I., Mouras, A., Horth, M., and Jacobs, M.**, Direct gene transfer to plants, present development and some future perspectives, *Plant Physiol. Biochem.*, 25, 493, 1987.
129. **Negrutiu, I., Shillito, R., Potrykus, I., Biasini, G., and Sala, F.**, Hybrid genes in the analysis of transformation conditions. I. Setting up a simple method for direct gene transfer in plant protoplasts, *Plant Mol. Biol.*, 8, 363, 1987.
130. **Tautorus, T. E., Bekkaoui, F., Pilon, M., Datla, R. S. S., Crosby, W. L., Fowke, L. C., and Dunstan, D. I.**, Factors affecting transient expression in electroporated black spruce (*Picea mariana*) and jack pine (*Pinus banksiana*) protoplasts, *Theor. Appl. Genet.*, 78, 531, 1989.
131. **Wilson, C. M.**, Plant nucleases. I. Separation and purification of two ribonucleases and one nuclease from corn, *Plant Physiol.*, 43, 1332, 1968.
132. **Sawicka, T.**, Membrane-bound nucleotic activity of corn root cells, *Phytochemistry*, 26, 59, 1987.

133. **Boston, R. S., Becwar, M. R., Ryan, R. D., Goldsbrough, P. B., Larkin, B. A., and Hodges, T. K.**, Expression from heterologous promoters in electroporated carrot protoplasts, *Plant Physiol.*, 83, 742, 1987.
134. **Tyagi, S., Spoerlein, B., Tyagi, A. K., Herrmann, R. G., and Koop, H. V.**, PEG and electroporation induced transformation in *Nicotiana tabacum*. Influence of genotype on transformation, *Theor. Appl. Genet.*, 78, 287, 1989.
135. **Huang, Y.-W. and Dennis, J.**, Factors affecting stable transformation of maize protoplasts by electroporation, *Plant Cell Tissue Organ Cult.*, 18, 281, 1989.
136. **Fry, S. C.**, *The Growing Plant Cell Wall: Chemical and Metabolic Analysis*, Longman Scientific and Technical, New York, 1988.
137. **Baron-Epel, O., Charyal, P. K., and Schindler, M.**, Pectins as mediators of wall porosity in soybean cells, *Planta*, 175, 389, 1988.
138. **Meyer, P., Walgenbach, E., Bussmann, K., Hombrecher, G., and Saedler, H.**, Synchronised tobacco protoplasts are efficiently transformed by DNA, *Mol. Gen. Genet.*, 201, 513, 1985.
139. **Okada, K., Nagata, T., and Takebe, I.**, Introduction of functional RNA into plant protoplasts by electroporation, *Plant Cell Physiol.*, 27, 619, 1986.
140. **Okada, K., Nagata, T., and Takebe, I.**, Co-electroporation of rice protoplasts with RNAs of cucumber mosaic and tobacco mosaic viruses, *Plant Cell Rep.*, 7, 333, 1988.
141. **Okada, K., Takebe, I., and Nagata, T.**, Expression and integration of genes introduced into highly synchronised plant protoplasts, *Mol. Gen. Genet.*, 205, 398, 1986.
142. **Koehler, F., Cardon, G., Pohlman, M., Gill, R., and Schieder, O.**, Enhancement of transformation rates in higher plants by low-dose irradiation: are DNA repair systems involved in the incorporation of exogenous DNA into the plant genome?, *Plant Mol. Biol.*, 12, 189, 1989.
143. **Hauptmann, R. M., Ozias-Akins, P., Vasil, V., Tabaeizadeh, Z., Rogers, S. G., Horsch, R. B., Vasil, I. K., and Fralay, R. T.**, Transient expression of electroporated DNA in monocotyledonous species, *Plant Cell Rep.*, 6, 265, 1987.
144. **Oard, J. H., Paige, D., and Dvorak, J.**, Chimeric gene expression using maize intron in cultured cells of bread-wheat, *Plant Cell Rep.*, 8, 156, 1989.
145. **Tsukada, M., Kusano, T. T., and Kitagawa, Y.**, Introduction of foreign genes into tomato protoplasts by electroporation, *Plant Cell Physiol.*, 30, 599, 1989.
146. **Jones, H., Ooms, G., and Jones, M. G. K.**, Transient gene expression in electroporated *Solanum* protoplasts, *Plant Mol. Biol.*, 13, 503, 1989.
147. **Topfer, R., Gronenborn, B., Schell, J., and Steinbiss, H. H.**, Uptake and transient expression of chimeric genes in seed-derived embryos, *Plant Cell*, 1, 133, 1989.
148. **Topfer, F., Prols, M., Schell, J., and Steinbiss, H. H.**, Transient gene expression in tobacco protoplasts: II. Comparison of the reporter gene systems for CAT, NPTII, and GUS, *Plant Cell Rep.*, 7, 225, 1988.
149. **Ecker, J. R. and Davis, R. W.**, Inhibition of gene expression in plant cell by expression of antisense RNA, *Proc. Natl. Acad. Sci. U.S.A.*, 83, 5372, 1986.
150. **Horn, M. E., Shillito, R. D., Conger, B. V., and Harms, C. T.**, Transgenic plants of orchardgrass (*Dactylis glomerata* L.) from protoplasts, *Plant Cell Rep.*, 7, 469, 1988.
151. **Chupeau, M.-C., Bellini, C., Guerche, P., Maisonneuve, B., Vastra, G., and Chupeau, Y.**, Transgenic plants of lettuce (*Lactuca sativa*) obtained through electroporation of protoplasts, *Bio/Technology*, 7, 503, 1989.
152. **Toriyama, K., Arimoto, Y., Uchimiya, H., and Hinata, K.**, Transgenic rice plants after direct gene transfer into protoplasts, *Bio/Technology*, 6, 1072, 1988.
153. **Shimamoto, K., Terada, R., Izawa, T., and Fujimoto, H.**, Fertile transgenic rice plants regenerated from transformed protoplasts, *Nature*, 338, 274, 1989.
154. **Christou, P., Murphy, J. E., and Swain, W. F.**, Stable transformation of soybean by electroporation and root formation from transformed callus, *Proc. Natl. Acad. Sci. U.S.A.*, 84, 3962, 1987.

155. **Christou, P., McCabe, D. E., and Swain, W. F.,** Stable transformation of soybean callus by DNA coated gold particles, *Plant Physiol.,* 87, 671, 1988.
156. **Christou, P., Swain, W. F., Yang, N., and McCabe, D. E.,** Inheritance and expression of foreign genes in transgenic soybean plants, *Proc. Natl. Acad. Sci. U.S.A.,* 86, 7600, 1989.
157. **Uchimya, H., Hirochika, H., Hashimoto, H., Hara, A., Masuda, T., Kasuimimoto, T., Harada, H., Ikeda, J. E., and Yoshicka, M.,** Co-expression and inheritance of foreign genes in transformants obtained by direct DNA transformation of tobacco protoplasts, *Mol. Gen. Genet.,* 205, 1, 1986.
158. **Yang, H., Zhang, H. M., Davey, M. R., Mulligan, B. J., and Cocking, E. C.,** Production of kanamycin-resistant rice tissues following DNA uptake into protoplasts, *Plant Cell Rep.,* 7, 421, 1988.
159. **Chen, W. H., Gartland, K. M. A., Davey, M. R., Sotak, R., Gartland, J. S., Mulligan, B. J., Power, J. P., and Coking, E. C.,** Transformation of sugarcane protoplasts by direct uptake of a selectable chimaeric gene, *Plant Cell Rep.,* 6, 297, 1987.
160. **Rhodes, C. A., Pierce, D. A., Mettler, I. J., Mascarenhas, D., and Detmer, J. J.,** Genetically transformed maize plants from protoplasts, *Science,* 240, 204, 1988.
161. **Prioli, L. and Saondahl, M. R.,** Plant regeneration recovery of fertile plants from protoplasts of maize (*Zea mays* L.), *Bio/Technology,* 7, 589, 1989.
162. **Vasil, I. K.,** Developing cell and tissue culture for the improvement of cereal and grass crops, *J. Plant Physiol.,* 128, 193, 1987.
163. **Zhang, W. and Wu, R.,** Efficient regeneration of transgenic plants from rice protoplasts and correctly regulated expression of the foreign gene in the plants, *Theor. Appl. Genet.,* 76, 835, 1988.
164. **Zhang, H. M., Yang, H., Rech, E. L., Golds, T. J., Davis, A. S., Mulligan, B. J., Cocking, E. C., and Davey, M. R.,** Transgenic rice plants produced from electroporation-mediated plasmid uptake into protoplasts, *Plant Cell Rep.,* 7, 379, 1988.
165. **Miki, L. A., Reich, T. J., and Iyer, V. N.,** Microinjection: an experimental tool for studying and modifying plant cells, in *Plant Gene Research: Plant DNA Infectious Agents,* Hohn, T. and Schell, J., Eds., Springer, Vienna, 1987, 249.
166. **Neuhaus, G. and Spangenberg, G.,** Plant transformation by microinjection techniques, *Physiol. Plant.,* 79, 213, 1990.
167. **Neuhaus, G., Spangenberg, G., Scheid, O. M., and Schweiger, H. G.,** Transgenic rape seed plants obtained by microinjection of DNA into microspore-derived proembryoids, *Theor. Appl. Genet.,* 75, 30, 1987.
168. **Crossway, A., Oakes, J. V., Irvine, J. M., Ward, B., Knauf, B. C., and Shewmaker, K.,** Integration of foreign DNA following micro-injection of tobacco mesophyll protoplasts, *Mol. Gen. Genet.,* 202, 179, 1986.
169. **Crossway, A., Hauptili, H., Houck, C., Irvine, J., Oakes, J., and Perani, L.,** Micromanipulation techniques in plant biotechnology. Biotechniques in plant biotechnology, *Biotechniques,* 4, 320, 1986.
170. **Neuhaus, G. and Schweiger, H. G.,** Expression of foreign genes in higher plants after electrofusion-mediated cell reconstruction of a microprojected karyoplast and a cytoplast, *Eur. J. Cell. Biol.,* 42, 236, 1986.
171. **Toyodo, H., Matsuda, Y., Utsumi, R., and Ouchi, S.,** Internuclear microprojection for transformation of tomato callus cells, *Plant Cell Rep.,* 7, 292, 1988.
172. **Greisbach, R. J.,** Protoplasts microinjection, *Plant Mol. Biol. Rep.,* 1, 32, 1983.
173. **Lawrence, W. A. and Davies, D. R.,** A method for microinjection and culture of protoplasts at very low densities, *Plant Cell Rep.,* 4, 33, 1985.
174. **Klein, T. M., Gradziel, T., Fromm, M. E., and Sanford, J. C.,** Factors influencing gene delivery into *Zea mays* by high-velocity microprojectiles, *Bio/Technology,* 6, 559, 1988.
175. **Klein, T. M., Wolf, E. D., Wu, R., and Sanford, J. C.,** High velocity microprojectiles for delivering nucleic acids into living cells, *Nature,* 327, 70, 1987.

176. **Klein, T. M., Fromm, M. E., Weissinger, A., Tomes, D., Schaaf, S., Sletten, M., and Sanford, J. C.**, Transfer of foreign genes into intact maize cells using high velocity microprojectiles, *Proc. Natl. Acad. Sci. U.S.A.*, 85, 4305, 1988.
177. **Klein, T. M., Harper, E. C., Svab, Z., Sanford, J. C., Fromm, M. E., and Maliga, P.**, Stable genetic transformation of intact *Nicotiana* cells by the particle bombardment process, *Proc. Natl. Acad. Sci. U.S.A.*, 85, 8502, 1988.
178. **Klein, T. M., Roth, B. A., and Fromm, E. M.**, Advances in direct gene transfer into cereals, in *Genetic Engineering, Principles and Methods II*, Setlov, J. K., Ed., Plenum, New York, 1989, 13.
179. **Stephen, A. G., Klein, T. M., Roth, B. A., Fromm, M. E., Cone, K. C., Radicells, J. P., and Chandler, V. L.**, Transactivation of anthocyanin biosynthetic genes following transfer of regulatory genes into maize tissues, *EMBO J.*, 9, 2517, 1990.
180. **Nash, J., Luehrsen, K. R., and Wallbot, V.**, Bronze-2 gene of maize: reconstruction of wild-type allele and analysis of transcription and splicing, *Plant Cell*, 2, 1039, 1990.
181. **Hanson, D. D., Hamilton, D. A., Travis, J. L., Bashe, D. M., and Mascarenhas, J. P.**, Characterization of a pollen-specific cDNA clone from *Zea mays* and its expression, *Plant Cell*, 1, 173, 1989.
182. **Twell, D., Klein, T. M., Fromm, M. E., and McCormick, S.**, Transient expression of chimeric genes delivered into pollen by microprojectile bombardment, *Plant Physiol.*, 91, 1270, 1989.
183. **Twell, D., Wing, R. A., Yamaguchi, J., and MacCormick, S.**, Isolation and expression of an anther-specific gene from tomato. *Mol. Gen. Genet.*, 217, 240, 1989.
184. **Schell, J.**, Transgenic plants as tools to study the molecular organization of plant genes, *Science*, 237, 1176, 1987.
185. **Callis, J., Fromm, M. V., and Walbot, V.**, Introns increase gene expression in cultured maize cells, *Genes Dev.*, 1, 1183, 1987.
186. **Quail, P. H., Christensen, A. H., Jones, A. M., Lissemore, J. L., Parks, B. M., and Sharrock, R. A.**, The phytochrome molecule and the regulation of its genes, in *Integration and Control of Metabolic Processes: Pure and Applied Aspects*, Kon, O.L., Ed., ISCU Press and Cambridge University Press, Cambridge, U.K., 1987, 41.
187. **Wang, Y.-C., Klein, T. M., Fromm, M., Cao, J., Sanford, J. C., and Wu, R.**, Transient expression of foreign gene in rice, wheat and soybean cells following particle bombardment, *Plant Mol. Biol.*, 11, 433, 1988.
188. **McCabe, D. E., Swain, W. F., Martinell, B. J., and Christou, P.**, Stable transformation of soybean (*Glycin max*) by particle acceleration, *Bio/Technology*, 6, 923, 1988.
189. **Tomes, D., Weissinger, A. K., Ross, M., Higgins, R., Drimmond, B. J., Schaaf, S., Malone-Schomesberg, J., Staebell, M., Flynn, P., Anderson, J., and Howard, J.**, Transgenic tobacco plants and their progeny by microprojectile bombardment of tobacco leaves, *Plant Mol. Biol.*, 14, 261, 1990.
190. **Boynton, J. E., Gillham, N. W., Harris, E. H., Hosler, J. P., Johnson, A. M., Jones, A. R., Randolph-Anderson, B. L., Robertson, D., Klein, T. M., Shark, K. B., and Sanford, J. C.**, Chloroplast transformation in *Chlamydomonas* with high velocity microprojectiles, *Science*, 240, 1534, 1989.
191. **Gordon-Kamm, W. M., Spencer, T. M., Mangano, M. L., Adams, T. R., Daines, R. J., Start, W. G., O'Brien, J. V., Chambers, S. A., Adams, W. R., Willets, N. G., Rice, T. B., Mackey, C. J., Kreuger, R. W., Kausch, A. P., and Lemaux, P. G.**, Transformation of maize cells and regeneration of fertile transgenic plants, *Plant Cell*, 2, 603, 1990.
192. **Finer, J. J. and McMullen, M. D.**, Transformation of cotton via particle bombardment, *Plant Cell Rep.*, 8, 586, 1990.
193. **Svab, Z., Hajdukiewcz, P., and Maliga, P.**, Stable transformation of plastids in higher plants, *Proc. Natl. Acad. Sci. U.S.A.*, 87, 8526, 1990.
194. **Vasil, V., Brown, S. M., Re, D., Fromm, M. E., and Vasil, I. K.**, Stably transformed callus lines from microprojectile bombardment of cell suspension culture of wheat, *Bio/Technology*, 9, 743, 1991.

195. **Christou, P., Ford, T. L., and Kofron, M.**, Production of transgenic rice (*Oryza sativa*) plants from agronomically important indica and japonica varieties via electric discharge particle acceleration of exogenous DNA into immature zygotic embryos, *Bio/Technology*, 9, 957, 1991.
196. **Bruce, W. B.**, Photoregulation of a phytochrome gene promoter from coat transferred into rice by particle bombardment, *Proc. Natl. Acad. Sci. U.S.A.*, 86, 9692, 1990.
197. **Damm, B., Schmidt, R., and Willmitzer, L.**, Efficient transformation of *Arabidopsis thaliana* using direct gene transfer to protoplasts, *Mol. Gen. Genet.*, 217, 6, 1989.
198. **Fearson, E. M.**, Beta-glucuronidase. *Adv. Enzymol.*, 16, 361, 1955.
199. **Hookyas, P. J. J.**, Transformation of plant cells via *Agrobacterium*, *Plant Mol. Biol.*, 13, 327, 1989.
200. **Ludwig, S. R., Bowem, B., Beach, L., and Wessler, S. R.**, A regulatory gene as a novel visible marker for maize transformation, *Science*, 247, 449, 1990.
201. **Luo, Z.-X. and Wu, R.**, A simple method for the transformation of rice via the pollen-tube pathway, *Plant Mol. Biol. Rep.*, 6, 165, 1988.
202. **Ledoux, L., Huart, R., and Jacobs, M.**, DNA-mediated genetic correction of thiaminless *Arabidopsis thaliana*, *Nature*, 249, 17, 1974.
203. **Luo, Z. and Wu, R.**, A simple method for the transformation of rice via pollen-tube pathway, *Plant Mol. Biol. Rep.*, 7, 69, 1989.
204. **Mouras, A., Saul, M. W., Essad, S., and Potrykus, I.**, Localization *in situ* hybridisation of a low copy chimaeric resistance gene introduced into plants by direct gene transfer, *Mol. Gen. Genet.*, 207, 204, 1987.
205. **Murray, E. E., Buchholz, W. G., Bowen, B., and Direct, B.**, Direct analysis of RNA transcripts in electroporated carrot protoplasts, *Plant Cell Rep.*, 9, 129, 1990.
206. **Nishiguchi, M., Langiridge, W. H. R., Szalay, A. A., and Zaitlin, M.**, Electroporation-mediated infection of tobacco leaf protoplasts with tobacco mosaic virus RNA and cucumber mosaic RNA, *Plant Cell Rep.*, 5, 57, 1986.
207. **Nishiguchi, M., Sato, T., and Motoyoshi, F.**, An improved method for electroporation in plant protoplasts: infection of tobacco protoplasts by tobacco mosaic virus particles, *Plant Cell Rep.*, 6, 90, 1987.
208. **Oard, J. H., Paige, D. F., Simmonds, J. A., and Gradziel, T. M.**, Transient gene expression of maize, rice, and wheat cells using an airgun apparatus, *Plant Physiol.*, 92, 334, 1990.
209. **Ooms, G., Burrel, M. M., Karp, A., Bevan, M., and Hille, J.**, Genetic transformation in two potato cultivars with T-DNA from disarmed *Agrobacterium*, *Theor. Appl. Genet.*, 73, 744, 1987.
210. **Pietrzak, Z. M., Shillito, R. D., Hohn, T., and Potrykus, I.**, Expression in plants of two bacterial antibiotic resistance genes after protoplast transformation with a new plant expression vector, *Nucleic Acids Res.*, 14, 5857, 1986.
211. **Prols, M., Topfer, R., Schell, J., and Steinbiss, H. H.**, Transient gene expression in tobacco protoplasts: time course of CAT appearance, *Plant Cell Rep.*, 7, 221, 1988.
212. **Saghai-Maroof, M. A., Soliman, K. M., Jorengsen, R. A., and Allard, R. W.**, Ribosomal DNA spacer-length polymorphism in barley: Mendelian inheritance, chromosomal location, and population dynamics, *Proc. Natl. Acad. Sci. U.S.A.*, 81, 8014, 1984.
213. **Sanford, J. C.**, The biolistic process, *Trends Biotechnol.*, 6, 299, 1988.
214. **Fillati, J. J., Sellemer, J., McCown, B., Hassig, B., and Comai, L.**, *Agrobacterium* mediated transformation and regeneration of *Populus*, *Mol. Gen. Genet.*, 206, 192, 1987.
215. **Spangenberg, G., Koop, H.-U., Lichter, R., and Schweiger, H. C.**, Microculture of single protoplasts of *Brassica napus*, *Physiol. Plant.*, 66, 1, 1986.
216. **Stiekema, W. J., Heidekamp, F., Louwerse, J. D., Verhoeven, H. A., and Dijkhuis, P.**, Introduction of foreign genes into potato cultivars 'Bintje' and 'Desiree' using an *Agrobacterium tumefaciens* binary vector, *Plant Cell Rep.*, 7, 47, 1988.

217. **Weber, G., Monajembashi, S., Greulisch, K. O., and Wolfrum, J.**, Injection of DNA into plant cells with a UV laser microbeam, *Naturewiss,* 75, 35, 1988.
218. **de la Pena, A., Lorz, H., and Schell, J.**, Transgenic rye plants obtained by injecting DNA into young floral tillers, *Nature,* 325, 274, 1987.
219. **Duan, X. and Chen, S.**, Variation of the characters in rice (*Oryza sativa*) induced by foreign DNA uptake, *China Agric. Sci.,* 3, 6, 1985.
220. **Feeman, J. P., Draper, J., Davey, M. R., Cocking, E. C., Gartland, K. M. A., Harding, K., and Pental, D.**, A comparison of methods for plasmid delivery into plant protoplasts, *Plant Cell Physiol.,* 25, 1353, 1984.
221. **Joersbo, M. and Brunstedt, J.**, Direct transfer to plant protoplasts by mild sonication, *Plant Cell Rep.,* 9, 207, 1990.
222. **Reich, T. J., Iyer, V. N., and Miki, B. L.**, Efficient transformation of alfalfa protoplasts by the intranuclear microinjection of Ti plasmids. *Bio/Technology,* 4, 1001, 1986.
223. **Sherman, S. and Bevan, M. W.**, A rapid transformation method for *Solanum tuberosum* using binary *Agrobacterium tumefaciens* vectors, *Plant Cell Rep.,* 7, 13, 1988.
224. **Vasil, I. K.**, Developing cell and tissue culture systems for the improvement of cereal and grass crops, *J. Plant Physiol.,* 128, 193, 1987.

Chapter 2

GENETIC ENGINEERING OF INSECT CONTROL AGENTS AND PRODUCTION OF INSECT-RESISTANT PLANTS

I. INTRODUCTION

Recent advances in recombinant DNA technology have made it possible to develop two types of insect control agents following introduction of genes into plants: (1) proteinase inhibitors, a class of proteins which, when present at relatively high levels in the diet are effective against certain insects, and; (2) endotoxins from *Bacillus thuringiensis* (Bt). In addition, the combination of Bt endotoxins with proteinase inhibitors is also being considered.

II. PROTEINASE INHIBITORS

Natural defense mechanisms against herbivorous insects appear to have evolved in many plants. One such mechanism is the synthesis of proteinase inhibitors.[1-3] In contrast to the Bt endotoxins, these proteins have antimetabolic activity against a wide range of insects. These protein inhibitors are present in the tissues of some plants at relatively high concentrations where they participate in a complex defense response with other molecules produced by the plant. The introduction of such specific proteinase inhibitors into plants is an alternative approach for obtaining crops that are resistant to insect attack.

Serine, thiol, metallo-, and aspartyl-proteinase are four classes of serine proteinase inhibitors and are further divided into 13 families. However, they all inhibit serine proteinases using the same competitive mechanism. On the other hand, the presence of trypsin inhibitor in the diet of insects reduces the effective concentration of trypsin available for digestion. A series of events including the inability to obtain amino acids from ingested food[3] is triggered as a result of reduction in trypsin concentration. Plants containing high levels of serine proteinase inhibitors can still be consumed by these insects. Colorado potato beetles, for example, readily consume potato tubers which contain serine proteinase inhibitors. Certain insects use thiol proteinases as primary digestive enzymes. Thiol (or cysteine) proteinase inhibitors have a major role in the plant-defense mechanism. Transformation of plants lacking thiol proteinase inhibitors with genes encoding these proteins could protect them against such insects.

It is advantageous to use proteinase inhibitors as insect-control agents because they are active against a wide range of insects, and they also act as insect control agents for insects that are resistant to the Bt endotoxin. In addition, they are inactivated during cooking, and are not consumed through food by human beings and animals. However, high levels of protein inhibitors are required for insect killing, and there is potential need to regulate expression of such inhibitors to specific plant organs.

In spite of all these limitations, the proteinase inhibitor proteins and their genes have received much attention during recent years. When leaves of, for example, potato or tomato are damaged by insect attack or by other forms of mechanical damage such as crushing (which perhaps mimics the chewing action of insects), serine proteinase inhibitors accumulate both in the damaged leaves (the local reaction) and also in undamaged leaves at a distance from the original wound site (systemic reaction). In nonwounded plants these proteins are generally restricted to storage tissues, such as seeds (the chymotrypsin inhibitors of barley) or tubers (patatin of potato), and in this situation their physiological role is not clearly understood. However, they are believed to play a role in defense against insects by virtue of their inhibitory action against proteases of insect, but not of plant origin and may interfere with the insect digestive process.

The systemic reaction requires that a signal be transmitted from the site of local damage to the more distant leaves and such a signal or "wound hormone" has been called the "proteinase inhibitor inducing factor", or PIIF. It has been demonstrated that pectinaceous plant cell wall fragments, or "oligosaccharins", can influence various aspects of plant metabolism and development, and fragments released by endopolygalacturonase activity can induce the accumulation of proteinase inhibitors. It is possible, therefore, that oligosaccharins released by mechanical damage may act as the signal to induce the systemic reaction, although the transmissibility of such fragments through plant tissues has yet to be demonstrated.

A very common proteinase inhibitor enzyme, known as inhibitor 1 (monomer M_r 8300) is one of the two serine proteinase inhibitors that accumulate in leaves of tomato and potato plants in response to wounding by predator attacks or by other mechanical damage.[4] The inhibitor is thought to be one of the inducible defensive proteins produced by plants.[1]

A wound-inducible inhibitor I gene has been isolated from tomato[5] and its biochemistry and regulatory mechanisms have been studied.[2,6] Genes containing the cauliflower mosaic virus 35S promoter fused to open reading frames coding for tomato proteinase inhibitor I, tomato inhibitor II, and potato inhibitor II were expressed in transgenic tobacco plants. Growth of *Manduca sexta* larvae (tobacco hornworms) feeding on leaves of transgenic plants containing inhibitor II, a powerful inhibitor of both trypsin and chymotrypsin, was significantly retarded, compared to growth of larvae fed on untransformed leaves. The presence of tomato inhibitor I protein, a potent inhibitor of chymotrypsin but a weak inhibitor of trypsin, in transgenic tobacco leaves had little effect on the growth of the larvae. Ryan and co-workers[7,8] incorporated stably a wound-inducible proteinase inhibitor I gene from tomato containing 725 bp of the 5' region and 2.5 kb of the 3' region into the genome of black

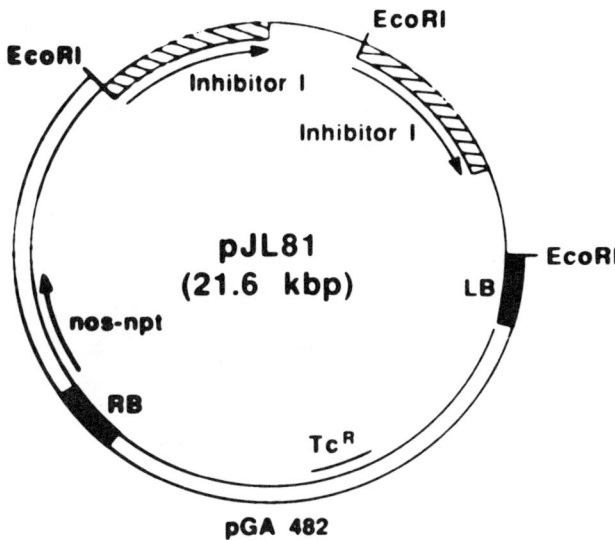

FIGURE 1. Plasmid map of pJL 81. Hatched regions are inhibitor I. RB, T-DNA right border; LB, T-DNA left border; *nos-npt*, a chimeric *nos-npt* fusion serves as a selectable marker in plants; Tc^R, tetracycline resistance. (From Johnson, R., Lee, J. S., and Ryan, C., *Plant Mol. Biol.*, 14, 349, 1990. With permission.)

nightshade plants (*Solanum nigrum*) using an *Agrobacterium* Ti plasmid-derived vector (Figures 1 and 2).[8] The tomato PI-1 gene was expressed in wounded leaves of transgenic *Solanum nigrum* (Figure 3). The protein inhibitor 1 in *S. nigrum* was identical to the native tomato inhibitor I in its immunological reactivity and in its inhibitory activity against chymotrypsin. The protein also exhibited the same M_r of 8 kDa as the native tomato inhibitor I, and its N-terminal amino acid sequence was identical to that of native tomato. This indicates that the factors that regulate both tissue specificity and wound induction of this gene are present in both these solanaceous species. However, unlike its natural environment in tomato leaves, the gene was expressed in unknown *Solanum nigrum* leaves at moderate levels and the expression was increased by a factor of the two. The PI-I expressed in *Solanum nigrum* was identical in every respect to tomato inhibitor I.

In the nonwounded potato plant the protein inhibitor II is localized to the tuber, but wounding of leaf tissue leads to its synthesis and accumulation in aerial parts. In addition, proteinase inhibitor II gene (PI-II) of potato has been cloned and expressed in transgenic tobacco plants.[9-12] This genomic clone was shown to be active in a systemic, wound-inducible manner in transgenic tobacco plants.[13] Sanchez-Serrano et al.[13] have also used *Agrobacterium tumefaciens* to transfer the PI-II gene to tobacco plants, which do not normally possess homologous DNA sequences, to study the aspects of regulation in transgenic plants. Using a cDNA of the gene as probe for PI-II mRNA, it

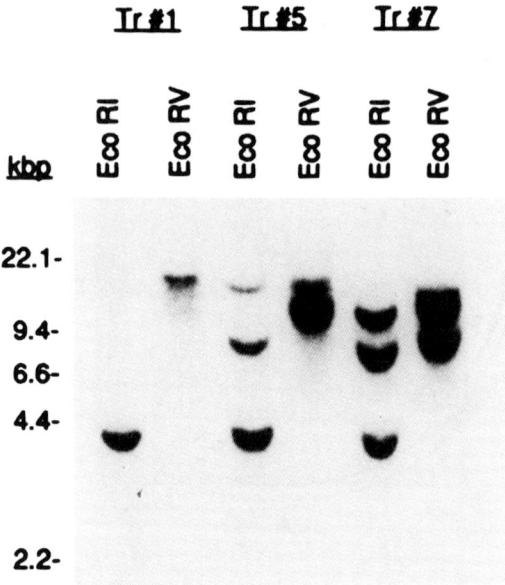

FIGURE 2. Southern blot analysis of three nightshade plants transformed with pJL81. 10 µg of genomic DNA from transformants 1, 5, or 7 was digested with *Eco*RI or *Eco*RV. After electrophoretic separation, the DNA was transferred to nitrocellulose and hybridized with an inhibitor I probe. (From Johnson, R., Lee, J. S., and Ryan, C., *Plant Mol. Biol.*, 14, 349, 1990. With permission.)

was found that, whereas little or no expression of the gene was detected in nonwounded leaves, high levels of the mRNA were detectable in wounded leaves. The level of expression was as high in some of the transgenic tobacco plants as in wounded leaves of potato plants, and the gene was transcribed from the same initiation site in both species. This indicated that, not only the general regulatory mechanisms, but also mRNA processing, were the same in both species. Furthermore, the systemic induction of the gene was demonstrated in the tobacco plants, and induction could also be achieved using the "artificial" oligosaccharins chitsan and polygalacturonic acid instead of wound-induced damage.

Thornburg et al.[14] have also used transgenic tobacco plants to study the regulation of the PI-II gene. A 1000-base pair restriction fragment of the 5' flanking region of the gene was linked to the chloramphenicol acetyltransferase (CAT) gene-coding region and to one of the two terminal sequences, either from the PI-II gene or from a Ti plasmid gene. It was found that the 5' flanking region of PI-II possessed a sequence that determined both the local and systemic expression of the *cat* gene in response to mechanical damage, but only if the PI-II terminator was present. This result shows that the 3' flanking region as well as the 5' regions are essential in determining the tissue and environmentally induced regulation of the gene. Interestingly,

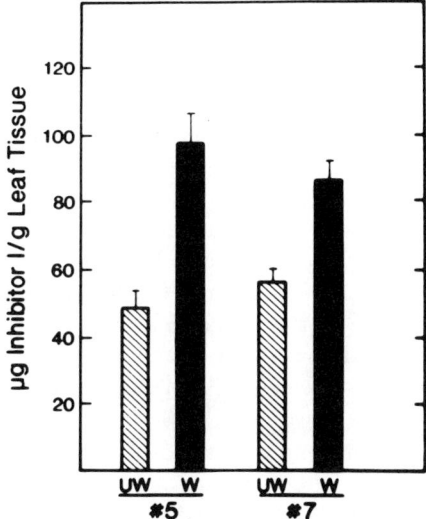

FIGURE 3. Expression of tomato inhibitor I (+SE) in unwounded (UW) and wounded (W) leaves of transgenic nightshade plants. (From Johnson, R., Lee, J. S., and Ryan, C., *Plant Mol. Biol.*, 14, 349, 1990. With permission.)

the 3' region of the PI-II bears some homology to that of the extensin gene, which is also induced by damage. The time-course CAT expression after wounding was also similar to the time-course of proteinase inhibitor accumulation in wounded tomato plants.

The mechanisms of defense reactions as a result of insect attack or wounding are not completely understood. In response to insect attack, tomatoes and potatoes activate quiescent genes, which leads to the accumulation of at least two small protein families. The activation of these genes is complex and incompletely understood. Initially, wounding disrupts cell and specific enzymes, and polyaglacturonases are released to the extracellular millieu. These polygalacturonases are then made free to act on their substrates, the plant cell walls to release small molecular weight oligosaccharides.[15-17] These oligosaccharides have been termed PIIF. They are believed to be the signals which interact with unwounded tissue in an undetermined manner to induce *de novo* synthesis of proteinase inhibitor mRNA and protein.[18]

These genes have been used in whole plants to express the native inhibitors[13] or foreign proteins[14] under wound-inducible control to understand the mechanism of proteinase inhibitor I (*pm2*) gene activation. Thornberg et al.[19] also revealed that the proteinase inhibitor II promoter directs gene expression in those portions of the leaf which are consumed preferentially by insects. The use of such promoters to direct chimeric gene expression in tissue preferentially consumed may have significant utility in engineeering plants to produce novel gene products, thus facilitating an integrated management program.

Several wound-inducible plant genes have been isolated and the activity of their respective promoters analyzed in transgenic plants[13,20,21] but deletion analysis of these promoter regions has not been carried out. The wound-inducible genes analyzed so far display different expression patterns with respect to the timing of induction and are tissue or organ specific. Thus the proteinase inhibitor II genes of potato are induced in a systemic fashion in both wounded and nonwounded aerial parts of the potato plant.[22] The induction is initiated in parenchymal cells close to vascular tissues and later extends to vascular tissue and further to distal tissues.[10] In contrast, wound induction of the TR2' promoter[21] as well as the promoter of the gene encoding the glycine-rich cell wall protein are in the immediate vicinity of the wounding site. Expression reaches a maximum in 30 to 45 min after wounding in tobacco, whereas in the case of potato proteinase inhibitor II genes, maximal expression is observed after 24 h.

A chimeric gene consisting of 1.3 kb of the 5' regulatory region, the coding region of the bacterial beta-glucuronidase (GUS) gene, and 260 bp of the proteinase inhibitor II 3'-untranslated region containing the poly(A) addition site, was introduced into potato and tobacco by *Agrobacterium tumefaciens*-mediated transformation.[10] Analysis of transgenic plants demostrated systemic, wound-inducible expression of this gene in the stem and leaves of potato and tobacco. Constitutive expression was found in stolons and tubers of nonwounded potato plants. Keil et al.[11] also described analysis of the expression after wounding of chimeric genes consisting of different parts of the proteinase inhibitor II gene fused to the CAT reporter gene in transgenic tobacco plants. They have shown that the promoter of the proteinase inhibitor II gene is rather complex, and consists of far upstream elements for high level wound induced gene expression and wound responsive element(s), probably in the region between positions -514 and -210. The promoter sequence was not homologous to other wound-inducible genes such as the genes encoding carrot extensin and two related wound-inducible genes of potato.[23]

Little progress has been made in this direction but two possible approaches for engineering plants for resistance to insects are apparent. First, it is possible to transfer proteinase inhibitor genes to plants which do not normally synthesize them, and their expression has been shown to be inducible. This raises one question, do these proteins really play a protective role? Plant transformation techniques provide a direct way of testing this. Second, the identification of regulatory sequences of wound inducible genes may prove valuable for engineering the inducible expression of genes encoding insecticidal products.

Hilder et al.[24] provided direct evidence for the effectiveness of proteinase inhibitors in reducing insect damage. A gene encoding a trypsin inhibitor of cowpea (*Vigna unguiculata*) was transferred, via *Agrobacterium tumefaciens*, into tobacco plants, and synthesis of the protein in the transformed plant was confirmed immunologically by Western blotting. The cowpea trypsin inhibitors had previously been correlated with resistance of cowpeas to the bruchid beetle *Callosobruchus maculatus,* and feeding experiments demonstrated their toxicity to a range of insect genera, including *Heliothis, Spodoptera, Diabrotica,* and *Tribolium,* all of which cause significant economic losses to crops. The expression of CAT protein driven by the wound-inducible promoter from the proteinase inhibitor II K (*pin2*) gene was examined in whole tobacco (*Nicotiana tabacum* L.) plants under field conditions.[19] Mechanical wounding of the field-grown leaves caused an accumulation of CAT protein in these leaves which began several hours after wounding and continued to accumulate for about 36 h. When sections of leaves were assayed for accumulation of CAT protein following wounding, the CAT protein was found to accumulate in the apical portions of the leaves. When endogenous insects attacked the leaves of transgenic plants grown in the field, the plants responded by inducing CAT protein. There is no evidence, however, of the toxicity of these proteins to either rats or humans. To determine any introduced resistance in the transgenic tobacco plants, untransformed and transformed individuals were infested with newly emerged larvae of the lepidopteran tobacco budworm, *Heliothis virescens,* a natural pest of tobacco.[19] While the untransformed plants were almost completely devoured, the larvae on the transformed plants either died or failed to develop normally, producing only limited damage to the plants.

III. *BACILLUS THURINGIENSIS* AND ITS ENDOTOXIN

Bacillus thuringiensis has been by far the most commercially successful biological pest control agent, accounting for more than 90% of sales, with a market value of 50 to 55 million per annum. Current products are from three major sources: *B. thuringiensis* var. *kurstaki* directed at lepidopteran pests of forests and agriculture; *B. thuringiensis* var. *berliner,* aimed at the wax moths, scourges of beekeepers; and, more recently, *B. thuringiensis* var. *israelensis,* as a control agent for dipteran vectors of human disease. Several strains of *Bacillus* have been isolated and their toxins or spores have been reported to be effective against several insect larvae.[25-30]

A. CHARACTERIZATION OF Bt ENDOTOXIN

B. thuringiensis is a common spore-forming Gram-positive soil bacterium, unique in its possession of a crystalline parasporal inclusion body. This crystal is proteinaceous, and usually bipyramidal. *B. thuringiensis* produces several types of toxins but it is the crystal toxins (called the δ-endotoxins) which are important commercially. The organisms are grown in batch culture to high cell densities, and the spores are collected and sprayed. Most δ-endotoxins are protoxins and must be proteolytically activated in the midgut of a susceptible insect to become toxic.

The protein which has the insecticidal property is a crystalline inclusion comprised of peptides of 134 to 140 KDa (Table 1). There are variations in

TABLE 1
Three Types of Lepidoptera-Specific Bt Toxins

	Molecular weight		Toxic to		
	Protoxin	Toxin	P. brassicae	M. sexta	S. littoralis
Type 1	130,000	60,000	+	+	−
Type 2	140,000	55,000	+	−	−
Type 3	130,000	63,000	−	−	+

number, shape, and composition of these inclusions. The protein, which is also known as delta endotoxin or P1, is the predominant parasporal component of most *B. thuringiensis*.

The most prevalent δ-endotoxin is comprised primarily of one species of polypeptide of ca. 135 kDa.[31-34] Several workers have contributed towards understanding the morphology and biochemistry of crystalline inclusion.[31,35-47] Toxicity of these proteinaceous crystals for lepidopteran larvae was established by Angus.[48,49] Of the 14 *B. thuringiensis flagella* (H) serotypes, all but 1 (H-14) contain the P1 type of crystalline inclusion.[42] There is usually one inclusion per cell, but there may be two or more.[50-52] Shape is characteristically bipyramidal, but occasional irregular shapes have been observed.[34,41,53]

Scherrer et al.[54] found that growth of *B. thuringiensis* in media containing more than 0.8% glucose resulted in amorphous crystals. The crystal is generally found outside the exosporium, exceptions being subspecies *finitimus*, *B. cereus* subsp. *fowler* and *lewin,* and *B. popilliae*.[51] There are also divergent opinions regarding the relationship between the exosporium and crystals. It has been suggested that the exosporium and crystal may be closely associated,[55-57] but others have found that crystal formation did not necessarily occur close to the exosporium.[51]

Several subspecies form an inclusion within the exosporium. Among all *B. thuringiensis*, subsp. *finitimus* has been most extensively studied.[58,59] However, the endotoxin of this subspecies is not toxic for the larvae of *Manduca sexta*, but is significantly toxic to the larvae of the cotton bollworm *Pectinophora gossypiella*. Subspecies *finitimus* was also found to produce both attached (within the exosporium) and unattached (outside the exosporium) inclusions, each containing predominantly 135-kDa polypeptides.[58] Gene(s) necessary for formation of the larger, attached inclusion are encoded in a 98-MDa plasmid, while the gene(s) encoding the unattached body are chromosomally located. When the 98-MDa plasmid was transferred to *B. cereus* by cell mating, the transcipient produced an inclusion within the exosporium.

In *B. thuringiensis* subsp. *israelensis* there are usually two to four inclusions per cell, which vary in shape from cuboidal to bipyramidal, ovoid, or amorphous.[60-62] They are relatively small (0.1 to 0.5 μm), and the toxic component is active against dipteran but not lepidopteran larvae.[63] Several polypeptides are extractable from these inclusions which also includes a major

26- to 28-kDa protein which is the actual toxin.[52,62,64,65] In addition, a 65-kDa polypeptide has also been implicated.[66,67] There is also confusion regarding the association of plasmid in the formation of inclusions and polypeptides.

The active form of P1 is produced in the midgut of susceptible insect larvae, where it is converted by gut proteases to a toxic moiety.[48,68,69] The precise size of active toxin is not available but it is reported to vary from 5 to 80 kDa.[39,41] Based on studies of crystal preparations digested with larval gut juice, Huber et al.[41] concluded that the molecular weights of active polypeptides could vary. Fragments smaller than 30,000 lose insecticidal property. However, the protoxins of *B. thuringiensis* subsp. *kurstaki* HD1 and HD73 were digested by gut juice proteinases to a 52-kDa "resistant core" which retained full toxicity for test larvae.[62]

The amino-terminal half of the polypeptide contains the active portion of P1 protoxin. A 58-kDa toxic fragment (T-fragment) from trypsin digests of the *B. thuringiensis* subsp. *dendrolimus* crystal protein and similar-sized toxic fragments from subspecies *kurstaki* HD263 were isolated.[44] The aminoterminal amino acid sequence of the T-fragment was Ile-Glu-X-Gly-Tyr-Thr, which matched that of the *kurstaki* HD1 (Dipel) gene.[70] HD1-Dipel is a commercially available derivative of *kurstaki* HD1 and its plasmid profile differs from that of HD1.[71] On the contrary, Chestunikha et al.[32] found that aminoterminal methionine was common in both the protoxin and the aminoterminal portion of protoxins.

The protoxin gene of *kurstaki* HD73 that contained the first 611 amino acids produced toxin which was effective against *M. sexta* larvae, whereas a subclone encoding the first 596 amino acids produced no toxin in *E. coli*.[50] Although there appears to be considerable conservation at the amino acid level, at least for the four genes sequenced to date, the function of the carboxyl half of the prototoxin molecule is not known. Surprisingly, the high-pressure liquid chromatography profiles of trypsin peptide digests of P1 from *kurstaki* HD73 and HD1 differed significantly.[62] Since the toxic portion is resistant to trypsin, most of the peptides should have been derived from the carboxyl half, although the interpretation of the profiles may be complicated because of partial digestion. In addition, there is some evidence that restriction enzyme maps of this portion of protoxin genes from different isolates are more variable than the amino-terminal halves.[72] There is, however, extensive conservation of the amino acid sequence of the carboxyl half of these protoxins.

In addition, the activities of the protease-derived toxins from *kurstaki* HD1 and HD73 P1 protoxins against cabbage (*Trichoplusia ni*) and tobacco budworm (*H. virescens*) larvae were virtually the same as the intact protoxins.[62] These two *kurstaki* strains have different 50% lethal concentrations for these particular Lepidoptera,[62] and since the relative activity was not altered in activated P1's, the carboxyl-terminal regions probably played no role in specificity.

Cloned genes and their subclones should be most useful for further defining the functions of various regions of protoxin molecules. A proteolytic digest of crystals of subspecies *berliner* 1715 was toxic to *Anagasta kuehniella* larvae, whereas the intact crystal had relatively little effect, implying that either solubilization or removal of the carboxyl-terminal portion by proteases may enhance or broaden specificity.[73] This indicates that the carboxyl-terminal half of the protoxin may have some function in determining specificity. In addition, this portion of the molecule has been reported to protect the toxin (in the bacterium, soil, etc.), and acts in the attachment to gut epithelial cell receptors,[40,74] or deposits the protoxin as an inclusion.

Two genes encoding insecticidal crystal proteins from *Bacillus thuringiensis* subsp. *kurstaki* HD-1 were cloned and sequenced. Both genes, designated *cryB1* and *cryB2*, encode polypeptides of 633 amino acids having a molecular mass displayed 87% identity in amino acid sequence, but exhibited different toxin specificities. The *cryB1* gene product is toxic to both dipteran (*Aedes aegypti*) and lepidopteran (*Manduca sexta*) larvae, whereas the *cryB2* gene product is toxic only to the latter. DNA sequence analysis indicates that *cryB1* is the distal gene of an operon which is comprised of three open reading frames (designated *orf1*, *orf2*, and *cryB1*). The amino acid sequence of *B. thuringiensis aizawai* ICI was found to differ by only four residues from the sequence of monospecific lepidopteran toxins (Bt *berliner* and Bt *kurstaki*).[75]

The time of appearance of insecticidal crystal protein (ICP) transcript and the corresponding crystal antigen of the coleopteran-toxic *Bacillus thuringiensis* var. *tenebrionis*, unlike other *B. thuringiensis* varieties, is dependent on the vegetative cell cycle. The expression of the ICP gene increases dramatically and reaches a plateau in early stationary cells. In addition, results from immunoblotting analysis also suggest that the ICP antigen is made in vegetative cells as a 73-kDa peptide, which by proteolytic processing is converted to a 65-kDa peptide to some extent in stationary cells.

B. PLASMID CODED GENES

Most *B. thuringiensis* plasmids are cryptic, but a few functions have been assigned, including bacteriocin production[50] and protoxin synthesis.[71,76] Initially, protoxin synthetic capacity was correlated with the presence of plasmids, in complete[56,77] or partial[78-80] curing studies and later via plasmid transfer in cell mating experiments. The cloning of protoxin genes from *B. thuringiensis* subsp. *berliner* 1715 and *kurstaki* HD1,[81-84] made it possible to develop DNA probes for the presence of gene in related species.

The strain H1 of *Bacillus thuringiensis* var. *thuringiensis* harbors three small cryptic plasmids, pGI1, pGI2, and pGI3 (8.2, 9.2, and 10.6 kb, respectively).[85,86] Two of these plasmids (i.e., pGI2 and pGI3) were successfully cloned into the vector pBR322, whereas only overlapping DNA fragments of pGI1 were cloned in *Escherichia coli*. A curing-hybridization technique was used to obtain isolates of *B. thuringiensis* missing one or another small cryptic plasmid.

DNA probes and cross-hybridization, revealed that the protoxin gene is present primarily in large plasmids and, in some cases, on more than one plasmid in a given subspecies.[81,82] In general, *B. thuringiensis* subspecies contain a substantial portion of their potential genetic information on plasmids. The number and sizes of the plasmids vary considerably. In subspecies with a larger number of plasmids such as *berliner* 1715 or *kurstaki* HD1, plasmids have a broad range, vary in size, (several larger than 30 MDa, with some greater than 100 MDa). In other subspecies, only a few plasmids have been reported, although in such subspecies as *subtoxicus* or *finitimus*, they tend to be larger. There is some variation in plasmid profiles of the same subspecies in data from different laboratories. These variations probably reflect the following: Different procedures were used for lysing cells resulting in variation, in particular, in the recovery of large plasmids (>100 MDa). It appears that the protoplast lysis procedure of Eckhardt as modified by Gonzalez and Carlton[78] is best suited for displaying the entire array of plasmids. Differences may also arise because of variations in plasmid profiles in separate isolates of the same subspecies.[71] There is also evidence for IS sequences, especially in large plasmids, and many of these embrace protoxin genes.[76] These factors may contribute to variations in plasmid profiles for different isolates with the same flagella serotype or for a given strain kept under laboratory conditions for a prolonged period.

Although protoxin genes seem to be in plasmids of various sizes, in all cases the plasmids are >30 MDa. The variety of locations may be related to the presence of inverted repeat elements in the vicinity of protoxin genes,[76] possibly providing a mobilizing capacity. In addition, Gonzalez and Carlton[79] provided evidence for the mobility of a protoxin gene from a study of the plasmid profiles and toxicity of *B. thuringiensis* subsp. *israelensis* derivatives. Some spontaneous derivatives had lost the protoxin-encoding plasmid of 75 MDa, but sequences hybridizing with this plasmid were present in the chromosomal DNA. These derivatives were still toxic for mosquito larvae, indicating that the toxin gene from the 75-MDa plasmid had presumably integrated in such a way that expression was still possible. Derivatives of this strain were found which were still toxic but contained no 75-MDa plasmid sequences in new plasmids of 65 and 80 MDa. Among the 75-MDa plasmid sequences, toxin-producing sequences were also thought to be present because the loss of the 65- and 80-MDa plasmids resulted in nontoxic derivatives containing only a portion of the 75-MDa plasmid sequences in the chromosome. These results have been confirmed by using a clone of the *israelensis* "protoxin" gene on a transposable element. The presumptive movement into and out of the chromosome may serve as a model to account for the presence of protoxin sequences in the chromosome of some subspecies.

C. CHROMOSOMAL CODED GENES

Not all the subspecies have prototoxin genes which are plasmid coded. Preliminary work from a DNA probe derived from a *B. thuringiensis* subsp.

kurstaki HD1 cloned protoxin gene revealed that protoxin genes are exclusively chromosomally encoded in *B. thuringiensis* subsp. *dendrolimus*[71,81,87,88] and subsp. *wuhanensis*.[71] These experiments, however, do not reveal the gene of *B. thuringiensis* subsp. *finitimus,* which contains a chromosomal gene encoding a 135-kDa parasporal protein that is related to any other genes.[58] The protein of this gene is immunologically distinct from the other major 135-kDa parasporal proteins present in inclusions which are enclosed within the exosporium. In addition, prototoxin DNA probes from the *B. thuringiensis* subsp. *kurstaki* HD1 or HD73 hybridized with a *Bam*HI-digested 98-MDa plasmid but not with chromosomal DNA. The gene for the chromosomally encoded inclusion is therefore different. The chromosomally encoded parasporal protein is synthesized later in sporulation than the plasmid-encoded protein. This reflects different modes of regulation of these genes.

A cryptic chromosomal gene that has sequence homology to an active plasmid gene as well as to the *kurstaki* HD1 gene was reported in *berliner* 1715.[81] This chromosomal DNA, when cloned into *kurstaki* HD1, produced toxic protein effective against *M. sexta* larvae,[89] even though plasmid-cured derivatives of this subspecies (still containing this chromosomal toxin gene) were nontoxic. This active toxin is produced by chromosomal genes in association with a plasmid gene.

D. CHARACTERIZATION AND CLONING OF Bt ENDOTOXIN GENES

Several subspecies of *B. thuringiensis* including *israelensis* and several subspecies of *B. sphaericus* have been extensively studied and their toxin genes have been cloned. These clones produced some toxin in *E. coli* (about 1 to 10% of an equivalent number of the bacilli from which the clone was derived). The chromosomal *kurstaki* HD1 gene was originally detected by production of antigen and its toxicity for *M. sexta* larvae.[89] However, a 4.6-kb *Eco*RI subclone was not reproducibly toxic, nor did it hybridize to a DNA fragment containing a plasmid-derived *kurstaki* HD1 protoxin gene or to the protoxin-encoding plasmid of subspecies *kurstaki* HD73, *berliner,* or *galleriae*.[50] Apparently, the *kurstaki* HD1 chromosomal clone contains a minor toxin which is not closely related to the major δ-endotoxin. The chromosomal gene from subspecies *berliner* 1715 did not produce toxin detectable by immunological procedures or bioassay. This clone was used as a hybridization probe, for the isolation of a plasmid gene copy that did produce toxin in *E. coli* and *B. subtilis* indicating a partial sequence homology to an active gene. In fact, a 114-base pair (bp) sequence of this chromosomal gene is present in an HD1 gene.[83,84] This chromosomal gene was not transcribed in *E. coli*; and in *B. subtilis* it was transcribed (only in sporulating cells) but not translated.[81] In addition, transfer of this clone to an acrystalliferous derivative of *kurstaki* HD1 by cell mating resulted in integration of the plasmid but no protoxin synthesis. In contrast, a clone of the plasmid-encoded *berliner* 1715 protoxin was stably maintained and produced protoxin when transferred to this *kurstaki* strain.[81]

The complete nucleotide sequence of plasmid pGI2, a *Bacillus thuringiensis* plasmid, has been studied.[85] Two types of 130-kDa insecticidal protein genes from a large plasmid of *israelensis* HD522 were cloned into an *E. coli* plasmid vector and their expression was identical to the host strain.[90]

The promoter regions of five of the cloned δ-endotoxin genes have been sequenced. This includes the protoxin genes of *B. thuringiensis* subsp. *kurstaki* HD1 (Dipel)[91] and HD73, a plasmid-encoded gene from subspecies *sotto*, another plasmid-encoded gene from subsp. *kurstaki* HD1, and the chromosomal gene from *B. thuringiensis* subsp. *berliner*1715.[87,88] This *berliner* gene is only transcribed and not translated in sporulating *B. subtilis* and produced no toxin when introduced into an acrystalliferous derivative of subspecies *kurstaki* HD1. There are three different "promoter" sites for active *kurstaki* HD1 (Dipel) gene,[70] only one recognized by *E. coli* and two by *B. thuringiensis* RNA polymerases.

Similar altered polymerases have been found in sporulating *B. thuringiensis*.[87,88] For example, it is known that certain *B. thuringiensis* δ-endotoxin genes can be transcribed by a unique sporulation polymerase referred to as the form II enzyme, but it has not yet been determined which promoters this enzyme recognizes. Another feature of the DNA sequence of the upstream region of the *kurstaki* HD1 gene is the presence of two regions of hyphenated dyad symmetry. These are located in such a way that they include transcription start sites and therefore could be sites where regulatory elements bind. Finally, the tandem promoters may contribute to regulation since they could ensure continued transcription by two distinct, altered polymerases present at different stages of sporulation.

Transcription studies have been done with only one of the *kurstaki* HD1 genes,[70] but it is likely that similar promoters exist for the four genes given the identical nucleotide sequences in the regulatory regions. Despite the identities of the kurstaki HD1 and HD73 genes, these toxins have different 50% lethal concentrations for certain Lepidopteran larvae,[62] and the plasmids containing these protoxin genes are regulated differently when transferred to *B. cereus*.[92] Clearly factors other than promoter sequences and unique forms of RNA polymerase are involved in regulation.

E. REGULATION OF Bt ENDOTOXIN GENES

At about stage II or III of sporulation, parasporal protein synthesis begins.[69,70,93] By stage V, the inclusion reaches maximum size which can approximate the size of the spore.[44] However, in *B. medusa* inclusion formation begins at the end of exponential growth, prior to the formation of the forespore septum.[51] The bipyramidal crystal formed in this way accounts for 20 to 30% of the total protein of the sporangium.[69] Activation of transcription and translation is thus closely coupled to protoxin accumulation. During the inclusion formation there is an increase in parasporal antigen, as well as of messenger RNA hybridizing with a cloned probe derived from the *B. thuringiensis* subsp. *kurstaki* HD1 protoxin gene.[70] In addition, the amount of protoxin is substantial and represents about 33 to 43% of the overall rate of protein synthesis.[94-96] In most subspecies, the major protoxin gene is on a large plasmid and is present in low copy number but there is dosage effect. Recently the nucleotide sequencing, gene expressions and subsequent manipulation of the Bt endotoxin gene to increase the production of toxin have been extensively worked out.[26,97-101]

Regulation of this gene also involves factors other than gene dosage, unique promoter sequences, and sigma subunits. Expression of cloned protoxin genes in *E. coli* and even in sporulating *B. subtilis* was poor. The clone of the *berliner* 1715 plasmid-encoded protoxin gene produced only about 10% as much protoxin antigen as the parental strain. In addition, plasmid curing and transfer experiments indicated a role for several *kurstaki* HD1 cryptic plasmids in regulation, and a *kurstaki* HD1 derivative (HD1-9) produced protoxin only at 25°C but sporulated well at both 25°C and 32°C.[92] This conditional phenotype could be suppressed, implicating cellular, metabolic processes in regulation. Cloned protoxin genes from subspecies *kurstaki* HD1 and HD73 have identical nucleotide sequences both for the initial part of the coding region and for several hundred base pairs upstream, through the regulatory region. Yet, when protoxin-encoding plasmids from HD1 or HD73 were transferred to *B. cereus*, only the latter expressed well, i.e., 30 to 50% of the parental strain. The 44-MDa plasmid form *kurstaki* HD1 produced little protoxin in *B. cereus* unless another HD1 plasmid of 4.9 MDa was present.[92] Similar results were found with a *B. cereus* transcipient containing the 98-MDa protoxin-encoding plasmid from subsp. *finitimus*. There were small inclusions (but enclosed within the exosporium) and much less 135-kDa parasporal antigen than an equivalent number of parental cells. In this case, the plasmid-encoded protoxin gene has not been sequenced. The very different expression of protoxin-encoding plasmids in *B. cereus* transcipients, even when the nucleotide sequences of the regulatory regions were identical, indicates that other factors, probably plasmid encoded, are involved.

Regulation of protoxin in derivatives of subsp. *kurstaki* HD1 synthesis may involve unique physiological factors that are not required for sporulation. It has been known for some time that variations in media influenced the relative yields of spores or inclusions.[102,103] The conditional strain, HD1-9, sporulated well at either 25 or 32°C, but produced protoxin only at the former temperature.[92] This regulation probably occurred at the level of transcription since there was no detectable protoxin messenger RNA in sporulating cultures of HD1-9 grown at 32°C. Conditional synthesis was also found in *B. cereus* transcipients containing the 29-MDa plus 44-MDa, or the 29-, 44-, and 4.9-MDa plasmids from *kurstaki* HD1. Thus, the 29-MDa plasmid (in the absence of one of 110 MDa) seemed to be responsible for the conditional phenotype.

The conditional phenotype of strain HD1-9 could be specifically suppressed by a subinhibitory concentration of D-cycloserine or in a D-cycloserine-resistant derivative of HD1-9.[50] Since neither the conditional phenotype nor its suppression affected sporulation, there must be regulatory signals specific for protoxin synthesis.

The regulation of the time of expression of protoxin genes has not been studied. In all of the *B. cereus* transcipients and in one case for the cloned protoxin gene from *berliner* 1715 in *B. subtilis*,[81] expression was confined to postexponential (sporulating) cells. As discussed above, the promoter regions seem to be unique and probably require special forms of RNA polymerase that are functional only in sporulating cells. The temperature-sensitive factor involved in the conditional *kurstaki* strain HD1-9 appeared to be synthesized during exponential growth,[92] but here again protoxin synthesis at the permissive temperature occurred only in sporulating cells.

In *B. subtilis,* the presence of unique forms of RNA polymerase is necessary but not sufficient for the expression of sporulation genes. Several but not all of the unique sigma factors seem to be present prior to the time they are functional and to require products of at least some of the *Spo*0 loci. For example, Zuber and Losick[104] found that fusions of the promoter for the *Spo*VG gene to β-galactosidase were not transcribed in most *B. subtilis Spo*0 mutants. These *Spo*0 mutations are pleiotropic, affecting the function not only of some 0(29) and 0(37) promoters but also of 0(28) promoters (68). The regulatory action of the *Spo*0 gene products may or may not be direct, but clearly other components are involved in regulating the transcription of these sporulation genes.

This complex regulatory pattern does not seem to apply to the closely related subspecies, *kurstaki* HD73, since transfer of the 50-MDa plasmid to *B. cereus* seems to be sufficient for protoxin synthesis at both 25 and 30°C and in amounts approaching (30 to 50%) the parental strain. Subspecies *kurstaki* HD73 has a simpler plasmid profile so that regulatory genes may have been incorporated into the same plasmid that carries the protoxin gene. On the contrary, transfer of the 98-MDa protoxin-encoding plasmid from subspecies *finitimus* to *B. cereus* resulted in the formation of small inclusions enclosed within the exosporium (as in the parental strain). These transcipients produced much less 135-kDa parasporal protein (shown by either protein stain or antigen in immunoblots) than the donor strain. Presumably there are regulatory factors encoded by the other *finitimus* plasmid (77 MDa) or by the chromosome which are not present in the *B. cereus* transcipients. Inclusions were found only in *B. cereus* transcipients grown at 25°C. In this case, however, growth at 30°C resulted in loss of the 98-MDa plasmid from transcipients (in contrast to the parental strain of *B. cereus* transcipients containing *kurstaki* HD1 plasmids), so some chromosomal function of *finitimus* may be involved in plasmid maintenance.

F. TRANSGENIC PLANTS EXPRESSING THE δ-ENDOTOXIN GENE

The choice of the Bt endotoxin gene for introduction into plants was based on the extensive knowledge of its gene regulation and protein as described previously. Thus the endotoxin gene from *Bacillus thuringiensis* has been the choice in a number of pesticide resistance experiments. The Bt toxin, believed to be an environmentally safe insecticide, is active against number of caterpillars including the tobacco hornworm and gypsy moth. The strategy, as will be described later, has been to link the toxin gene to a constructive promoter that will express the toxin in all plant tissues. T-DNA transformation has been used to move the gene into tobacco and tomato plants where it appears to express strongly enough to kill a large percentage of caterpillars. An indirect approach to pest management bypasses the problem of plant transformation altogether. This involves inserting a toxin gene into the genome of leaf- or root-colonizing bacteria, which synthesize and secrete the pesticide *in situ* on the leaf surface. For example, Bt endotoxin and biotoxin for root cutworm have been inserted into strains of *Pseudomonas fluorescens*. The advantage of this system is the ease of transforming bacteria compared with plants. Although Bt toxin genes have been transferred to *Pseudomonas fluorescens* strains able to colonize maize roots, permission for field trials has not yet been granted.[105]

A Bt endotoxin gene with insecticidal activity against lepidopteran larvae was reported to be introduced successfully into tobacco plants in July 1987[106] by Plant Genetic Systems (PGS), a Belgian biotechnology company. Truncation of the full-length gene to the size encoding the toxic core did not reduce insecticidal activity. The success of PGS was attributed to the choice of promoter (wound-inducible upon insect feeding) and to fusion of the Bt endotoxin with the gene for kanamycin resistance, which allowed them to use resistance to kanamycin to select plants expressing high levels of the fusion gene product. The levels of Bt endotoxin produced were sufficient to kill first-instar *Manduca sexta* larvae, and this insecticidal property was heritable.

Immediately thereafter, a second report of successful transformation with Bt endotoxin gene followed in August 1987, with the expression of Bt in tomato by researchers at the Monsanto Co.[107] 35S CaMV promoter was used to direct expression of the insecticidal activity in plants expressing a truncated Bt endotoxin gene. Later in 1987, the Agracetus Co. reported expression of the Bt endotoxin in tobacco with the 35S CaMV promoter linked to an AMV (alfalfa mosaic virus) leader sequence giving enough Bt mRNA to be easily detectable on Northern blot analysis. Using immunoblot techniques, they identified a peptide in resistant plants corresponding in size to that expected for a truncated Bt endotoxin. In these three initial reports (above), the Bt endotoxin genes chosen generated insecticidal activity against lepidopteran insects.

Since 1987, Bt endotoxin genes have been reported to be introduced into additional plant species including potato[106] and cotton.[108] Efforts are under way to engineer a variety of other crops (such as sunflower and vegetables) to resist lepidopteran and some coleopteran insects. Modifications to the bacterial gene sequence of the Bt endotoxin to make it more readily expressed in plants have been reported[109] and have been shown to increase concentration (up to 500-fold) and, consequently, host range to include less sensitive pests such as the tomato fruitworm (*Heliothis zea*) and tomato pinworm (*Keiferia lycopersicella*).

Generally *Agrobacterium*-mediated T-DNA transfer is used to express chimeric *B. thuringiensis* genes. The toxin sequences are inserted in Ti plasmid-derived expression vectors behind the promoter of the *Agrobacterium* TR2′ gene.[106] This promoter directs expression of mannopine synthase in plant cells transformed with the TR DNA of plasmid pTiA6.[110] Two divergent promoters, TR1′ and TR2′, are generally present on the TR DNA fragments inserted in the expression vectors.

In the *B. thuringiensis* genes, either the entire bt2 coding sequence or truncated fragments comprising the toxic part, are cloned behind the TR2′ promoter.[106] The neomycin phosphotransferase gene (*neo*), coding for the NPTII enzyme, which confers resistance to aminoglycoside antibiotics such as kanamycin, was used as a selectable marker and inserted behind TR1′. In some constructs, *neo* was fused to the bt2 coding sequence to generate a transcriptional and translational fusion. Fusion proteins encoded by such hybrid genes were stably expressed in *Escherichia coli* and exhibited both insect toxicity and phosphotransferase activity.

The expression vectors were mobilized into the *Agrobacterium* recipient, C58C1 Rif^R pGV2260.[106] Recombination between pG2260 and the expression vector through the homologous pBR322 sequences produced Ti plasmids pG1161, pG1163, pGS1151, and pGS1152 containing *bt2, bt884, bt: neo23*, and *bt: neo860* genes, respectively. Both tobacco (var. Petit Havana SR1) and potato (var. Berolina) produced transformed shoots upon cocultivation of leaf disks with *Agrobacterium* containing these constructs. Tobacco shoots were selected on 50 to 100 mg/ml kanamycin and rooted on nonselective medium.[106] Potato shoots were selected on 100 mg/ml antibiotic. Because Vaeck et al.[106] expected the expression of *bt2* and NPTII to be linked, they selected transformed plants expressing high levels of antibiotic resistance for further testing in insect bioassays. Insecticidal activity in transgenic plants was determined in feeding assays with first instar *Manduca sexta* larvae. Transgenic plants inducing efficient insect killing were obtained for both tobacco and potato.[106] For tobacco, high toxicity resulting in 80 to 100 percent mortality of the larvae was observed in 14% of the plants expressing the longer fusion protein Bt: NPT23 and 56% of those with the shorter Bt:NPT860. This suggests that the shorter *bt:neo860* gene provides higher levels of biologically active protein than does the longer *bt:neo23*. High insecticidal activity was also detected in plants expressing the truncated *bt884* gene. None of the plants transformed with the full-length *bt2* gene produced insect-killing activity above levels obtained in NPTII-expressing control plants.[106] Potato plants transformed with a fusion gene or truncated *bt2* gene also exhibited significant insecticidal activity.

Further greenhouse experiments demonstrated that the transgenic plants were protected against insect feeding damage.[106] The *B. thuringiensis* protein-containing plants showed very limited damage that was restricted to feeding areas of a few square millimeters, whereas control plants were entirely consumed within 10 days. *B. thuringiensis* protein levels in the insect-resistant plants ranged from 7 to 40 ng/mg total plant protein, or between 4 and 25 mg/cm of leaf. This compares to the general commercial application rate of *B. thuringiensis*, which is approximately 0.5 kg of dry powder per hectare, equivalent to 15 mg of active toxin per square centimeter of plant leaf when sprayed in optimal conditions. Vaeck et al.[106] also analyzed inheritance of the new trait in plants with one or two T-DNA inserts. Kanamycin resistance and insecticidal activity were coupled and the inheritance patterns were fully compatible with simple Mendelian rules for a dominant trait.

Although plants expressing low levels of Bt insect control protein are afforded some protection from insect damage caused by lepidopteran insects, field tests with tomato plants have indicated that higher levels of expression are required to control the agronomically important lepidopteran insects.[111] Insect-tolerant plants will be of value to farmers only if they produce sufficient Bt insect control protein to provide consistent plant protection. In order to achieve increased insecticidal efficacy, the Bt coding sequence has been the target of specific alterations.

It was clear from the experiments of Vaeck et al.[106] that the original complete gene(s) of *B. thuringiensis* was not fully expressed in plants, suggesting that it needs modification and should be made more or less similar to the plant genome to improve overall translational efficiency. In an attempt to improve the efficiency of these genes at levels that provided effective control of agronomically important lepidopteran insect pests, Perlack et al.[108] have initially attempted to express truncated forms of the insect control protein genes of *Bacillus thuringiensis* var. *kurstaki* HD-1 (*cry IA[b]*) and HD-73 (*cry IA[c]*) in cotton plants. Total protection from insect damage of leaf tissue from these plants was observed in laboratory assays when tested with two lepidoteran insects: an insect relatively sensitive to Bt insect control protein, *Trichoplusia ni* (cabbage looper) and an insect that is 100-fold less sensitive, *Spodoptera exigna* (beet army worm). Whole plants, assayed under conditions of high insect pressure with *Heliothis zea* (cotton bollworm) showed effective square and boll protection (Table 2). The number of plants that contained highly modified *cry*A(b) insect control protein genes (pMON 5377) were toxic to cabbage looper larvae and sustained no leaf damage. Similarly these cotton plants containing the modified gene sustained little damage from beet army worm larvae under laboratory as well as greenhouse conditions.

TABLE 2
Results of Cotton, Bollworm Bioassays on Whole Plants

Plant line	B.t.k. protein	Mean instar	Percent protected (Bolls/squares)
81	CryIA(b)	3.4	70
Control		4.6	00
81	CryIA(b)	1.9	76
Control		2.7	16
249	CryIA(c)	1.8	87
Control		3.1	12

From Perlak F. J. et al., *Bio/Technology*, 8, 939, 1990. With permission.

Immunological analysis by Western blotting indicated that cotton plants containing the modified genes produced Bt insecticidal protein at up to 1% of their total soluble protein.

Truncating the gene, keeping essentially the N-terminal half of the protein intact, resulted in improved expression of the gene in barley to detectable levels.[94,95] The use of different promoters, fusion proteins, and leader sequences has not significantly increased insect control protein gene expression.[51,94] However, Perlak et al.[109] were interested in the modification of the B.t.k. gene to make the production of B.t.k. protein still higher. They hypothesized that a gene with a sequence adapted for a Gram-positive prokaryote may not have the appropriate coding sequence for efficient plant expression. Examination of the insect control protein gene coding sequence indicated that it differs significantly from plant genes in G+C content. Multiple sequence motifs that are not common in the coding region of plant genes were found to be common in the wild-type (WT) *cry*IA(b) sequence. These included localized regions of A+T richness resembling plant introns, potential plant polyadenylation signal sequences, ATTTA sequences, which have been shown to destabilize mRNA in other systems, and codons rarely used in plants.[112]

Perlak et al.[109] initiated two approaches to increase the level of *cry*IA(b) and *cry*IA(c) insect control proteins in genetically modified plants. First, DNA sequences predicted to inhibit efficient plant gene expression at both the translational and mRNA level were selectively removed throughout the coding sequence by site-directed mutagenesis to partially modify the gene (PM gene) without changing the amino acid sequence (Figure 4). Wholesale changes in the DNA characterized the second approach, which required the use of a fully modified (FM) synthetic gene (Figure 4). The FM genes encoded proteins nearly identical in amino acid sequence to the wild type (WT) gene.

Perlak et al.[109] compared the expression of the *cry*IA(b) gene to partially modified (3% nucleotide difference) and to fully modified (21% nucleotide

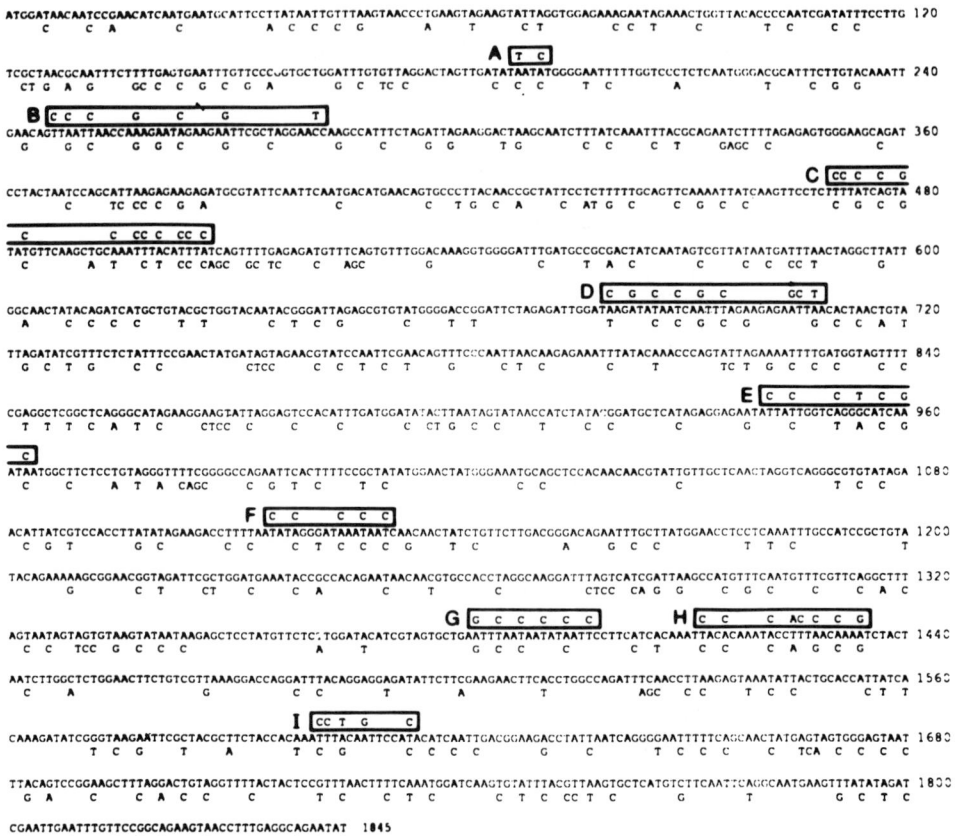

FIGURE 4. DNA sequence of the *cry*IA(b) gene on the numbered line with the modifications found in the PM gene on the line above and FM gene on the line below. The differences between the WT and PM genes are within the labeled boxed areas (A-I). (From Perlak, F. J. et al., *Proc. Natl. Acad. Sci. U.S.A.*, 88, 3324, 1991. With permission.)

difference) *cry*IA(b) and *cry*IA(c) genes in tobacco and tomato (Table 3). The modified genes increased the frequency of plants that produced the proteins at quantities sufficient to control insects and dramatically increased the levels of these proteins. Among the most highly expressing transformed plants for each gene, the plants with the partially modified *cry*IA(b) gene had a 10-fold higher level of insect control protein and plants with the fully modified *cry*IA(b) had a 100-fold higher level of *cry*IA(b) protein compared with the wild-type gene (Figure 5). Similar results were obtained with the fully modified *cry*IA(c) gene in plants. In addition, specific sequences of the partially modified *cry*IA(b) gene were analyzed for their ability to affect *cry*IA(b) gene expression in tobacco. The modified genes (PM and FM) were routinely expressed above the biological threshold set for insect control in tomato and

TABLE 3
Summary of Changes Made in the PM and FM
cryIA(b) Genes Compared to the WT cryIA(b) Gene

	WT	PM	FM
No. of bases different from WT	—	62 of 1743	390 of 1845
No. of codons different from WT	—	55 of 579 (9.5%)	365 of 615 (60%)
G+C content	37%	41%	49%
No. of potential PPSS[a]	18	7	1
No. of ATTTA sequences	13	7	0

[a] PPSS — plant polyadenylylation signal sequences.

From Perlak, F. J. et al., *Proc. Natl. Acad. Sci. U.S.A.*, 88, 3324, 1991. With permission.

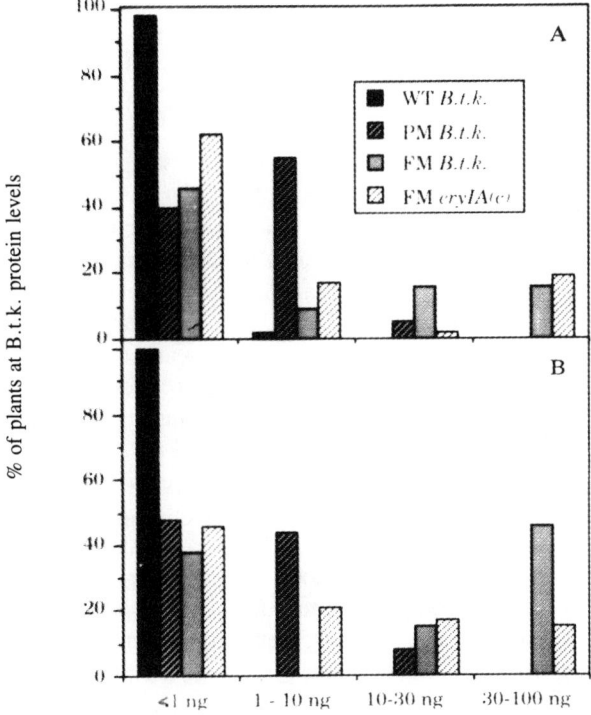

FIGURE 5. Distribution of tomato (A) and tobacco (B) plants that exhibited insect control. The numbers of tomato plants that exhibited insect control, used to calculate the percentage of B.t.k. protein levels, are as follows: WT gene, 53; PM gene, 63; FM gene, 55; FM cryIA(c), 97. The numbers of tobacco plants are as follows: WT gene, 3; PM gene, 42; FM gene, 22; FM cryIA(c), 95. Levels of insect control protein are per 50 μg of total plant protein. (From Perlak, F. J. et al., *Proc. Natl. Acad. Sci. U.S.A.*, 88, 3324, 1991. With permission.)

TABLE 4
Comparison of the Fraction of Plants Exhibiting Insect Control with the WT, PM, and FM

	Plants exhibiting insect control/total kanamycin-selected plants	
Gene	Tobacco	Tomato
WT cryIA(b)	3/54 (6%)	53/204 (26%)
PM cryIA(b)	25/42 (60%)	40/63 (63%)
FM cryIA(b)	12/22 (55%)	32/55 (58%)
FM cryIA(c)	—	52/97 (54%)
FM cryIA(c)[a]	48/95 (51%)	—

[a] The coding region of the N-terminal amino acid s of this FM *cry*IA(c) gene has been deleted as in the WT PM *cry*IA(b) genes.

From Perlak, F. J. et al., *Natl. Acad. Sci. U.S.A.*, 88, 3334, 1991. With permission.

tobacco (Table 4). This occurrence was less frequently observed with the WT gene, especially in tobacco. The DNA sequence of a single region was identified as important to the improvement of plant expression of the *cry*IA(b) gene. The increased levels of *cry*IA (b) mRNA were not directly proportional to the increased levels of *cry*IA(b) protein in plants transformed with the modified *cry*IA(b) genes, indicating that the nucleotide sequence of these genes had an effect in improving their translational efficiency in plants.

Analysis of the mRNA levels of some of the PM and FM plants indicated that comparable amounts of FM mRNA led to the production of more insect control protein than PM mRNA, suggesting that the presence of predominantly plant preferred codons in the FM gene has improved its overall translational efficiency. The impact of the exceptional levels of expression that the PM and FM genes provide is most apparent in the efficacy these plants exhibit to withstand insect attack.

Insect control protein levels in plants increased as a consequence of modification of the DNA sequence, up to 100-fold over levels seen with the wild-type gene in plants.[109] The increased gene expression appeared to be the result of increased translational efficiency. The higher level of insect control protein in these plants was directly correlated to increased insecticidal activity. The demonstration of the utility of these genes to provide protection from insects has far-reaching implications for the future of insect-resistant plants and for the application of these gene modification principles to the design of heterologous genes for high-level expression in plants. The use of these plants in an integrated pest management system will expand the options of farmers to protect their harvest from insect destruction.

IV. GENETICALLY ENGINEERED INSECT-RESISTANT PLANTS AND ECOLOGICAL CONSIDERATIONS

Traditional techniques of selecting plants resistant to insects have offered farmers an economically and ecologically reasonable means of suppressing insect and microbial pests.[113] Unfortunately, some pathogens and insects have evolutionarily circumvented resistance factors that were initially very effective.[114-122] Concern about pest adaptation to resistant crops has led to theoretical and empirical research on the dynamics of pest adaptation and strategies for delaying such adaptation.[119,123,124] To circumvent these problems, scientists were in search of alternatives which could provide a more meaningful kind of insect control. The development of recombinant DNA techniques brought new hope, and it appears that soon we may have the ability to routinely produce cultivars with the capacity to kill over 95% of the pest populations.[106,120] Thus, genetically engineered crops will have partial resistance to insects. Genetic engineering techniques could also produce plants that kill a variable proportion of a pest population, slow the pest's growth rate, and/or decrease the pest's reproductive capacity.[12,120] In some cases this "partial resistance" would be sufficient in and of itself to decrease yield loss. In other cases this partial resistance, coupled with the effects of natural enemies, could offer sufficient pest suppression.[121,125]

However, the release of genetically engineered plants is also not as safe as it has been considered to be. Although some engineered characteristics, such as delayed fruit ripening and genetic markers appear to have no obvious competitive value, others such as tolerance to insects would seem to confer a competitive advantage over plants without such characteristics. Thus a critical analysis and ecological consideration of such plants must be made before their release into the enviroment. Already reports have started appearing in the literature of engineered insect-resistant plants becoming susceptible to insects in the course of time.[126]

Van Emden[127] used a straightforward mathematical model to demonstrate that partial host plant resistance interacting with natural enemies of pests could prevent pest outbreaks in certain situations. This model has received support from several subsequent studies.[125-134] The use of partial resistance and natural enemies should prove to be of long-term benefit because pests would not adapt to such a system as rapidly as they would to a system where a strong resistance factor was used as a unilateral approach to pest suppression.[125] It is believed that adaptation of some pests to mixtures of resistant and susceptible plants might be slower in the presence of natural enemies than without them because of the stochastic nature of host-plant colonization and interactions between the predator and pest population dynamics.[135] In contrast, herbivores would adapt more rapidly to plant defenses in the presence of natural enemies than without them.[136]

The effects of natural enemies on pest adaptation to a resistance factor depend on the details of the specific tritrophic interactions in the system of interest. Gould et al.[122] therefore examined interactions between a variety of resistance factors and natural enemies for effects on the rate of adaptation. In their examination, they used both qualitative and quantitative models, specific ecological interactions between a resistance factor, a pest, and its natural enemies as outlined by several workers.[127,137,138]

A major difficulty in the release of genetically engineered plants may be the adaptation of insects to such insecticides as happened with several chemical pesticides. A pest can adapt to resistance factors in many ways. These include behavioral, physiological, and biochemical alterations in the insect. These alterations could be controlled by alleles at one or more genetic loci. The specific types of adaptions and their genetic control could certainly affect rates of adaptation. The interactive effects of partial resistance and natural enemies on the population dynamics and genetics of pests will almost always have intra- and intergenerational components.[122]

The emphasis placed on the development of insect resistance as one of the many goals in a crop improvement program is determined in large part by the benefits expected from the insect resistance compared with other crop characteristics. As mentioned earlier the development of insect-resistant crop cultivars, whether through biotechnology or conventional plant breeding, is a time-consuming and costly endeavor. It will be worthwhile if we consider the problems of insects becoming resistant to transgenic plants more seriously. Appraisal of these benefits must include consideration of both the level of crop protection needed and the potential durability of the resistance once deployed. A variety of factors influence both the levels of pest control provided by a particular resistance and its durability.[139,140]

Considerable attention has been given recently to the contribution of natural enemies to pest population suppression on resistant plants[141] and to the selection for adaptation to resistance. Several authors have argued that the rate of pest adaptation to host-plant defenses would be increased,[136] or decreased[135] by the occurrence of mortality due to natural enemies. Gould et al.[122] indicated that there is a need for a detailed understanding of the ecological interactions of a pest and its natural enemies before predictions can made about rates of pest adaptation. Through the use of qualitative and quantitative models, Gould et al.[122] have illustrated that even in an extremely simple system (monogenic pest adaptation, a monophagous pest, and 100% resistance in the host-plant biomass) the magnitude of the differential fitness, and hence the rate of adaptation in the presence of natural enemies may be influenced by the details of the pest response (behavioral or physiological) to the plant defense, biology of the natural enemy, and dynamics of the pest/natural enemy interaction.

In some cases of host-plant resistance, the natural enemy complex may be less likely to change significantly the rate of pest adaptation than in other cases of crop resistance (Mexican bean beetle/soybean). However, in many cases of partial resistance, general predictions cannot be made. Pest/natural enemy interactions vary greatly among different areas for a pest on a given crop and among crops for a pest in a given area. In addition, the level of population suppression attributable to a particular natural enemy species in a larger natural enemy complex is uusually varied. Many of the important interactions determining the contribution of natural enemies to pest adaptation to resistant crops and to overall pest suppression on resistant crops, are system and site specific. As a result, case by case, detailed empirical studies of plant/pest/natural enemy interactions are necessary to develop meaningful predictions of the impact of natural enemies on pest adaptation to resistance in a particular system. In a few large, uniform crop/pest systems such as those associated with corn, soybean, or wheat production in the midwestern U.S., a relatively few local and in-depth studies might provide information common to the majority of the production areas. However, for most crop/pest interactions, site- or area-specific research is needed to generate the requisite information. Such research is expensive and large numbers of site-specific studies on how to design and deploy resistant crop cultivars in order to provide durable crop protection cannot be justified in most crops. Thus, if ecologists and evolutionary biologists are to contribute meaningfully to the design and deployment of resistant germplasm for durable crop protection, general principles must be developed which describe the likely impact of a variety of common, important system parameters on levels of pest suppression and rate of adaptation. In the absence of such principles, inputs of these scientists are likely to be ignored.[142]

Results of the model proposed by Gould et al.[122] suggest that, in general, selection for adaptation to a resistant plant type is lower at a given level of pest population suppression when that suppression is achieved by the combined action of plant resistance and natural enemies than by strong resistance alone. If additive genetic variance in the pest for adaptation to strong resistance and partial resistance were equal, this would result in more rapid pest adaptation to the strong resistance. If there was more additive genetic variance or a higher rate of mutation to alleles conferring full adaptation to the partial resistance than the strong resistance, the expectation could be reversed. The estimates of such mutation rates or additive genetic variance are lacking.

V. CONCLUSIONS

With the advent of biotechnology, interest in *B. thuringiensis* and production of insect resistant plants has greatly increased.[143-174] By using the *Agrobacterium* vector system, a researcher can transfer foreign genes that are expressed in many dicotyledonous plants. This method can be used to express *B. thuringiensis* genes in plant varieties which are protected from insect feeding damage.

Most *B. thuringiensis* δ-endotoxins have a molecular weight of 130×10^3 to 140×10^3. They are in fact "protoxins" which, after solubilization in the insect midgut, are proteolytically processed to smaller, toxic polypeptides of approximately 60×10^3 daltons.[166] For Bt2, a recombinant Bt toxin cloned from *B. thuringiensis berliner* 1715, the complete DNA sequence of the protoxin gene was determined.[158-160] In addition, based on the amino-terminal amino acid sequence, Hofte and co-workers[158-160] localized the toxic polypeptide to residues 29 to 607 on the protoxin sequence. A similar processing step is involved in the activation of a dipteran-specific δ-endotoxin called Bt8. The gene has been cloned from *B. thuringiensis israelensis*.[158-160] Interestingly, the coleopteran toxin from *B. thuringiensis tenebrionis*, B13 appears in the crystal as a naturally processed toxin.

Despite their clearly distinct spectrum of insecticidal activity, Bt2, Bt8, and B13 are members of a family of structurally related proteins. Apart from some amino acid sequence similarities, these proteins exhibit a strikingly similar pattern of hydrophobicity in the biologically active part of the molecule. Such observations indicate that these toxins may be functionally related and use similar mechanisms to interact with the insect cells.

The availability of different cloned toxin genes and detailed knowledge about their structural properties has led to an understanding of parameters that determine their specificity. Ultimately this information can be used to construct new types of insecticidal proteins with improved activity or a broader host range. Such insect toxins are becoming extremely useful in the development of new crop varieties that are resistant to a whole range of pest.

Progress in engineering insect resistance in transgenic plants has been achieved through the use of the insect control protein gene of *Bacillus thuringiensis* (Bt). Most strains of Bt are toxic to lepidopteran (moth and butterfly) larvae, although some strains with toxicity to coleopteran or dipteran larvae have been described. Bt toxins disrupt ion transport across brush border membranes of susceptible insects.

Insect control proteins from a prokaryotic source, *Bacillus thuringiensis* var. *kurstaki* are specific for lepidopteran insects and exhibit no activity against humans, other vertebrates, and beneficial insects. These properties have made the genes of these insect-specific proteins attractive candidates for genetic modification of crops for protection against lepidopteran pests. Genes encoding lepidopteran-specific insect control proteins have been cloned and sequenced. Truncated genes, which produced insecticidal active protein, have been expressed in tomato, tobacco, and cotton. Field tests of these plants revealed that higher levels of insect control protein in the plant tissue would be required to obtain commercially useful plants.

The insect control proteins are highly expressed in their natural host, *B. thuringiensis*. Up to 50% of the total protein in sporulated cultures of B.t.k. are the insect control proteins deposited as crystals within the cell. Insect control protein genes are expressed well in *Escherichia coli*[64] or *Pseudo-*

monas.[52] Poor expression in plants is a well-reported characteristic of the B.t.k. insect control proteins.

Although in the present transgenic plants *B. thuringiensis* protein levels are sufficiently high to control *M. sexta,* control of other insect species, which are less sensitive to Bt toxin, may require higher levels of expression. Alternatively, one may consider the use of more active *B. thuringiensis* toxins. Even within the lepidopteran pathotype, differences exist in the insecticidal spectra of a large collection of *B. thuringiensis* toxins. Strains have been identified that are highly toxic to *Spodoptera littoralis,* an insect which is insensitive to most commonly known *B. thuringiensis* strains. A recent approach has been the use of the PM or FM Bt gene in order to make them compatible to the plant system. This has increased the toxin level by several fold in plants.

On the basis of the insecticidal spectrum and size of the toxic polypeptide fragment, Vaeck et al.[106] defined three types of lepidopteran-active toxins. Bt2 is an example of type 1 δ-endotoxin; it has a relatively broad spectrum of insecticidal activity, although it is nontoxic to *S. littoralis*. A type 2 toxin is found in pure form in the crystals of *B. thuringiensis* strain 4412 and is toxic to *Pieris brassicae* but not to *Manduca sexta* and *S. littoralis*. An example of a type 3 toxin is presented together with a type 1 toxin in *B. thuringiensis aizawai* strain HD 127; it is characterized by a high toxicity to *S. littoralis*.[159]

Specific monoclonal antibodies have been generated against these three types of δ-endotoxins. These antibodies provide an efficient tool for very rapid immunological screening of large numbers of *B. thuringiensis* strains. Among 28 strains tested with the monoclonals against type 3 toxin, only eight showed positive reactions. In a bioassay, the same eight strains, and only these, exhibited toxicity to *S. littoralis*.[159] This approach can be extended to the selection and cloning of *B. thuringiensis* genes and is an essential step in the engineering of plant resistance against diverse insect species in agriculturally important crop cultivars.

Transgenic tomato, tobacco, and cotton plants containing the Bt gene exhibited tolerance to caterpillar pests in laboratory tests. The level of insect control observed in the field tests with tobacco and tomato plants has been excellent. The level of expression of the protein encoded by these genes in the plants is low (typically about 0.001% of total soluble protein) compared to many other heterologous genes. However, plants expressing low levels of B.t.k. insect control protein are afforded some protection from insect damage caused by lepidopteran insects. In one such test, tomato plants containing the Bt gene suffered no agronomic damage under conditions that led to total defoliation of control plants. The excellent insect control observed under field conditions indicates that this technology may have commercial applications in the near future.

As the progress in the production of insect-resistant transgenic plants is tremendous, such plants may be released into the market very soon. However, the impact of such engineered plants on the natural population of insect predators and other harmful effects are yet to be clearly visualized. Scientists are also beginning to realize the problem of insects becoming resistant to transgenic plants.

REFERENCES

1. **Ryan, C.**, Quantitative determination of soluble cellular proteins by radial diffusion in agar gels containing antibodies, *Trends Biochem.*, 12, 434, 1967.
2. **Ryan, C.**, Proteinase inhibitors in plant leaves, a biochemical model for pest-induced natural plant protection, *Trends Biochem. Sci.*, 5, 148, 1978.
3. **Ryan, C.**, Insect control with genetically engineered crops, *BioEssays*, 10, 20, 1989.
4. **Plunkett, G., Senear, D., Zuroske, G., and Ryan, C.**, Proteinase inhibitor I and II from leaves of wounded tomato plants, *Arch. Biochem. Biophys.*, 213, 456, 1982.
5. **Lee, J. S., Brown, W., Graham, J., Pearce, G., Fox, E., Dreher, T., Ahern, K., Pearson, G., and Ryan, C.**, Molecular characterization and phylogenetic studies of a wound-inducible proteinase inhibitor I gene in *Lycopersicon* species, *Proc. Natl. Acad. Sci. U.S.A.*, 83, 7277, 1986.
6. **Thornberg, R. N., An, G., Cleveland, T. E., Johnson, R., and Ryan, C. A.**, Wound-inducible expression of a potato inhibitor II-chloramphenicol acetyltransferase gene fusion in transgenic tobacco plants, *Proc. Natl. Acad., Sci. U.S.A.*, 84, 744, 1987.
7. **Johnson, R., Narvaez, J., An, G., and Ryan, C.**, Expression of proteinase inhibitor I and II in transgenic tobacco plants: effect on natural defense against *Manduca sexta* larvae, *Proc. Natl. Acad. Sci. U.S.A.*, 86, 9871, 1989.
8. **Johnson, R., Lee, J. S., and Ryan, C.**, Regulation and expression of wound inducible tomato inihibitor I gene in trangenic nightshade plants, *Plant Mol. Biol.*, 14, 349, 1990.
9. **Keil, M., Sanchez-Serrano, J., Schell, J., and Willmaitzer, L.**, Primary structure of a proteinase inhibitor II gene from potato (*Solanum tuberosum*), *Nucleic Acids Res.*, 14, 5641, 1986.
10. **Keil, M., Sanchez-Serrano, J., Schell, J., and Willmitzer, L.**, Both wound-inducible and tuber-specific expression are mediated by the promoter in a single member of the potato proteinase inhibitor II gene family, *EMBO J.*, 8, 1323, 1989.
11. **Keil, M., Sanchez-Serrano, J., Schell, J., and Welmitzer, L.**, Localization of elements important for wound inducible expression of chimeric potato proteinase inhibitor II-CAT gene in transgenic tobacco plants, *Plant Cell*, 2, 61, 1990.
12. **Sanchez-Serrano, J., Schmidt, R., Schell, J., and Willmitzer, L.**, Nucleotide sequence of a proteinase inhibitor II encoding cDNA of potato (*Solanum tuberosum*) and its mode of expression, *Mol. Gen. Genet.*, 203, 15, 1986.
13. **Sanchez-Serrano, J. J., Keil, M., O'Conner, A., Schell, J., and Willmitzer, L.**, Wound-induced expression of a potato proteinase inhibitor II gene in transgenic plants, *EMBO J.*, 6, 303, 1987.
14. **Thornberg, R. W., An, G., Cleveland, T. E., Johnson, R., Ryan, C. A.**, Wound inducible expression of potato inhibotor II — CAT gene fusion in transgenic tobacco plants, *Proc. Natl. Acad. Sci. U.S.A.*, 84, 744, 1987.

15. **Bishop, P. D., Makus, D. J., Pearce, G., and Ryan, C. A.**, Proteinase inhibtor-inducing factor activity in tomato leaves residues in oligosaccharides enzymatically release from cell walls, *Proc. Natl. Acad. Sci. U.S.A.*, 78, 35, 1981.
16. **Bishop, P. D., Pearce, G., Bryant, J. E., and Ryan, C. A.**, Isolation and characterization of the proteinase inhibitor-inducing factor from tomato leaves: identity and activity of poly- and oligolacturonide fragments, *J. Biol. Chem.*, 259, 13172, 1984.
17. **Ryan, C. A., Bishop, P. D., Graham, J. S., Broadway, R. M., and Duffey, S. S.**, Plant cell wall fragments activate proteinase inhibitor genes for plant defense, *J. Chem. Ecol.*, 12, 1025, 1986.
18. **Graham, J. S., Hall, G., Pearce, G., and Ryan, C. A.**, Regulation of synthesis of proteinase inhibitors I and II mRNAs in leaves of wounded tomato plants, *Planta*, 169, 399, 1986.
19. **Thornberg, R. W., Kerman, A., and Molin, L.**, Chloramphenicol acetyl transferase (CAT) protein is expressed in transgenic tobacco in field tests following attack by insects, *Plant Physiol.*, 92, 500, 1990.
20. **Keller, B., Schmid, J., and Lamb, C. J.**, Vascular expression of bean cell wall glycinrich protein-β-glucuronidase gene fusion in transgenic tobacco, *EMBO J.*, 8, 1309, 1989.
21. **Teeri, T. H., Lehvaslaiho, H., Franck, M., Uotila, J., Heino, P., Palva, T. H., Van Montagu, M., and Herrera-Estrella, L.**, Gene fusions to LacZ reveal new expression patterns of chimeric genes in transgenic plants, *EMBO J.*, 8, 343, 1989.
22. **Pena-Cortes, H., Sanchez-Serrano, J., Rocha-Sosa, M., and Willmitzer, L.**, Systemic induction of proteinase inhibitor II gene expression in potato plants by wounding, *Planta*, 174, 84, 1988.
23. **Chen, J. and Varner, J. E.**, An extracellular matrix protein in plants, characterization of genomic clone for carrot extensin, *EMBO J.*, 4, 2145, 1985.
24. **Hilder, V. A., Gatehouse, A. M. R., Sheerman, S. E., Barker, R. F., and Boulter, D.**, A novel mechanism for insect resistance engineered in tobacco, *Nature*, 330, 160, 1987.
25. **Deschle, W. E., Hagen, H. E., Rutschke, J., Stamer, M., and Meyer, T.**, Black fly (Diptera: simuliidae) in West Germany with the biological larvicide *Bacillus thuringiensis* var. *israelensis*. *Bull. Soc. Vector. Ecol.*, 13, 280, 1988.
26. **Donovan, W. P., Gonzalez, J. M., Jr., Gilbert, M. P., and Dankocsik, C.**, Isolation and characterization of EG2158, a new strain of *Bacillus thuringiensis* toxic to coleopteran larvae, and nucleotide sequence of the toxin gene, *Mol. Gen. Genet.*, 214, 365, 1988.
27. **Dubois, N. R., Reardon, R. C., and Kolodny-Hirsch, D. M.**, Field efficiency of the NRD-12 strain of *Bacillus thuringiensis* against gypsy moth (Lepidoptera: Lymantriidae), *J. Econ. Entomol.*, 81, 1672, 1988.
28. **Jaques, R. P. and Laing, D. R.**, Effectiveness of microbial insecticides in control of the Colorado potato beetle (Coleoptera: Chrysomelidae) on potatoes and tomatoes, *Can. Entomol.*, 120, 1123, 1988.
29. **Nicoli, G., Corazza, L., Cornale, R., and Marzoochchi, L.**, Survey of predatory insects in pear orchards using various pest-control strategies, Proc. 15th Italian National Entomology Congress, L'Aquila, June 13th to 17th, 1988, p. 489, 1988.
30. **Srivastva, K. L.**, Consumption of leaves by *Achaea janata* in *Bacillus thuringiensis* and calcium arsenate treatments, *J. Environ. Biol.*, 9, 387, 1988.
31. **Calabrese, D. M., Nickerson, F. W. and Lane, L. C.**, A comparison of protein crystal subunit sizes in *Bacillus thuringiensis*, *Can. J. Microbiol.*, 26, 1006, 1980.
32. **Chestukhina, G. G., Zalumin, J. A., Kostina, L. I., Kotova, T. S., Katturukha, S. P., and Stepanov, V. M.**, Crystal-forming proteins of *Bacillus thuringiensis*, *Biochem. J.*, 187, 457, 1980.
33. **Chestukhina, G. G., Kostina, L. I., Mikhailova, A. L., Tyuria, S. A., Klepikoca, F. S., and Stepanov, V. M.**, The main features of *Bacillus thuringiensis* δ-endotoxin molecular structure, *Arch. Microbiol.*, 132, 159, 1982.

34. **Huber, H. E. and Luthy, P.**, *Bacillus thuringiensis* delta endotoxin: composition and activation, in *Pathogenesis of Invertebrate Microbial Diseases*, Davidson, E.W., Ed., Allenheld, Osmun, Totowa, NJ, 1981, 209.
35. **Bulla, L. A., Jr., Bechtel, D. B., Kramer, K. J., Shethna, Y. I., Aronson, A. I., and Fitz-James, P. C.**, Ultrastructure, physiology, and biochemistry of *Bacillus thuringiensis*, *Crit. Rev. Microbiol.*, 8, 147, 1980.
36. **Bulla, L. A., Jr., Costilow, R. N., and Sharpe, E. S.**, Biology of *Bacillus popilliae*, *Adv. Appl. Microbiol.*, 22, 1, 1978.
37. **Bulla, L. A., Jr., Kramer, K. J., Cox, D. J., Jones, B. L., Davidson, L. I., and Lockhart, G. L.**, Purification and characterization of the entomocidal protein of *Bacillus thuringiensis*, *J. Biol. Chem.*, 256, 3000, 1981.
38. **Fast, P. G.**, The crystal toxin of *Bacillus thuringiensis*, in *Microbial Control of Pests in Plant Diseases*, Burges, H. D., Ed., Academic Press, New York, 1981, 223.
39. **Fast, P. G. and Martin, W. C.**, *Bacillus thuringiensis* crystal toxin: dissociation into toxic low molecular weight peptides, *Biochem. Biophys. Res. Commun.*, 95, 1314, 1980.
40. **Fast, P. C., Murphy, D. W., and Sohi, S. S.**, *Bacillus thuringiensis* δ-endotoxin: evidence that toxin acts at the surface of suscpetible cells, *Experentia*, 34, 762, 1978.
41. **Huber, H. E., Luthy, P., Ebersold, H.-R., and Cordier, J.-L.**, The subunits of the parasporal crystal of *Bacillus thuringiensis*: size, linkage, and toxicity, *Arch. Microbiol.*, 129, 14, 1981.
42. **Lüthy, P.**, Insecticidal toxins of *Bacillus thuringiensis*, *FEMS Microbiol. Lett.*, 8, 1, 1980.
43. **Lüthy, P., Jaquet, F., Huber-Lukac, H. E., and Huber-Lukac, M.**, Physiology of the delta endotoxin of *Bacillus thuringiensis* including the ultra-structure and histopathological studies, in *Basic Biology of Microbial Larvicides of Vectors of Human Diseases*, Michal, F., Ed., UNDP/World Bank/WHO, Geneva, 1981, 29.
44. **Nagamatsu, Y., Itai, Y., Hatanaka, C., Funatsu, G., and Hayashi, K.**, A toxic fragment from the entomocidal crystal protein of *Bacillus thuringiensis*, *Agric. Biol. Chem.*, 48, 611, 1984.
45. **Nagamatsu, Y., Tsutsui, R., Ichimaru, T., Nagamatsu, M., Koga, K., and Hayashi, K.**, Subunit structure and toxic component of delta endotoxin from *Bacillus thuringiensis*, *J. Invertebr. Pathol.*, 32, 103, 1978.
46. **Nickerson, K. W.**, Structure and function of the *Bacillus thuringiensis* protein crystal, *Biotechnol. Bioeng.*, 12, 1305, 1980.
47. **Tyrell, D. J., Bulla, L. A., Andrews, R. E., Kramer, K. J., Davidson, L. I., and Nordin, P.**, Comparative biochemistry of entomocidal parasporal crystals of selected *Bacillus thuringiensis* strains, *J. Bacteriol.*, 145, 1052, 1981.
48. **Angus, T. A.**, A bacterial toxin paralyzing silkworm larvae, *Nature (London)*, 173, 545, 1954.
49. **Angus, T. A.**, Extraction, purification, and properties of *Bacillus sotto* toxin, *Can. J. Microbiol.*, 2, 416, 1956.
50. **Aronson, A. I. and Dunn, P. E.**, Regulation of protoxin synthesis in *Bacillus thuringiensis*: conditional synthesis in a variant is supressed by D-cycloserine, *FEMS Microbiol. Lett.*, 27, 237, 1985.
51. **Aronson, A. I. and Fitz-James, P.**, Structure and morphogenesis of the bacterial spore coat, *Bacteriol. Rev.*, 40, 360, 1976.
52. **Aronson, A. I., Tyrell, D. J., Fitz-James, P. C., and Bulla, L. A., Jr.**, Relationship of the synthesis of spore coat protein and parasporal crystal protein in *Bacillus thuringiensis*, *J. Bacteriol.*, 151, 399, 1982.
53. **Hannay, C. L. and Fitz-James, P. C.**, The protein crystals of *Bacillus thuringiensis berliner*, *Can. J. Microbiol.*, 1, 694, 1955.
54. **Scherrer, P., Luthy, P., and Trumpi, B.**, Production of delta endotoxin by *Bacillus thuringiensis* as a function of glucose concentration, *Appl. Microbiol.*, 25, 644, 1973.

55. **Somerville, H. J.**, Formation of the parasporal inclusion of *Bacillus thuringiensis*, *Eur. J. Biochem.*, 18, 226, 1971.
56. **Somerville, H. J.**, Microbial toxins, *Ann. N.Y. Acad. Sci.*, 217, 93, 1973.
57. **Somerville, H. J. and James, C. R.**, Association of the crystalline inclusion of *Bacillus thuringiensis* with the exosporium, *J. Bacteriol*, 102, 580, 1970.
58. **Debro, L., Fitz-James, P. C., and Aronson, A.**, Two different parasporal inclusions are produced by *Bacillus thuringiensis* subsp. *finitimus*, *J. Bacteriol.*, 165, 258, 1986.
59. **Short, J. A., Walker, P. D., Thompson, R. O., and Somerville, H. J.**, The fine structure of *Bacillus finitimus* and *Bacillus thuringiensis* spores with special reference to the location of the crystal antigen, *J. Gen. Microbiol.*, 84, 261, 1974.
60. **Charles, J. F. and deBarjac, H.**, Sporulation of crystalogenese de *Bacillus thuringiensis* var. *israelensis* em microscopie electronique, *Ann. Microbiol (Inst. Pasteur)*, 133A, 425, 1982.
61. **Mikkola, A. R., Carlberg, G. A., Vaara, T., and Gylenberg, H. G.**, Comparison of inclusions in different *Bacillus thuringiensis* strains. An electron microscope study, *FEMS Microbiol. Lett.*, 13, 401, 1982.
62. **Yamamoto, T., Tizuku, T., and Aronson, J. N.**, Mosquitocidal protein of *Bacillus thuringiensis* subsp. *israelensis*: identification and partial isolation of the protein, *Curr. Microbiol.*, 9, 279, 1983.
63. **Goldberg, L. J. and Margalit, J.**, A bacterial spore demonstrating rapid larvicidal activity against *Anopheles sergentii*, *Uranotaenia unguiculata*, *Culex univitattus*, *Aedes aegypti*, *Culex pipiens*, *Mosquito News*, 37, 355, 1977.
64. **Armstrong, J. L., Rohermann, G. F., and Beaudeau, G. S.**, Delta endotoxin of *Bacillus thuringiensis* subsp. *israelensis*, *J. Bacteriol.*, 161, 39, 1985.
65. **Yamamoto, T. and McLaughlin, R. E.**, Isolation of protein from the parasporal crystal of *Bacillus thuringiensis* var. *kurstaki* toxic to the mosquito larva, *Aedes taeniorhychus*, *Biochem. Biophys. Res. Commun.*, 103, 414, 1981.
66. **Hurley, J. M., Lee, S. G., Andrews, R. E., Jr., Klowden, M. J., and Bulla, L. A., Jr.**, Separation of the cytolytic and mosquitocidal proteins of *Bacillus thuringiensis* subsp. *israelensis.*, *Biochem. Biophys. Res. Commun.*, 126, 961, 1985.
67. **Lee, S. G., Eckbad, W., and Bulla, L. A., Jr.**, Diversity of protein inclusion bodies and identification of mosquitocidal protein in *Bacillus thuringiensis* subsp. *israelensis*, *Biochem. Biophys. Res. Commun.*, 126, 953, 1985.
68. **Cooksey, K. E.**, The protein crystal toxin of *Bacillus thuringiensis*: biochemistry and mode of action, in *Microbial Control of Insects and Mites*, Burges, H. D. and Hussey, N. W., Eds., Academic Press, New York, 1971, 247.
69. **Lecadet, M.-M. and de Barjac, H.**, *Bacillus thuringiensis* during sporulation, *Eur. J. Biochem.*, 23, 282, 1971.
70. **Wong, H. C., Schnepf, H. E., and Whiteley, H. R.**, Transcriptional and translational start sites for the *Bacillus thuringiensis* crystal protein gene, *J. Biol. Chem.*, 258, 1960, 1983.
71. **Kronstad, J. W., Schnepf, H. E., and Whiteley, H. R.**, Diversity of locations for *Bacillus thuringiensis* crystal protein genes, *J. Bacteriol.*, 134, 419, 1983.
72. **Klier, A., Lereclus, D., Fibier, J., Bourgouin, C., Menou, G., Lecadet, M.-M., and Rapoport, G.**, Cloning and expression in *Escherichia coli* of the crystal protein gene from *Bacillus thuringiensis* strain *aizawal* 7-29 and comparison of the structural organization of the genes from different serotypes, in *Molecular Biology of Microbial Differentiation*, American Society for Microbiology, Washington, D.C., 1985, 217.
73. **Yamvrias, C.**, Contribution a letude du mode d'action de *Bacillus thuringiensis berliner* vis-a-vis de la teigne de la forine flower moth *Anagaste* (Ephestia) *kuniella*-source *berliner*, *Entomophaga*, 7, 101, 1962.
74. **Knowels, B. H., Thomas, W. E., and Ellar, D. J.**, Lectin-like binding of *Bacillus thuringiensis* var. *kurstaki* lepidopteran-specific toxin is an initial step in insecticidal action, *FEBS Microbiol. Lett.*, 168, 197, 1984.

75. Hiader, M. Z. and Ellar, D. J., Nucleotide sequence of *Baccilus thuringiensis aizawai* IC1 entomocidal crystal protein gene, *Nucleic Acids Res.*, 16, 10927, 1988.
76. Kronstad, J. W. and Whiteley, H. R., Inverted repeat seqeunces flank a *Bacillus thuringiensis* crystal protein gene, *J. Bacteriol.*, 160, 95, 1984.
77. Debabov, V. G., Azizbekyan, R. R., Khlebalina, O. I., Dyachenko, V. V., Galushka, F. P., and Belykh, R. A., Isolation and preliminary characterization of extra chromosomal elements of *Bacillus thuringiensis* DNA, *Genetika*, 13, 496, 1977.
78. Gonzalez, J. M., Jr. and Carlton, B. C., Patterns of plasmid DNA in crystalliferous and acrystalliferous strains of *Bacillus thuringiensis*, *Plasmid*, 3, 92, 1980.
79. Gonzalez, J. M., Jr. and Carlton, B. C., A large transmissible plasmid is required for crystal toxin production in *Bacillus thuringiensis* variety *israelensis*, *Plasmid*, 11, 28, 1984.
80. Gonzalez, J. M., Jr., Dulmage, H. T., and Carlton, B. C., Correlation between specific plasmids and δ-endotoxin production in *Bacillus thuringiensis*, *Plasmid*, 5, 351, 1981.
81. Klier, A., Fargette, F., Riber, J., and Rapaport, G., Cloning and expression of the crystal proteins genes from *Bacillus thuringiensis* strain *berliner* 1715, *EMBO J.*, 1, 791, 1982.
82. Schnepf, H. E. and Whiteley, H. R., Cloning and expression of the *Bacillus thuringiensis* crystal protein gene in *Escherichia coli*, *Proc. Natl. Acad. Sci. U.S.A.*, 78, 2893, 1981.
83. Schnepf, H. E. and Whiteley, H. R., Delineation of a toxin-encoding segment of *Bacillus thuringiensis* crystal protein gene, *J. Biol. Chem.*, 260, 6273, 1985.
84. Schnepf, H. E., Wong, H. C., and Whiteley, H. R., The amino acid sequence of a crystal protein from *Bacillus thuringiensis* deduced from the DNA base sequence, *J. Biol. Chem.*, 260, 6264, 1985.
85. Mahillon, J. and Seurinck, J., Complete nucleotide sequence of pG12, a *Bacillus thuringiensis* plasmid containing Tn4430, *Nucleic Acids Res.*, 16, 11827, 1988.
86. Mahillon, J., Hespel, F., Perssens A.-M., and Delcour, J., Cloning and partial characterization of three small cryptic plasmids from *Bacillus thuringiensis*, *Plasmid*, 19, 169, 1988.
87. Klier, A., Parsot, C., and Rapaport, G., *In vitro* transcription of the cloned chromosomal gene from *Bacillus thuringiensis*, *Nucleic Acids Res.*, 11, 3975, 1983.
88. Klier, A., Bourgouinm C., and Rapport, G., Mating between *bacillus subtilis* and *Bacillus thuringiensis* and transfer of cloned crystal genes, *Mol. Gen. Genet.*, 191, 257, 1983.
89. Held, G. A., Bulla, L. A., Jr., Ferrari, E., Hoch, J. A., Aronson, A. I., and Minnich, S. A., Cloning and localization of the lepidopteran protoxin gene of *Bacillus thuringiensis* subsp. *kurstaki*, *Proc. Natl. Acad. Sci. U.S.A.*, 79, 6065, 1982.
90. Sen, K., Honda, G., Koyama, N., Nishida, M., Neki, A., Sakai, H., Himeno, M., and Komano, T., Cloning and nucleotide sequence of the two 130 KDa insecticidal protein genes of *Bacillus thuringiensis* var. *israelensis*, *Agric. Biol. Chem.*, 52, 873, 1988.
91. Wie, S. I., Andrews, R. E., Hammock, B. D., Faust, R. M., and Bulla, L. A., Jr., Enzyme-linked immunosorbent assay for detection of the entomocidal parasporal crystalline protein of *Bacillus thuringiensis* subsp. *kurstaki* and *israelensis*, *Appl. Environ. Microbiol.*, 43, 891, 1982.
92. Minnich, S. A. and Aronson, A. I., Regulation of protoxin synthesis in *Bacillus thuringiensis*, *J. Bacteriol.*, 158, 447, 1984.
93. Fitz-James, P. and Young, E., Morphology of sporulation, in *The Bacterial Spore*, Could, G. W. and Hurst, A., Eds., Academic Press, New York, 1969, 39.
94. Andrews, R. E., Jr., Bechtel, D. B., Campbell, B. S., Davidson, L. I., and Bulla, L. A., Jr., Solubility of parasporal crystals of *Bacillus thuringiensis* and presence of toxic protein during sporulation, germination and outgrowth, in *Sporulation and Germination*, Levinson, H. S., Soneshein, A. L., and Tipper, D. J., Eds., American Society for Microbiology, Washington, D.C., 1981, 174.

95. **Andrews, R. E., Jr., Iandolo, J. J., Campbell, B. S., Davidson, and Bulla, L. A., Jr.**, Rocket immunoelectrophoresis of the entomocidal parasporal crystal of *Bacillus thuringiensis* subsp. *kurstaki*, *Appl. Environ. Microbiol.*, 40, 897, 1980.
96. **Andrews, R. E., Jr, Faust, R. M., Wabiko, H., and Raymond, K. C.**, Insect control with genetically engineered crops, *Crit. Rev. Biotechnol.*, 6, 163, 1987.
97. **Ghosh, D., Majumdar, T., and Bhattacharya, T.**, Efficiency of *Bacillus thuringiensis israelensis* as larvicide of *Culex* at laboratory condition, *Environ. Ecol.*, 6, 509, 1988.
98. **Green, B. D., Battisti, L., and Thorne, C. B.**, Involvement of Tn4430 in transfer of *Bacillus anthracis* plasmids mediated by *Bacillus thuringiensis* plasmid pXO12, *J. Bacteriol.*, 17, 104, 1989.
99. **Adam, L. F., Visick, J. E., and Whiteley, H. R.**, A 20 KDa protein is required for efficient production of the *Bacillus thuringiensis* subsp. *israelensis* 27 KDa crystal protein in *Escherichia coli*, *J. Bacteriol.*, 171, 521, 1989.
100. **Sekar, V.**, The insecticidal crystal protein gene is expressed in vegetative cells of *Bacillus thuringiensis* var. *tenebrionis*, *Curr. Microbiol.*, 17, 347, 1988.
101. **Li, J., Henderson, R., Carroll, J., and Ellar, D.**, X-ray analysis of the crystalline parasporal inclusion in *Bacillus thuringiensis* var. *tenebrionis*, *J. Mol. Biol.*, 199, 543, 1988.
102. **Dulmage, H. T.**, Insecticidal activity of HD-1, a new isolate of *Bacillus thuringiensis* var. *alesti*, *J. Invertebr. Pathol.*, 15, 232, 1970.
103. **Dulmage, H. T.**, Production of δ-endotoxin by eighteen isolates of *Bacillus thuringiensis*, serotype 3, in fermentation media, *J. Invertebr. Pathol.*, 18, 355, 1971.
104. **Zuber, P. and Losick, R.**, Use of a *lacZ* fusion to study the role of the *spoO* genes of *Bacillus subtilis* in development regulation, *Cell*, 35, 275, 1983.
105. **Davison, J.**, Plant beneficial bacteria, *Bio/Technology*, 6, 282, 1988.
106. **Vaeck, M., Reynaerts, A., Hofte, H., Jansens, S., De Beuckeleer, M., Dean, C., Zabeau, M., Van Montagu, M., and Leemans, J.**, Transgenic plants protected from insect attack, *Nature*, 327, 33, 1987.
107. **Fischoff, D. A., Bowdish, K. S., Perlak, F. J., Marrone, P. G., McCormick, S. M., Niedermeyer, J. G., Dean, D. A., Kusano-Kretzmer, K., Mayer, E. J., Rechester, D. E., Rogers, S. G., and Fraley, R. T.**, Insect tolerant transgenic plants, *Bio/Technology*, 5, 807, 1987.
108. **Perlak, F. J., Deaton, R. W., Armstrong, T. A., Fuchs, R. L., Sims, S. R., Greenplate, J. T., and Fischhoff, D. A.**, Insect resistant cotton plants, *Bio/Technology* 8, 939, 1990.
109. **Perlak, F. J., Fuchs, R. L., Dean, D. A., McPherson, S. L., and Fischhoff, D. A.**, Modification of the coding sequence enhances plant expression of insect control protein genes, *Proc. Natl. Acad. Sci. U.S.A.*, 88, 3324, 1991.
110. **Velten, J., Velten, L., Hain, R., and Schell, J.**, Isolation of a dual plant promoter fragment from the Ti plasmid of *Agrobacterium tumifaciens*, *EMBO J.*, 3, 2723, 1984.
111. **Delannay, X., LaVallee, B. J., Proksch, R. K., Fuchs, R. L., Sims, S. R., Greenplate, J. T., Marohe, P. G., Dudson, R. B., Augustine, J. J., Layton, J. G., and Fischhoff, D. A.**, Field performance of transgenic tomato plants expressing the *Bacillus thuringiensis* var. *kurstaki* insect control protein, *Bio/Technology*, 7, 1265, 1989.
112. **Murray, E. E., Lotzer, J., and Eberle, M.**, Codon usage in plant genes, *Nucleic Acids Res.*, 17, 477, 1989.
113. **Harris, M. K.**, Biology and breeding for resistance to arthropods and pathogens in agricultural plants, Proc. Int. Short Course in Host Plant Resistance, July–August, 1979, Texas A & M University, College Station, 1980.
114. **Lamberti, F., Waller, J. M., and Van der Graaf, N. A.**, *Durable Resistance in Crops*, Plenum Press, New York, 1983.
115. **Gould, F.**, The Evolution of Adaptation to Host Plants and Pesticides in a Polyphagous Herbivore, *Tetranychus urticae* Kochk., Ph.D. thesis, S.U.N.Y. at Stoney Brook, New York, 1977.

116. **Gould, F.**, Resistance of cucumber varieties to *Tetranychus urticae*: Genetic and environmental determinants, *J. Econ. Entomol.*, 71, 680, 1978.
117. **Gould, F.**, Rapid host range evolution in a population of the phytophagous mite, *Tetranychus urticae* Koch., *Evolution*, 33, 791, 1979.
118. **Gould, F.**, Genetics of plant-herbivore systems: interactions between applied and basic study, in Variable Plants and Herbivores in *Nature and Managed Systems*, Denno, R. and McClure, B., Eds., Academic Press, New York, 1983, 599.
119. **Gould, F.**, Simulation models for predicting durability of insect-resistant germplasm: a deterministic diploid, two-locus model, *Environ. Entomol.*, 15, 1, 1986.
120. **Gould, F.**, Evolutionary biology and genetically engineered crops, *BioScience*, 38, 26, 1988.
121. **Gould, F.**, Ecological genetic approaches for the design of genetically engineered crops, in *Biotechnology, Biological Pesticides, and Novel Plant-Pest Resistance for Insect Pest Management*, Roberts, D. W. and Granados, R. R., Eds., Boyce Thompson Institute, Cornell University, Ithaca, NY, 1989, 146.
122. **Gould, F., Kennedy, G., and Johnson, M. T.**, Effect of natural enemies on the rate of herbivore adaptation to resistant host plants, *Entomol. Exp. Appl.*, 58, 1, 1991.
123. **Browning, J. A.**, Genetic protective mechanisms of plant-pathogen populations: their coevolution and use in breeding for resistance, *Biology and Breeding for Resistance to Arthropods and Pathogens in Agricultural Plants*, Harris, M. K., Ed., Texas A & M University, College Station, 1980, 52.
124. **Leonard, K. J.**, The host population as a selective factor, in *Populations of Plant Pathogens: Their Dynamics and Genetics*, Wolfe, M. S. and Caten, C. E., Eds., Blackwell Scientific, Oxford, U.K., 1986, 163.
125. **Bergman, J. M. and Tingey, W. M.**, Aspects of the interaction between plant genotypes and biological control, *Bull. Entomol. Soc. Am.*, 25, 275, 1979.
126. **Wrubel, R. P., Krimsky, and Wetzler, R. E.**, Field testing transgenic plants, *BioScience*, 42, 280, 1992.
127. **Van Emden, H. F.**, Plant insect relationships and pest control, *World Rev. Pest Cont.*, 5, 115, 1966.
128. **Kartohardjono, A. and Heinrichs, E. A.**, Populations of the brown planthopper, *Nilaparvata lugens* (Stal.) (Homoptera: Delphacidae), and its predators on rice varieties with different levels of resistance, *Environ. Entomol.*, 13, 359, 1984.
129. **Obrycki, J. J.**, The influence of foliar pubescence on entomophagous species, in *Interactions of Plant Resistance and Parasitoids and Predators of Insects*, Boethel, D. J. and Eikenbary, E. D., Eds., Ellis Horwood, Chichester, England, 1986, 61.
130. **Obrycki, J. J., Tauber, M. J., and Tingey, W. M.**, Predator and parasitoid interaction with aphid-resistant potatoes to reduce aphid densities: a two-year field study, *J. Econ. Entomol.*, 76, 456, 1983.
131. **Price, P. W.**, *Insect Ecology*, 2nd ed., J. Wiley-Interscience, New York, 1984.
132. **Price, P. W., Bouton, C. E., Gross, P., McPheron, B. A., Thompson J. N., and Weiss, W. E.**, Interactions between insect herbivores and natural enemies, *Annu. Rev. Ecol. Syst.*, 11, 41, 1980.
133. **Trichilo, P. J. and Leigh, T. F.**, The impact of cotton plant resistance on spider mites and their natural enemies, *Hilgardia*, 54, 1, 1986.
134. **Starks, K. J., Muniappan, R., and Eikenbary, R. D.**, Interaction between plant resistance and parasitism against greenbug on barley and sorghum, *Ann. Entomol. Soc. Am.*, 65, 650, 1972.
135. **Wilhoit, L. R.**, The Effects of Plant Varieties on the Evolution of the Aphid, *Schizaphis graminum* (Rondani), Ph.D. dissertation, University of California, Berkeley, 1988.
136. **Schultz, J. C.**, Impact of variable plant defensive chemistry on susceptibility of insects to natural enemies, in *Plant Resistance to Insects*, Hedin, A., Ed., ACS Symp. Ser. Vol. 208, American Chemical Society, Washington, D.C., 37, 1983.

137. **Van Emden, H. F.**, The interaction of plant resistance and natural enemies: effects on populations of sucking insects, in *Interactions of Plant Resistance and Parasitoids and Predators of Insects*, Boethel, D. J. and Eikenbary, R. D., Eds., Ellis Horwood, Chichester, England, 1986, 138.
138. **Bergelson, J. M. and Lawton, J. H.**, Does foliage damage influence predation on the insect herbivores of birch, *Ecology*, 69, 434, 1988.
139. **Kennedy, G. G.**, 2-tridecanone, tomatoes and *Heliothis zea*: potential incompatibility of plant antibiosis with insecticidal control (*Lycopersicon hirsutum f. glabratum*), *Entomol. Exp. Appl.*, 35, 305, 1984.
140. **Kennedy, G. G., Gould, F., dePonti, O. M. B., and Stinner, R. E.**, Ecological, agricultural, genetic, and commercial considerations in the deployment of insect-resistant germplasm. *Environ. Entomol*, 16, 327, 1987.
141. **Boethel, D. J. and Eikenbary, R. D.**, *Interactions of Plant Resistance and Parasitoids and Predators of Insects*, Ellis Horwood Ltd., Chichester, England, 1986, 224.
142. **Raffa, K. F.**, Genetic engineering of trees to enhance resistance to insects, *BioScience*, 39, 524, 1989.
143. **Abdul-Naser, S. F., Amman, E. D., Merdan, A. I., and Farrag, S. M.**, Infectivity tests on *Bacillus thuringiensis* and *B. Sypiella*, *Z. Angew. Entomol.*, 85, 60, 1979.
144. **Adang, M. J., Staver, M. J., Rocheeleau, T. A., Leighton, J., Barker, R. F., and Thompson, D. V.**, Characterized full-length and truncated plasmid clones of the crystal protein of *Bacillus thuringiensis* subsp. *kurstaki* HD-73 and their toxicity to *Manduca sexta*, *Gene*, 36, 289, 1985.
145. **Barsomian, G. D., Robilard, N. J., and Thorne, C. B.**, Chromosomal mapping of *Bacillus thuringiensis* by transduction, *J. Bacteriol.*, 157, 746, 1984.
146. **Barton, K. A., Whiteley, H. R., and Yang K.**, *Bacillus thuringiensis* δ-endotoxin expressed on transgenic *Nicotiana tabacum* provides resistance to lepidopteran insects, *Plant Physiol.*, 85, 1103, 1987.
147. **Battisti, L., Green, B. D., and Thorne, C. B.**, Mating system for transfer of plasmids among *Bacillus anthracis, Bacillus cereus,* and *Bacillus thuringiensis*, *J. Bacteriol.*, 162, 543, 1985.
148. **Baumann, P. B., Unterman, B. M., Baumann, L., Broadwell, A. H., Abbene, S. J., and Bowditch, R. D.**, Purification of the larvicidal toxin of *Bacillus sphaericus* and evidence for high molecular weight precursors, *J. Bacteriol.*, 163, 738, 1985.
149. **Bourgouin, C., Delecluse, A., Ribier, J., Klier, A., and Rapopri, G.**, *Bacillus thuringeinsis* subsp. *israelensis* inverted repeat sequences, *J. Bacteriol.*, 170, 3575, 1988.
150. **Bouton, C. E.**, Plant Defensive Traits: Translation of Their Effects on Herbivorous Insects into Reduced Plant Damage, Ph.D. dissertation, University of Illinois, Urbana, 1984.
151. **Brousseau, R. and Masson, L.**, *Bacillus thuringiensis* insecticidal crystal toxins: Gene structure and mode of action, *Biotech. Adv.*, 6, 697, 1988.
152. **Campos, F., Donskov, N., Arnason, J. T., Philogene, B. J. R., Atkinson, J., Morand, P., and Wertiuk, N. H. K.**, Biological effects of toxicokinetics of DIMBOA in *Diadegma terebrans* (Hymenoptera: Ichneumonidae), an endoparasitoid of *Ostrinia nubilalis* (Lepidoptera: Pyralidae), *J. Econ. Entomol.*, 83, 356, 1990.
153. **Casagrande, R. A. and Haynes, D. L.**, The impact of pubescent wheat on the population dynamics of the cereal leaf beetle, *Environ. Entomol.*, 5, 153, 1976.
154. **Danks, H. V., Rabb, R. L., and Southern, P. S.**, Biology of insect parasites of *Heliothis* larvae in North Carolina, *J. Ga. Entomol. Soc.*, 14, 36, 1979.
155. **Davidson, E. W. and Myers, P.**, Parasporal inclusions in *Bacillus sphaericus*, *FEMS Microbiol. Lett.*, 10, 261, 1981.
156. **Herrnstadt, C., Gilory, T. E., Sobieski, D. A., Bennett, B. D., and Gaertner, F. H.**, Nucleotide sequence and deduced amino acid sequence of a coleopteran-active delta-endotoxin gene from *Bacillus thuringiensis* subsp. *san diego*, *Gene*, 57, 37, 1987.

157. **Honee, G., van der Salm, T., and Visser, B.**, Nucleotide sequence of crystal protein gene isolated from *B. thuringiensis* subspecies *entomocidus* 60.5 coding for a toxin highly active against *Spodoptera* species, *Nucleic Acids Res.*, 16, 6240, 1988.
158. **Hofte, H. and Whiteley, H. R.**, Insect control with genetically engineered crops, *Micro. Rev.*, 53, 242, 1989.
159. **Hofte, H., van Rie, J., Jansens, S., van Houtven, A., Vanderbruggen, H., and Vaeck, M.**, Monoclonal antibody analysis and insecticidal spectrum of three types of lepidopteran-specific insecticidal crystal proteins of *Bacillus thuringiensis, Appl. Environ. Micobiol.*, 54, 2010, 1988.
160. **Hofte, H., De Greve, H., Seurinkck, J., Jansens, S., Mahillon, J., Ampe, C., Vandekerchhove, J., Vanderbruggen, H., Van Montagu, M., Zabeau, M., and Vaeck, M.**, Structural and functional analysis of a cloned delta endotoxin of *Bacillus thuringiensis berliner* 1715, *Eur. J. Biochem.*, 161, 273, 1986.
161. **Horsch, R. B., Fry, J. E., Hoffmann, N. L., Wallroth, M., Eicholt, Z. D., Rogers, S. J., and Fraley, R. T.**, A simple and general method for transferring genes into plants, *Science*, 227, 1229, 1985.
162. **Kalfon, A. R. and De Barjac, H.**, Screening of the insecticidal activity of *Bacillus thuringiensis* strains against the Egyptian cotton leaf worm *Spodoptera littoralis, Entomophaga*, 30, 177, 1985.
163. **King, E. G. and Coleman, R. J.**, Potential for biological control of *Heliothis* species, *Annu. Rev. Entomol.*, 34, 53, 1989.
164. **Krieg, A., Huger, A. M., Langenbruch, G. A., and Schnetter, W.**, *Bacillus thuringiensis* var. *tenebrionis*: ein neuer gegenuber Larven von Coleopteren wirksamer Pathotype, *Z. Angew. Entomol.*, 96, 500, 1983.
165. **Lerclus, D., Ribber, J., Klier, A., Menou, G., and Lecadet, M.-M.**, A transposon-like structure related to δ-endotoxin gene of *Bacillus thuringiensis, EMBO J.*, 3, 2561, 1984.
166. **Lilley, M., Ruffel, R. N., and Somerville, H. J.**, Purification of the insecticidal toxin in crystals of *Bacillus thuringiensis, J. Gen. Microbiol.*, 118, 1, 1980.
167. **Longmann, J., Lipphardt, S., Lorz, H., Hauser, I., Willimitzer, L., and Schell, J.**, 5′ upstream sequences from the *wun*1 gene are responsible for gene activation by wounding in transgenic plants. *Plant Cell*, 1, 151, 1989.
168. **Matur, A. and Ceber, K.**, The utilization of bacilli as larvicidal agents against anophenline and culicide mosquitoes in Turkey: I. Larvicidal activity of *Bacillus thuringiensis* serotype H-14, *J. Trop. Med. Hyg.*, 91, 229, 1988.
169. **Shibano, Y., Yamagata, Nakamura, N., Iizuku, T., Sugisaki, H., and Takanami, M.**, Nucleotide sequence coding for the insecticidal fragment of the *Bacillus thuringiensis* crystal protein, *Gene*, 34, 243, 1985.
170. **Stahly, D. P., Dingman, D. W., Irgens, R. L., Field, C. E., Feiss, M. G., and Smith, G. L.**, Multiple extrachromosomal deoxyribonucleic acid molecules in *Bacillus thuringiensis, FEMS Microbiol. Lett.*, 3, 139, 1978.
171. **van Frankenhuyzen, K. and Fast, P.G.**, Susceptibility of three coniferophagous *Choristoneura* species (Lepidoptera: Tortiricidae) to *Bacillus thuringiensis* var. *kurstaki, J. Econ. Entomol.*, 82, 193, 1989.
172. **Winstein, S. A., Bernheimer, A. W., and Oppenheim, J. D.**, Isolation of a hemolysin from a spore-crystal mixture of *Bacillus thuringiensis israelensis* (serotype H-14), *Toxicon.*, 26, 733, 1988.
173. **Widner, W. R. and Whiteley, H. R.**, Two highly related insecticidal crystal proteins of *Bacillus thuringiensis* subsp. *kurstaki* posses different host range specificities, *J. Bacteriol.*, 171, 965, 1989.
174. **Woods, S. A., Elkinton, J. S., and Shapiro, M.**, Effects of *Bacillus thuringiensis* treatments on the occurrence of nucelra polyhedrosis virus in gypsy moth (Lepidotera: Lymantriidae) populations, *J. Econ. Entomol.*, 81, 1706, 1988.

Chapter 3

ENGINEERING HERBICIDE RESISTANCE INTO PLANTS

I. INTRODUCTION

Herbicide treatments are an integral part of modern agriculture, since they provide cost-effective increases in agricultural productivity. Increased yields result from reduced weed competition for water, light, and nutrients. In addition, crop quality often improves in the absence of contaminating weed seeds, such as wild mustard seeds in harvested canola or wild garlic seeds in wheat. Herbicides can also aid soil conservation efforts through no-till agricultural practices, wherein herbicides rather than tillage are used to reduce weed populations prior to planting.

An ideal herbicide should combine the following properties: (1) control of all plant species except for the crop of interest; (2) a high degree of environmental safety; and (3) minimal persistence in soil. However, it is difficult to find a herbicide with all these properties. Most herbicides are designed to affect the photosynthetic process, which is common both to plants and weeds. Consequently, at present, selectivity is only based on differential herbicide uptake between crop and weed, controlled timing and site of application, or detoxification of the herbicide by crop plant.

Thus selective toxicity of herbicides to weeds but not to crops is one of the most difficult properties to achieve. However, selectivity is a function of the physicochemical properties of a compound, and of the biochemical interactions of the compound with the crop and the weeds, and these parameters have been exploited to achieve selectivity in a few cases. For example, herbicides that do not percolate beyond the top soil layer, and thus do not affect crop roots that extend below this layer, can provide selectivity. In addition, environmental conditions such as climate, soil pH, and soil organic content influence these interactions. For some compounds, management practices (e.g., timing and/or site of the application) can be used to impart selectivity. Glyphosate, a nonselective herbicide, is used prior to planting as a substitute for tillage, thereby taking advantage of its rapid and wide-spectrum weed-killing activity and short lifetime in the soil.

A number of important herbicides (e.g., the triazines, sulfonyl-ureas, and imidazolinones) are more toxic to weeds than to specific crops. In these examples, selectivity results from a unique or enhanced metabolic detoxification of the herbicide by the crop plant but not by the weed. In other cases, herbicide selectivity results from the sequestering of the herbicide within an internal compartment of the crop plant. Alternatively, external barriers such as plant cuticles can prevent penetration of the herbicide. In some cases it has been possible to achieve selectivity by seed coat applications of a "safener", a second chemical that reduces the toxicity of the herbicide to the crop.

In addition to this, several attempts have been made to produce crops resistant/tolerant to herbicides by conventional methods, although this, too, has not yielded very encouraging results. Subsequently the tissue culture methods and regeneration potential of plant tissues have been exploited for this purpose. It is interesting to note that in some instances resistant cells appeared in large numbers even in the absence of a mutagenic agent. This high variability appears to be triggered by the process of *in vitro* culture, particularly when cell cultures are cultivated over a prolonged period as undifferentiated cells. Under these circumstances, chromosomal mutations accumulate at a very high rate. This variability is known as "somaclonal variation"[1] and may possibly be accounted for by the mobilization of transposable elements due to conditions in the cell culture.[2]

In fact, manipulations of protoplasts have yielded cells showing somaclonal variation and herbicide-resistant plants from such cells have been raised. It is also not absolutely necessary to obtain protoplasts for the selection of mutants *in vitro*. Any plant cell culture that can be regenerated is suitable for this purpose. Application of these techniques permitted the selection and regeneration of plants with resistance to the herbicides chlorate,[3] picloram,[4,5] amitrole,[6] glyphosate,[6] chlorsulfuron, and sulfometuron methyl[5] and imazapyr.[7,8] Due to the randomness of mutations leading to resistance, the molecular mechanisms of the resistance in many of these cases remain unknown. As an example of this, the glyphosate-resistant tobacco plants described by Singer and McDaniel[6] exhibit cross-resistance with amitrole, a herbicide with a completely different chemical structure and mode of action. Some of the regenerated herbicide-resistant plants show poorer growth rates than the corresponding wild-type plants. This has been attributed to cell culture conditions,[9] in particular to chromosomal transpositions. However, the selected plants as a rule serve only as "donors" of the resistance gene in further crossing experiments, so that this would not appear to constitute a major problem for the development of herbicide-resistant plants by cell culture techniques.

With the development of recombinant DNA technology, considerable progress has been made to clone and insert foreign genes into plants conferring herbicide tolerance. Herbicide resistance, particularly where mode of action had been identified as a sensitive enzyme, was an obvious candidate for such manipulations. Such information is handy for the isolation of genes providing resistance/tolerance to several herbicides. The transgenic plants thus produced showed the expression of foreign genes and a high level of herbicide tolerance.

II. STRATEGIES FOR THE DEVELOPMENT OF HERBICIDE-RESISTANT PLANTS

In general, the production of herbicide-resistant plants is based on the utilization of enzymes which detoxify the herbicide. In fact, natural selectivity is primarily based on the fact that the resistant crop plant detoxifies the herbicide metabolically.[10] Resistant plants modify herbicides to inactive derivatives, often by conjugation with other compounds. This detoxification may take place via a whole series of separate metabolic steps. The biochemical processes underlying metabolism are very similar to the detoxification reactions familiar in animal tissues[11] and comprise oxidation, reduction, hydrolysis, and/or conjugate formation resulting in hydroxylation of aliphatic chains or aromatic residues in herbicides. The greatest advances in our knowledge have been made in connection with the detoxifying enzymes belonging to the group of glutathione-S-transferases (GST), which catalyze the conjugation of glutathione to an electrophilic center of hydrophobic herbicides by means of the -SH group. For example, GSTs are responsible for resistance in maize, millet, sugarcane, and a number of grasses (e.g., *Gigiratia sanguinalis, Panicum dichotomifolium, Setaria italica*).

At least three different GST isoenzymes have been detected in maize tissue, which together constitute about 1 to 2% of the total soluble protein and show different substrate specificities.[12] One of the isoenzymes (GST II) is induced by treatment of the plant with herbicide safeners.[12,13] This GST isoenzyme induced by the safener possesses a greater detoxification capacity than the other two isoenzymes.[14]

In addition to the conjugation of herbicides with glutathione, detoxification reactions are also known to occur via conjugations of glucose or amino acids with amino-, carboxyl-, or hydroxyl groups of herbicides. The literature contains description of glucose conjugates of metribuzin,[15,16] dichlofop-methyl,[17] 2,4-dichlorophenoxyacetic acid (2,4-D),[18-20] sulfonylurea herbicides,[21,22] imidazolinones,[23] and many others.[24] Conjugates with amino acids are known between 2,4-D and glutamic or aspartic acid.[25] Practically no biochemical data exist on the proteins which catalyze these conjugations. This is also true of the mixed-function oxidases which play an important role in the metabolism of 2,4-D,[26] chlorotoluron, sulfonylureas,[22] imidazolinones,[23] and other classes of herbicides.[17,27]

Nonselective herbicides, as a rule, cannot be detoxified by plants. A possible source of genes coding enzymes that metabolize nonselective herbicides is the wide spectrum of microbial organisms that detoxify herbicides in the soil. For example, in the case of the nonselective herbicide glyphosate, some microorganisms are able to degrade the herbicide.[28-31] The pathways of glyphosate metabolism have been extensively studied.[32-34] The use of microbial herbicide-metabolizing enzymes for conferring herbicide resistance to plants has been reviewed.[35] Indeed, it was shown that the expression of the bacterial nitrilase gene in plants under the control of a light-regulated tissue-specific promotor confers high levels of tolerance to the compound bromoxynil.[36,37] It would appear that the transfer of genes which code herbicide metabolizing enzymes is the only practicable way of producing herbicide-resistant plants via genetic engineering in cases where the target of the herbicide is unknown. Thus there are, in general, two strategies for engineering herbicide resistance in plants. These strategies are discribed in detail below.

A. ALTERING THE LEVEL AND SENSITIVITY OF THE TARGET OF THE HERBICIDE

The increasing knowledge of modes of action of herbicides, and rapid progress in molecular genetics have led to the isolation and modification of numerous genes encoding the target proteins for herbicide detoxification or inactivation (from both plants and microorganisms). Commercially important herbicides against which resistance has been engineered using this approach can be divided primarily into photosynthesis inhibitors or inhibitors of amino acid biosynthesis. In the former, several mutant organisms resistant to a particular herbicide have been generated and resistant genes utilized subsequently to spread resistance to different plants.

1. Mutant *psb*A Gene

Herbicides from many different chemical classes interfere with the reduction of plastoquinone on the acceptor site of PS II[38] (Table 1). Substances with totally different structures (triazines, triazinones, urea derivatives, carbonilides, uracils, and many others) bind to the thylakoid membrane of chloroplasts, thereby blocking photosynthetic electron transport. The fact that herbicides from different chemical classes were able to dislodge one another from the thylakoid membrane led to the idea of distinct but overlapping herbicide-binding sites. With the aid of a photoaffinity marker (azidoatrazine), it was possible to show that a 32-kDa protein in the thylakoid membrane is responsible for herbicide binding. This herbicide binding protein, which has since become known under several names (32-kDa protein, Q_B protein, D_1, herbicide-binding protein) has been studied in detail, at least in part of its uniquely high turnover when exposed to light. Molecular genetic investigations have identified the gene which codes a precursor of the Q_B protein. This plastid gene was given the name *psb*A. This gene codes a 34-kDa protein which is processed post-translationally at the C-terminus to yield the mature 32-kDa protein. *psb*A was the first gene identified for a herbicide target. The sequences of the *psb*A genes of a number of different plants show a very high degree of homology.[39-52]

The herbicide-resistant weeds, as well as resistant mutants selected in several microorganisms, such as *Chlamydomonas, Euglena,* and the cyanobacterium *Anacystis,* have provided a source of genes for studies of the molecular genetic basis for resistance. More than a dozen mutants of the *psb*A

TABLE 1
Herbicide, Primary Target, and Successful Steps towards Engineering Herbicide-Resistant Plants

Inhibited pathway	Herbicide	Primary target	Strategy for engineering resistance	
			Target modification	Detoxification
Photosynthesis	Atrazine	Q_B protein	Mutant *psbA* gene (plant)	GST gene (plant)
	Bromoxxynil	Q_B protein	—	*bxn* gene (bacterial)
Amino acid biosynthesis	Phosphinothricin	Glutamine synthase (GS)	GS gene amplification	*bar* gene (bacterial)
	Sulfonylurea and imidazolinone	Acetolactate synthase (ALS)	Mutant ALS genes (plant)	—
	Glyphosate	5-enolpyruvyl shikimate-3-phosphate synthase (EPSPS)	EPSPS gene amplification; mutant *AroA* gene (bacterial)	—
Growth regulation	2,4-D	Not known	—	*tfdA* (bacterial)

gene have been isolated and sequenced; at least six sites where amino acid substitutions result in herbicide resistance have been identified.[51] More than two thirds of the mutations result in a substitution of Ser-264 with Gly or Ala. A list of herbicide-resistant mutations identified in *psbA* genes is given in (Table 2).[52]

The mutated Q_B proteins show various degrees of cross-resistance with other PSII herbicide classes. Moreover, photoaffinity labeling with ^{14}C-azide-atrazine and amino acid Met_{214} of Q_B protein revealed that this amino acid is located in the vicinity of herbicide-binding site. Another decisive construction was made by the pioneering work of Michel and colleagues[53-55] who crystallized the photosynthetic reaction center of the photosynthesizing bacterium *Rhodopseudomonas viridis*. Models were developed which reproduced the exact three-dimensional arrangement of the amino acids involved in the binding of the herbicides.[56] It was shown that these amino acids are components of a Q_B protein niche into which both the secondary quinone Q_B, which is essential for photosynthetic electron transport, and the PSII herbicides are bound. Different amino acids of this niche are involved in the binding of the quinone and the individual herbicide classes.

As the PSII herbicides dislodge quinone Q_B from its binding site, its binding areas in the niche are of interest for an understanding of the herbicidal mode of action. The quinone binds to the protein via at least two hydrogen bridge linkages, one to His_{215} and the other to a peptide bond in the vicinity of Ser_{264}. This amino acid is also involved in the binding linkage to its -OH

TABLE 2
Amino Acid Changes in Herbicide-Resistant Q_B Proteins

Amino Acid	Wild type	Mutant	Relative level of resistance	
			Atrazine	DCMI
264	Ser	Ala	100	10
264	Ser	Gly	1000	1
219	Val	Ilw	2	15
251	Ala	Val	—	—
255	Phe	Tyr	15	0.5
275	Leu	Phe	—	—

group. Mutation of this serine to alanine or glycine in triazine-resistant organisms results in a loss of this triazine-binding site and thus in reduced affinity to this class of herbicide. However, these serine mutants also exhibit reduced photosynthetic electron transport activity, which can probably be accounted for by the fact that a peptide bond of Ser_{264} is involved in the binding of the quinone. The binding of other classes of herbicides to the Q_B is less affected by this mutation because these other classes interact mainly with amino acids of the binding niche.[39,53] The understanding of the involvement of amino acids in the binding of quinone and herbicides might allow identification of which amino acid exchanges are connected with resistance to a particular class of herbicides without, at the same time, impairing photosynthetic electron transport.

However, there are several difficulties in using this system for creating herbicide-resistant plants. The first is the *psb*A gene itself, which, as mentioned earlier, is located in the plastid genome. In addition, there is as yet no known method of transforming chloroplasts. The Q_B protein coded by the *psb*A gene is synthesized as precursor protein. For its functionally correct insertion into the thylakoid membrane and integration into the PSII complex, a posttranslational processing of the C-terminal end of the protein will be necessary. There is, however, no evidence that a nuclear-coded *psb*A gene product transported by the transit-peptide from the small subunit of ribulose biphosphate carboxylase/oxygenase of pea can be functionally integrated into the thylakoid membrane.[57] Furthermore, there are increasing indications that *psb*A mutations resulting in herbicide resistance are recessive.[58] This will make it absolutely necessary to eliminate endogenous sensitive copies of the *psb*A gene before any phenotypical expression of the resistance can be expected. This handicap may also complicate the transfer of resistance by conventional breeding methods.

To bypass this technical problem, the *psb*A mutant gene from the *Amaranthus* hybrid was converted into a nuclear gene by fusing its coding and transit-peptide-encoding sequences of a nuclear gene.[59] These chimeric constructs were included in the nuclear genome of tobacco, using *Agrobacterium* Ti transformation techniques. The protein product of nuclear *psb*A was identified in the chloroplasts of transgenic tobacco plants. These plants showed increased tolerance to atrazine, demonstrating the transported *psb*A product functions in photosynthesis.[59]

Studies in *Brassica* indicate that the introduction of complete chloroplast genomes from a different species may have major consequences for yield. Maternally inherited triazine resistance has been introduced into *Brassica napus* from *Brassica campestris* via sexual crossing.[10] After eight backcross generations to *Brassica napus*, the nuclear genome was almost biogenic, but by using the resistant plant as the female parent in each cross, the *Brassica campestris* chloroplast genome was retained. The resulting cultivar has been released commercially,[10] its yield is reduced by 30%. Thus, in this crop the combination of *Brassica campestris* chloroplasts with the *B. napus* nuclear background has a significant effect on plant growth. The extent to which this problem can be overcome by introducing a single chloroplast gene rather than the complete genome has yet to be tested. Despite extensive research and the availability of the number of resistant genes, engineering triazine resistance into crops via transformation methods is still problematic.

Engineering of atrazine resistance through the introduction of genes encoding detoxifying enzymes has been attempted but has also proved difficult. Genes encoding atrazine and alachlor detoxifying glutathione-*S*-transferases (GSTI and GSTIII) have been isolated from maize.[14,29,60] Preliminary studies have been reported indicating some increased tolerance to atrazine of transgenic tobacco plants expressing an atrazine-metabolizing GST gene.[61]

2. Acetolactate Synthase (ALS) Genes

Enzymes from biosynthetic pathways specific to plants are ideal herbicide targets. Essential amino acid biosynthetic enzymes have therefore a great potential as herbicide targets; approximately half of the amino acids used in proteins can be synthesized only by plants and microorganisms. The essential amino acids comprise the aromatic amino acids phenylalanine, tyrosine, tryptophan, histidine, lysine, threonine, and methionine. Three amino acid biosynthetic enzymes have been identified in the last few years as targets of important new herbicide classes (Figure 1). Since the metabolic pathways for amino acid biosynthesis are practically identical in plants and microorganisms, the extensive information available on microbial amino acid biosynthesis has proved useful in order to generate plants resistant to sulfonylureas, imidazolinones and glyphosate.

Selective sulfonylurea herbicide toxicity has been achieved by genetic modification of crops. The first example of a sulfonylurea herbicide-resistant mutant plant was obtained by selection of tobacco cells in tissue culture and the subsequent regeneration of mutant plants from the cell lines. Resistance was semidominant and segregated as a single nuclear gene.[5,62] It has also

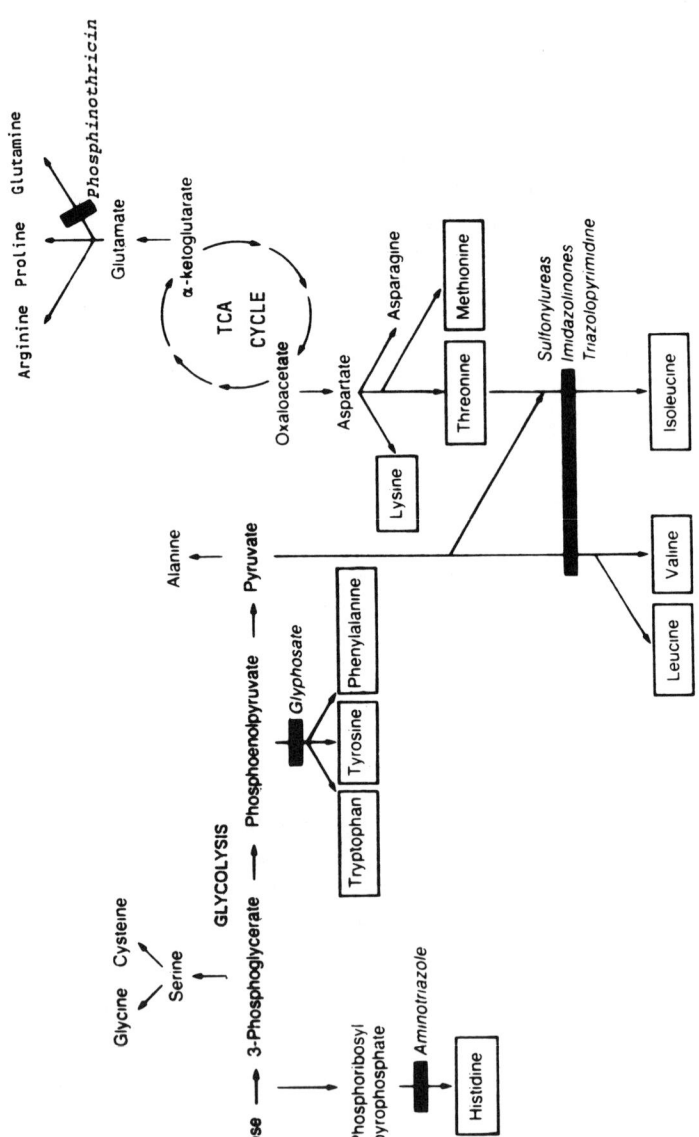

FIGURE 1. Herbicide targets in amino acid biosynthetic pathways. Boxes indicate essential amino acids.

been possible to isolate resistant mutants of other crop plants (soybeans) by seed mutagenesis, but considerable effort and time have been required.[63] A third approach has been to engineer sulfonylurea-resistant crops through plant transformation. For this, the gene(s) responsible for the resistance trait, either the native genes in the metabolic detoxification pathway or the mutant genes identified by selection, had to be isolated. The use of microbial species as models has played a central role in the identification and isolation of the necessary plant genes.

Studies on the mode of action of the sulfonylurea herbicide chlorsulfuron showed that inhibition of cell division in plant tissue was an early response to treatment.[64,65] This result, along with the ability to isolate resistant mutants in cultured plant cells, suggested that the herbicide antagonized a single basic cellular function and encouraged the use of microbial models to investigate herbicide action. Physiological studies in *Salmonella typhimurium* suggested that the target of the sulfonylurea herbicide sulfometuron methyl was the enzyme acetolactate synthase (ALS), also known as acetohydroxy acid synthase (AHAS). This enzyme is required for the synthesis of isoleucine, leucine, and valine.[66,67] Multiple ALS isozymes exist in the enterobacteria *S. typhimurium* and *E. coli,* and it was shown that the ALS II and ALS III, but not ALS I, are inhibited by sulfometuron methyl.[66] *In vitro* analyses of ALS activity from yeast, pea, tobacco, and *Chlamydomonas* demonstrated that the eukaryotic enzymes are very sensitive to sulfometuron methyl.[5,68]

a. Microbial ALS Mutant Genes

Indications that the sulfonylurea herbicides act by inhibition of ALS came from a combination of genetic and biochemical studies.[69,70] Sulfonylurea-resistant mutants of *S. typhimurium, Saccharomyces cerevisiae, Nicotiana tabacum,* and *Arabidopsis thaliana* were isolated. Most of the mutants from each organism produced ALS activity insensitive to the herbicide, and the resistant enzyme activity cosegregated with cellular resistance in genetic crosses.[62,68,71,72] The bacterial and yeast mutations were mapped to the loci of the ALS structural genes *ilv*G and *ilv*2, respectively.[66,68] The identification of ALS as the target of the sulfonylurea herbicides provided at least a partial explanation of their low toxicity to animals (animals lack this enzyme and must obtain the branched-chain amino acids from their diets).

ALS is also the target of two other structurally distinct classes of herbicides, the imidazolinones[56] and the triazolopyrimidines or sulfonanilides.[73] Thus, ALS may be a particularly susceptible target for herbicides. It has been demonstrated that the toxicity of sulfometuron methyl to bacteria is enhanced by the accumulation of an ALS substrate, 2-ketobutyrate, which is itself toxic. Therefore, the deficiency of branched-chain amino acids and the increase in concentration of the toxic intermediate combine to make ALS a particularly good target for herbicides.[67] However, no evidence that 2-ketobutyrate is accumulated in and is toxic to plant cells has been reported.

Genetic and biochemical studies have provided strong evidence that ALS enzymes from enteric bacteria are tetramers containing two subunits. ALS I purified from *E. coli* is composed of two large 60-kDa and two small 9.5-kDa subunits.[74,75] Molecular cloning and DNA sequencing have provided a physical characterization of the *ilv*BN operon that encodes these subunits.[76,77] Similarly, ALS II has been purified from *Salmonella typhimurium* and shown to be composed of two large 59-kDa and two small 9.7-kDa subunits. DNA sequence analysis of the cloned *E. coli ilv*GMEDA operon of *S. typhimurium*, along with amino acid sequence analysis of purified ALS II, has demonstrated that the *ilv*G and *ilv*M genes encode these subunits.[78,79] Genetic studies first suggested that ALS III was specified by two genes, designated *ilv*I and *ilv*H.[80] DNA sequence analysis again provided physical evidence for the existence of the two genes, organized as an operon.[81,82] Purification of ALS III provided confirmation of an analogous subunit structure for this isozyme.

The molecular cloning and DNA sequence analysis of the yeast ALS gene, designated *ilv*2, permitted a comparison of the ALS amino acid sequence from a eukaryotic source with those from *E. coli*.[68,69] The most striking structural difference between the deduced yeast and bacterial proteins was the presence of an approximately 90-amino acid long amino-terminal sequence extension on the former that was absent from the latter.[69] Subcellular fractionation experiments[83,84] had shown yeast ALS to be localized in mitochondria and the amino acid sequence in the region has characteristics common to other known mitochondrial transit sequences. It was, therefore, suggested that this region might include a mitochondrial transit sequence.[69]

The molecular basis for resistance to the sulfonylurea herbicides has been investigated by using cloned yeast and bacterial ALS genes. Overexpression of the yeast ALS enzyme by about 4-fold occurred when the gene was present on a high-copy-number plasmid; this overexpression resulted in a 4- to 5-fold increase in the minimal concentration necessary to inhibit growth.[68] An increase in the minimal inhibitory concentration of a sulfonylurea herbicide for *E. coli* was also observed when a functional *E. coli ilv*G gene was present on a high-copy-number plasmid.[85] Mutations that resulted in the production of sulfonylurea-resistant ALS were isolated in the cloned yeast *ilv*2 and *E. coli ilv*G genes by genetic selection. DNA sequencing showed that each mutant gene contained a single nucleotide change, resulting in a single amino acid substitution in the ALS protein. In yeast ALS, Pro_{192} was substituted with Ser and in *E. coli* ALS II Ala_{26} was changed to Val.[85]

Spontaneous mutations in the yeast *ilv*2 gene were also isolated and characterized by DNA sequencing to determine the amino acid substitutions responsible for herbicide resistance.[86] The analysis showed 24 different amino acid substitutions, at 10 different sites, ranging from the amino to the carboxy ends of the protein.[52]

The amino acid residues present in the wild-type yeast enzyme at these 10 sites were also present in the three wild-type *E. coli* ALS isozymes at most sites.[69] One of the exceptions to this generalization occurs in *E. coli* ALS II, where a serine residue is present at the site analogous to that where the proline-to-serine substitution had resulted in herbicide resistance in yeast. Since *E. coli* ALS II was known to be more resistant to the sulfonylurea herbicide sulfometuron methyl than the yeast or plant enzymes,[69] it was plausible that this serine residue contributed to the increased resistance. This hypothesis was tested by converting the serine codon in the ALS II gene of *E. coli* to a proline codon by site-directed mutagenesis; this change indeed increased the sensitivity of the enzyme to the herbicide.[69] In another example, *E. coli* ALS I, which is much more resistant to the sulfonylurea herbicides than is ALS II, had a glutamine residue at one site analogous to that where substitutions for the wild-type tryptophan residue had resulted in herbicide-resistant yeast ALS, and a serine residue at a second site where substitutions for the wild-type alanine residue had resulted in herbicide resistance in both yeast ALS and *E. coli* ALS II. Conversion of the glutamine residue in *E. coli* ALS I to tryptophan also resulted in increased sensitivity to the herbicides.[87] Thus, these residues appear to contribute to inhibition by sulfonylureas in all ALS enzymes. The herbicide-resistant plant enzymes for which sequence information is available also have amino acid substitutions at these same sites.

Site-directed mutagenesis was used to expand the spectrum of amino acid substitutions at the ten sites in yeast ALS.[52,88,89] At some of these sites, such as Ala_{117}, Pro_{192}, and Trp_{586}, nearly any substitution for the wild-type amino acid resulted in a herbicide-resistant enzyme. At other sites, only a few substitutions had that result.

b. *Plant Mutant ALS Genes*

Efforts to use microbial genes to generate herbicide resistant transgenic plants faced a number of difficulties and uncertainties. The two of the major difficulties are: (1) ALS is localized in the chloroplast,[90] and (2) bacterial ALS is composed of two different subunits. Thus the generation of herbicide-resistant plants using bacterial genes requires not only expression of two protein subunits of the enzyme, but also their translocation into and assembly in plant chloroplasts. Since ALS had not been purified from yeast, neither the subunit structure of the enzyme nor the amino terminus of the mature protein present in the mitochondria is known; this has added considerable uncertainty to any effort to express that enzyme in plants. For these reasons the isolation and use of plant ALS genes appeared to be a more attractive route for engineering sulfonylurea-resistance in plants.

Even before the conservation of amino acid sequences between yeast and bacterial ALS enzymes had been discovered, hybridization between the yeast and *Salmonella ilv*G genes had been detected under low-stringency conditions.[52] Together, these observations led to an attempt to detect ALS genes from other species, using heterologous DNA hybridization. A segment of the yeast ALS gene that spanned most of the coding region was used as a probe.[88,89]

Hybridization was detected between the yeast ALS gene and genomic DNA libraries from the cyanobacterium *Anabaena* 7120 and the higher plants *Arabidopsis thaliana* and *N. tabacum*; ALS genes were isolated from all three species.[88,89]

DNA sequence analysis of the *Arabidopsis* and *Nicotiana* ALS genes indicated that neither gene has introns, and that they code for proteins of 667 and 670 amino acids, respectively. The deduced ALS protein sequences are similar throughout most of their length; approximately 75% of the nucleotides and 85% of the encoded amino acids are identical.[88,89] The 5' ends of the coding sequences, which are the only regions that are not highly conserved between the two plant ALS sequences, appear to encode chloroplast transit sequences. The nucleotide sequences of these regions are more similar than are the deduced amino acid sequences, suggesting that there are few constraints on the amino acid sequences in the transit peptides. Comparison of the deduced amino acid sequences of the plant ALS genes with those of the yeast and the three *E. coli* ALS genes showed that all share the same three conserved domains. At the ten sites where substitutions result in sulfonylurea-resistant yeast ALS, the amino acid residues present in the plant ALS enzymes are identical to those found in the wild-type yeast enzyme.

The number of ALS genes present in *N. tabacum* and *A. thaliana* was determined by Southern blot hybridization analyses. Homologous cloned ALS genes were used as probes. A single ALS gene hybridized to the probe in *Arabidopsis*, while two hybridized in *N. tabacum* (an allotetraploid).[88,89] The identification of two ALS genes in tobacco was consistent with genetic data, which had indicated that tobacco mutations that code herbicide resistance have two loci.[5,62] Southern blot analysis of DNA from several other crop species has indicated that many of them, including corn and soybean, carry multiple ALS genes.

The cloned plant ALS genes were used as hybridization probes to isolate genes carrying ALS mutations from herbicide-resistant plants. In tobacco, the Hra line, which is mutated at the SURB locus and which is 1000-fold more resistant to sulfonylureas than are wild-type lines,[62] was used as one source of a mutant ALS gene. A second tobacco line, C3, which carries a mutation at the SURA locus,[5] was used as a source of a second mutant ALS gene. Four genes, representing all of the ALS loci from the two mutant lines, were isolated and sequenced. The molecular characterization of mutant and wild-type genes from each plant line permitted the assignment of the genes to the appropriate genetic locus and the determination of the amino acid substitutions in the mutant enzymes. The SURB-Hra gene, which was isolated by two successive rounds of genetic selection, contained two mutations, which resulted in Pro_{196} to Ala and Trp_{573} to Leu substitutions. The SURA-C3 gene carried a single mutation that resulted in a Pro_{196} to Gln substitution.[90,91] Similarly, a gene that conferred sulfonylurea herbicide resistance was isolated from *Arabidopsis* and sequenced; this gene carried a single mutation that resulted in a Pro_{197} to Ser substitution.[71] Both the Pro_{196} to Gln mutation in the tobacco gene and the analogous Pro_{197} to Ser mutation in the *Arabidopsis* gene conferred selective resistance to sulfonylurea herbicides, but not to imidazolinone herbicides. The double tobacco mutant that carried the Pro_{196} to Ala and Trp_{573} to Leu substitutions was cross-resistant to both classes of herbicides.[91]

The mutant plant ALS genes were introduced into *N. tabacum* by transformation and conferred useful levels of herbicide resistance in transgenic plants.[71] The tobacco SURB-Hra gene was also used to transform a number of heterologous species to sulfonylurea herbicide resistance at the cellular, and in some cases, whole plant level. These species include tomato, sugar beet, oilseed rape, alfalfa, lettuce, and melon.[87] In some of the heterologous transformants, such as tomato, expression of the resistant tobacco ALS gene was efficient, 20 to 60% of ALS activity being derived from the mutant gene.[88,89] In others, such as rape, only a low level of resistance was observed.[87] This suggested that the effectiveness of heterologous genes must be evaluated on a case-by-case basis.

Additional herbicide-resistance mutations in plant ALS genes were generated by site-directed mutagenesis, based upon the mutations identified in the yeast ALS gene. The mutant genes were introduced into tobacco by transformation, and their ability to confer herbicide resistance was monitored.[70] In addition, a bacterial expression assay system for the *Arabidopsis* ALS gene was developed to permit the rapid isolation and characterization of new mutations. In this assay system, the plant gene, including its chloroplast transit sequence, was expressed in an *E. coli* auxotroph that produced no ALS. The wild-type *Arabidopsis* ALS gene promoted growth only in the absence of herbicides, while mutant *Arabidopsis* ALS genes promoted growth both in the presence and absence of herbicides. This selection thus allowed assay for the efficacy of herbicide-resistance mutations in a plant ALS gene.[72]

As a more critical measure of agronomically useful herbicide resistance, some of the tobacco transformants were treated with sulfonylurea herbicides in field tests and evaluated for phytotoxic symptoms.[88,89] Foliar sprays were applied at rates corresponding to 0, 8, 16, and 32 grams of herbicide per hectare. Transformed plants showed no damage at the highest application rate. Wild-type plants showed damage at the 8 g/ha application rate.[88,89] Thus, expression of the SURB-Hra gene can provide an effective means of producing a sulfonylurea herbicide-resistant crop.

3. EPSPS and Mutant *Aro*A Genes

This strategy has proved useful in engineering glyphosate resistance in plants. Glyphosate is an exceptionally reliable, phloem-mobile, broad-spectrum herbicide with little residual soil activity. Because of its desirable properties, less toxicity to all major crops, and considerable commercial importance, extensive efforts have been aimed at the development of herbicide-resistant cultivars. A number of reviews have described this work in depth.[34,35,60,90-93] The primary target of glyphosate is 5-enolpyruvyl-shikimate-3 phosphate synthase, or EPSPS, an enzyme in the aromatic amino biosynthetic pathway. This pathway is found only in microbes and plants. EPSPS was originally inferred to be the target of glyphosate action through identification of shikimic acid as an intermediate that accumulated following glyphosate treatment, through experiments showing that glyphosate activity could be suppressed by the addition of aromatic amino acids[94] and by biochemical studies of glyphosate inhibition of EPSPS.[95,96,97] These findings were subsequently corroborated by the selection of glyphosate-resistant mutant strains of enteric bacteria. A resistant strain of *Salmonella* was identified following two cycles of chemical mutagenesis. The resistant mutation was shown to be in the *aro*A gene; the mutant gene was subsequently isolated and was shown to confer resistance when transferred to *E. coli*. DNA sequencing indicated that two distinct types of mutations had led to the glyphosate-resistance phenotype. The first round of mutagenesis had created a promoter mutation in the *aro*A gene, which conferred low levels of glyphosate tolerance by elevating the level of expression of the gene. The second cycle of mutagenesis had generated a point mutation in the *aro*A structural gene, which caused a proline to serine substitution at residue 101 of the protein.[37,92,98] Overexpression of the *E. coli aro*A gene, as a consequence of its presence on a high-copy-number plasmid, could also confer glyphosate tolerance,[99] providing additional evidence that EPSPS was the primary target of this compound.

Glyphosate-tolerant transgenic plants were first generated by transferring the mutant *Salmonella aro*A gene, linked to either an octopine or mannopine synthase promoter for plant cell expression, to tobacco.[98] The bacterial gene lacked a chloroplast transit sequence, and thus the herbicide-resistant EPSPS was expected to be localized in the cytoplasm; the plant EPSPS is predominantly localized in the chloroplast.[100] The transformed tobacco plants showed increased, but incomplete, tolerance to glyphosate. The same bacterial gene was also transferred to tomatoes. The gene again conferred glyphosate tolerance, but after foliar herbicide treatments the transformants were smaller than the unsprayed controls.[101] Subsequent to these studies, mutant bacterial genes were fused to plant EPSPS chloroplast transit sequences and then transferred to plants. In one series of experiments, the petunia EPSPS gene transit sequence, along with the first 27 codons for the mature protein, was fused to a mutant *E. coli aro*A gene. The chimeric preprotein was imported into petunia chloroplasts and conferred resistance in the transformed plants.[100]

III. ALTERNATIVE STRATEGIES FOR ACHIEVING HERBICIDE RESISTANCE

A. GENE AMPLIFICATION AND OVEREXPRESSION

Because the most agronomically important crops were not amenable to transformation, efforts were mounted to generate resistance to them using

Chapter 3

alternative methods.[101-108] Imidazolinone-resistant mutants of maize were isolated using genetic selection of cultured maize cells. Fertile plants exhibiting a greater than 100-fold increase in resistance to imidazolinones and cross resistance to the sulfonylureas were regenerated from one line. Homozygous progeny showed more than a 300-fold increase in resistance to the herbicides. Resistance was inherited as a single dominant nuclear gene, and ALS activity from the mutants was resistant to the herbicides *in vitro*.[109] This mutant, which did not have any associated growth or yield defects, is being developed by Pioneer Hi-Bred International, Inc.[52] Although a significant breeding program has been required to introduce the resistance trait into elite lines, it is likely to be the first ALS-targeted herbicide resistant crop to be commercialized, with release anticipated in the early 1990s.

Imidazolinone-resistant mutants of maize that show no cross-resistance to the sulfonylureas have also been isolated. In one case, the mutant showed tolerance to a much lower level of herbicide. Tolerance was inherited as a recessive trait, and no change in ALS activity was observed.[109] Sulfonylurea herbicide-tolerant mutants of soybean have been obtained by genetic selection from mutagenized soybean seeds. These mutants showed only 5- to 10-fold increase in tolerance, were recessive, and did not affect sensitivity of ALS to inhibition by sulfonylureas.[63] Recently, sulfonylurea-resistant mutant lines of soybean with significantly higher levels of tolerance have been obtained. These mutants have an altered ALS that showed reduced sensitivity to inhibition by sulfonylureas.[52]

Glyphosate resistance has also been imparted to plants through EPSPS gene amplification. The plant genes were isolated by taking advantage of the finding that plant lines tolerant to glyphosate arose through overproduction of EPSPS.[57,96,102] A glyphosate-tolerant petunia line was isolated by applying increasingly stringent stepwise selection conditions until a line was established in which EPSPS DNA, RNA, and protein levels were elevated approximately 20-fold.[14,60,102] This elevation in EPSPS-specific macromolecules facilitated the subsequent cloning of the EPSPS gene. EPSPS was purified from the line, and the N-terminal amino acid sequence of the protein was determined by microsequencing. Based on this sequence, three sets of potentially complementary oligonucleotide probes for the gene were synthesized and were used to screen messenger RNA populations from the amplified line. The set containing the oligonucleotide complementary to the gene was determined from Northern blot hybridizations, and was then used to identify a partial cDNA clone for EPSPS. The cDNA clone was in turn used to identify a genomic DNA clone; a complete cDNA clone was subsequently constructed.[14,60]

The petunia EPSPS gene was shown to be a nuclear gene that spans 9 kb of DNA and is interrupted by 7 introns.[103] The mature protein was predicted to have a molecular mass of approximately 48 kDa.[103] A comparison of the inferred petunia, *Arabidopsis,* and tomato EPSPS proteins showed that they are highly conserved, except in the region of the chloroplast transit peptide.[103]

The petunia gene codes for a 72-amino-acid chloroplast transit sequence, while the tomato chloroplast transit sequence is more than twice as long, with 148 codons.

Shah et al.[60] used petunia EPSPS gene to produce glyphosate-resistant transgenic plants. The petunia cDNA clone was linked to a CaMV 35S promoter, in order to obtain a high level of expression of the gene in plants. Expression of the wild-type gene from this promoter resulted in a 20-fold increase in EPSPS activity in transgenic petunia plants. The plants tolerated applications of glyphosate approximately four times greater than that needed to kill nontransformed plants.[60] In order to provide additional herbicide tolerance in the transgenic plants, particularly in the meristematic regions where glyphosate accumulates,[104] site-specific mutations were introduced into the wild-type petunia EPSPS gene. All of the substituted EPSPS proteins had reduced catalytic efficiencies.[93] The modified genes were again coupled to the CaMV 35S promoter, and were able to confer higher levels of resistance in transgenic plants. In a parallel series of experiments, mutated bacterial EPSPS genes were fused to plant EPSPS chloroplast transit sequences, and the chimeric genes introduced into plant cells. These constructions also conferred herbicide resistance.[93] Using such herbicide-resistance constructs, a wide variety of glyphosate-resistant transgenic plant species have been created. Some of these plants, including tomato and oilseed rape, were tested in field trials during 1987 and 1988.[52]

However, much success has been obtained in engineering phosphinothricin (PPT) resistance in plants. PPT acts as a herbicide by inhibiting the amino acid enzyme, glutamine synthase (GS) of plants and bacteria. In plants, GS is involved in the assimilation of nitrogen. A PPT-tolerant alfalfa cell line had a 10-fold increase in GS activity; this was consistent with the levels of GS-specific RNA and DNA.[110] Gene amplification of GS required enhanced GS expression, allowing sufficient enzyme activity even in the presence of the inhibitors. Expression of a mutant alfalfa gene in transgenic tobacco plants has also been reported to confer a low level of tolerance to PPT, but only when it is taken up by the roots.[52] However, converting more glutamate into glutamine might produce undesired effects on the nitrogen metabolism of the plants.

B. HERBICIDE-DETOXIFYING GENES

Manipulations involving introduction into plants of additional enzymes that are designed specifically to inactivate the relevant herbicide, are also considered superior to those aimed at changing the herbicide's target. If the target enzyme is altered, it may not retain full function, and phenotypic consequences may be detrimental to the engineered plant. Thus herbicide detoxifying enzymes appear to be suitable in such situations; however, the obvious setback with the detoxification approach is the lack of availability of suitable enzymes.

Chapter 3

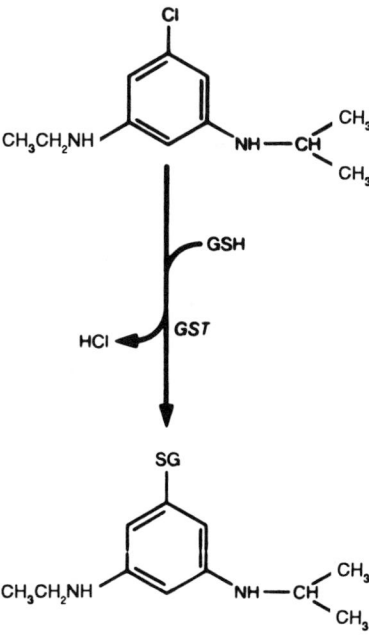

FIGURE 2. Detoxification of atrazine by GST. GSH, reduced glutathione; SG, conjugated glutathione; GST, glutathione-S-transferase.

1. Herbicide-Detoxifying Enzymes from Plants

Little information is available on the molecular basis of herbicide detoxification in plants. Consequently, the development of genetic engineering strategy based on plant herbicide-detoxifying enzymes is slower than that of microbial systems.

Herbicide-detoxifying enzymes such as mixed function oxidase, amidase, decarboxylases, and conjugative systems have been identified in tolerant plants.[110] The system most extensively studied is the conjugation of xenobiotic compounds with glutathione (Figure 2). This conjugation is catalyzed by multifunctional glutathione-S-transferase (*GSTs*). Genes encoding atrazine and alachlor detoxifying GSTs (GSTI-GSTIII) have been isolated from maize,[14,60] though the introduction and expression of these genes in other plants have met little success. This is perhaps not surprising; in maize the active atrazine GST is the product of more than one gene locus, and glutathione level, in contrast to many other crop species, is very high.

Another class of conjugative enzymes is represented by the N-glucosyl transferase present in metribuzin-tolerant tomato.[35] Detoxification of the herbicide is due to increased activity of N-glucosyl transferase. Metribuzin tolerance in tomato is controlled by one gene locus. A third class of plant enzyme is the mixed function oxidases, involved in the detoxification of 2,4-D in tolerant pea and of dicamba in tolerant barley.[35]

2. Herbicide-Detoxifying Genes from Bacteria

As mentioned earlier, many herbicides are rapidly degraded by soil microflora. Such microorganisms provide an alternative strategy for obtaining herbicide-tolerance genes.[105] Bacterial degradative systems have been extensively studied, and a number of detoxifying enzymes have been identified. Engineering herbicide resistance into plants has been primarily achieved by introducing bacterial genes encoding enzymes that inactivate the herbicide by nitryl hydrolysis, acetylation, or removal of an acetate chain.

Bacterial degradation genes have been proved extensively helpful in generating glyphosate-resistant plants. Initially *Pseudomonas* and *Arthrobacter* species capable of growing on glyphosate as a sole carbon source have been identified.[33,105] These strains metabolize glyphosate to phosphate, glycine, and a one-carbon unit.[32-34] In contrast, through soil organism metabolism, glyphosate is primarily degraded to aminomethylphosphonic acid.[28,58,106] The bacterial genes that carry out these metabolic degradations could theoretically be isolated and expressed in plants, to produce glyphosate-tolerant plants. No glyphosate-tolerant plants have yet been reported in which expression of cloned genes enables metabolism of glyphosate to nontoxic products.

A strain of soil bacterium *Klebsiella ozaenae* has also been isolated which can use 3,5-dibromo-4-hydroxy benzonitrile (an inhibitor of photosynthetic electron transport) as its sole nitrogen source. It possesses a plasmid-encoded nitralase gene (*bxn*) that is constitutively expressed and converts the herbicide to its inactive metabolite, 3,5-dibromo-4-hydroxybenzoic acid in a single step. The *bxn* gene has been cloned and introduced into plants.[36] Transgenic tobacco and tomato plants expressing the *bxn* gene were able to tolerate bromoxynil.[36]

De Block et al.[111] reported the engineering of plants resistant to the nonselective herbicide PPT and bialaphos. Bialaphos is a tripeptide produced by *Streptomyces hygroscopicus*. It consists of PPT and two L-alanine residues. GS inhibition leads to ammonia accumulation which causes rapid death of the plant.[112] At present, two products are available as commercial herbicides: glufosinate ammonium is the ammonium salt of the chemically synthesized PPT (Basta®, Hoechst AG, FRG), and bialaphos is produced by fermentation of *S. hygroscopicus* (Herbiacer, Meijii Seika, Japan). The gene conferring resistance to PPT (*bar*) was isolated from the bialaphos biosynthetic pathway of *S. hygroscopicus*.[113] It encodes the enzyme phosphinothricin acetyl transferase (PAT), which converts PPT with high efficiency into a nonherbicidal acetylated form.[49] Transgenic tobacco (*Nicotiana tabacum*) and potato (*Solanum tuberosum*) plants expressing the *bar* gene are resistant to application of glufosinate and bialaphos under greenhouse conditions.[111]

S. hygroscopicus is used for the commercial production of bialaphos. The bialaphos is synthesized from three carbon precursors (probably pyruvate or phosphoenolpyruvate) in a series of at least thirteen conversions.[114] Many of these conversions are positively regulated.[115] Cloned DNA fragments which either restore productivity to these mutants or confer resistance to bialaphos (*bar*) have been mapped to an 18-kb gene cluster.[113] Although enzyme assays for most of these steps have not been developed, it has been shown that *S. hygroscopicus* contains an acetyl-coenzyme A-dependent activity which can be modified by dimethylphosphinothricin (DMPT), an intermediate in the pathway, or PPT itself.

De Block et al.[111] described the cloning and characterization of the bialaphos resistance gene (*bar* or *pat*) from *S. hygroscopicus* which is involved in the bialaphos biosynthetic pathway. They placed the *bar* gene under the center of the CaMV 35S promoter and transferred to tobacco, tomato, and potato plants. PAT was used as a selectable marker in protoplast cocultivation. The chimeric *bar* gene was expressed in tobacco, potato, and tomato plants. Transgenic plants showed complete reistance towards high doses of the commercial formulations of phosphinothricin and bialaphos.

Pühler and co-workers[116] isolated and characterized the phosphinothricin tripeptide (*Pt*) resistance gene from *S. viridochromogenes* Tü94 mutant strain. A 4.0-kb *Bam*HI fragment coding resistance was isolated from *S. viridochromogenes* and cloned *S. lividans* TK23. Pt resistance was assigned to 0.8-kb *Bgl*II fragment which also included the Pt resistance promoter (Figure 3). Subcloning this fragment downstream from the *lac*Z promoter conferred PPT resistance in *E. coli* (Figure 4). This *Pt* or *pat* gene was modified and introduced into *Nicotiana tabacum* by *Agrobacterium*-mediated leaf disc transformation (Figure 5).[117] The GTG start codon of *pat* was replaced by ATG and the modified *pat*-coding region was fused to CaMV 35S promoter. Transgenic tobacco plants could directly be selected on PPT-containing medium.

Wohlleben et al.[117] also compared the sequence of the *bar* gene of *S. hygroscopicus* studied by Thompson et al.[149] with the *pat* gene of *S. viridochromogenes* and found significant homology between the two genes (Figure 6). However, variations were detected in the 5' noncoding region of the two resistance genes which may reflect differences in regulation.

Field experiments of herbicide resistance in transgenic crops have been published.[118] De Greef et al.[118] have analyzed two herbicide-resistant tobacco (*Nicotiana tabacum*) and four potato (*Solanum tuberosum*) lines under field conditions. These lines resulted from *Agrobacterium*-mediated transformation of tobacco and potato leaf discs with chimeric *bar* genes encoding phosphinothricin acetyl transferase. Greenhouse tests have demonstrated that this enzyme protects transformed plants against the broad-spectrum herbicides glufosinate and bialaphos. In field tests, the transformants revealed the same agronomic performance as untransformed controls. Complete resistance to field dose applications of glufosinate was observed, although *pat* expression in these lines varied by two orders of magnitude. These field data confirm that glufosinate can be applied as a selective postemergence herbicide on engineered crops.

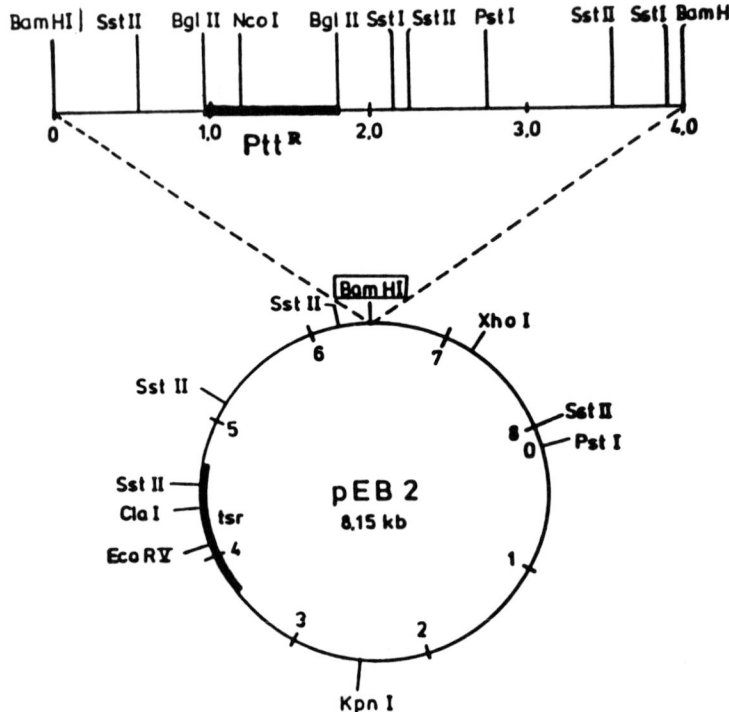

FIGURE 3. Restriction map of plasmid pES1.1. Plasmid pES1.1 consists of the *Streptomyces* vector pEB2 and 4.0-kb *Bam*HI fragment of *S. viridochromogenes* mutant ES1 DNA. Inserting the 4.0-kb *Bam*HI fragment in the opposite direction resulted in plasmid pES1.2 (not shown). The black bar in the insert indicates the location of the Ptt^R gene. The heavy line segment *tsr* outlines the position of the Thio^R gene from *S. azureus* in the vector. Note: *Sst*I and *Nco*I sites were not determined for the vector part of pES1.1. No sites for *Kpn*I, *Pvu*I, *Xho*I, *Cla*I, *Eco*RV, *Eco*RI, and *Hin*dIII were found in the 4.0-kb *Bam*HI fragment. (From Straunch, E., Wohlleben, W., and Pühler, A., *Gene*, 63, 65, 1988. With permission.)

IV. CONCLUSIONS

Modifying plants to become resistant to broad-spectrum herbicides would allow their selective use for crop protection. As a consequence, a major effort has been devoted in several laboratories to engineer herbicide-resistance in plants.[119-164] The development of herbicide-resistant varieties of crops has generated not only considerable interest but also considerable controversy over recent years. However, much of this controversy is based upon misconceptions of reasons for this work. In the initial transformation studies there has been an understandable, and inevitable, emphasis on plants which are simple, predictable, and easily selected and tested.

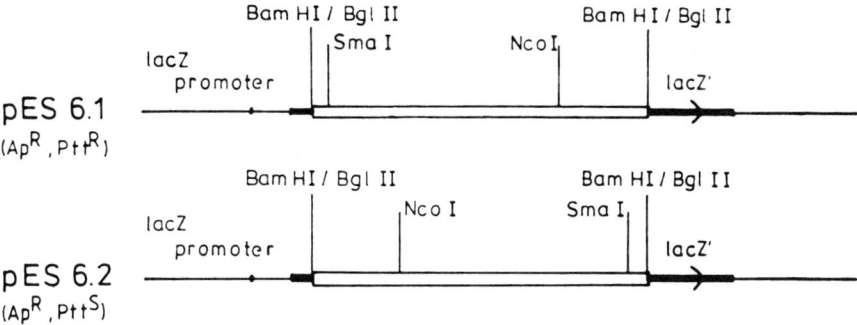

FIGURE 4. Construction of plasmids pES6.1 and pES6.2 and their use for the expression of Ptt^R gene in *E. coli*. The 0.8 *Bgl*II fragment carrying the Ptt^R gene (open box) was cloned in both orientations into *E. coli* plasmid vector pSVB20 giving pES6.1 and pES6.2. The heavy line represents the *lacZ'* gene; the direction of transcription is indicated by an arrow. A relevant part of the restriction maps of pES6.1 and pES6.2 is presented. The antibiotic resistance phenotype of *E coli* clones carrying these plasmids is indicated. (From Straunch, E., Wohlleben, W., and Pühler, A., *Gene*, 63, 65, 1988. With permission.)

Herbicide resistance, particularly where studies on the mode of action had identified a sensitive enzyme, was an obvious candidate for investigation. Such an interest has led to the isolation of genes providing resistance/tolerance to several herbicides. However, the development of such resistant/tolerant plants, useful as they might be for scientific study, will not lead to their automatic commercialization and is not evidence of any determined policy on behalf of chemical companies to increase herbicide usage. It is illogical to suppose that the availability of a crop variety resistant to a particular herbicide will force farmers to purchase supplies of that compound. They will make decisions on economic grounds and the herbicide regime chosen will be that which can provide effective control at the lowest possible cost. It will also be possible to replace older products with ecologically more favorable ones with fewer carrying problems. The potential problem of herbicide-resistant weeds will be made neither better nor worse by the availability of herbicide-resistant crops. Such contingencies will be prevented or managed, as now, by sensible rotation of crops and herbicides.

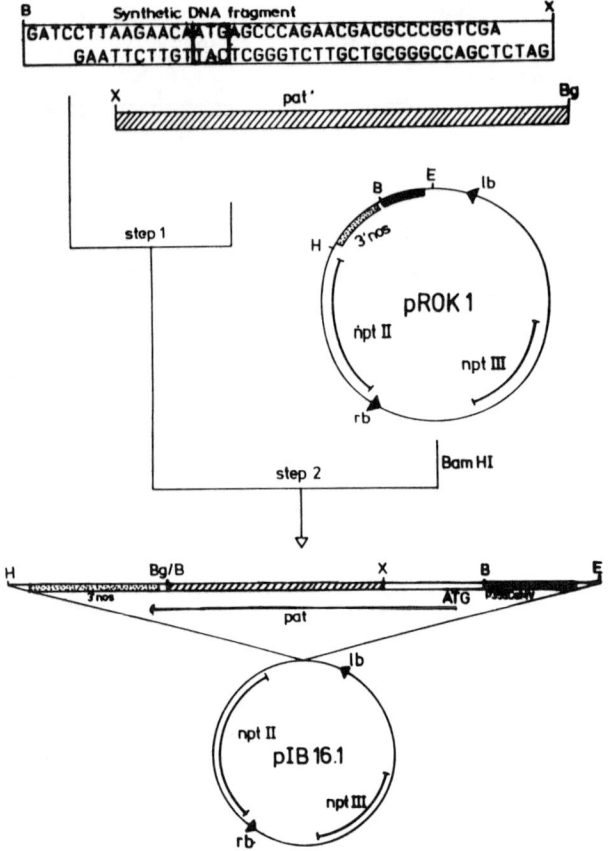

FIGURE 5. Strategy for the construction of a plant vector containing the modified *pat* gene. The cosntruction strategy consists of two steps. In step 1 the modifications of the *pat* gene are presented. A synthetic DNA fragment (open box) was fused to the *Xho*II-*Bgl*II fragment of the *pat* gene (hatched box). The fusion provides the *pat* gene with an ATG start codon (box) instead of the original GTG. In step 2, cloning of the modified *pat* gene into the vector pROK1 is shown. A *Bam*HI-*Bgl*II fragment carrying the modified *pat* gene was inserted into the single *Bam*HI cloning site of pROK1 between the 35S CaMV promoter (blackened box) and the 3' end of the *nos* gene (stippled box) to give pIB16.1. Plasmid pIB16.1 carries the *npt*II gene containing the Tn5 *aph*II coding region inserted between the nos promoter and terminator between the right (rb) and left (lb) borders of the T-DNA. The *npt*III gene from *Streptococcus faecalis* is present on plasmid pIB16.1 as selection marker in bacteria. The drawings are not to scale. Abbreviations: B, *Bam*HI; Bg, *Bgl*II; E, *Hind*III; X, *Xho*II. (From Wohlleben, W.A., Arnold, W., Broer, I., Hillemann, D., Straunch, E., and Pühler, A., *Gene*, 70, 25, 1988. With permission.)

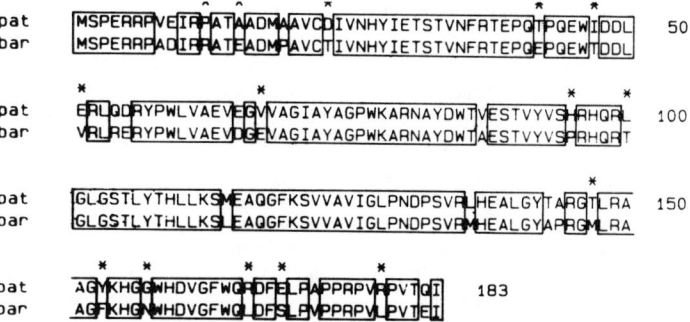

FIGURE 6. Comparison of the amino acid sequence specified by the *pat* gene of *S. viridochromogenes* Tüugy and the *bar* gene of *S. hydroscopicus*. The amino acid sequences of *pat* (top lines) and *bar* (bottom lines) are compared. Identical aa are boxed in and nonconservative changes are marked by asterisks. (From Wohlleben, W. A., Arnold, W., Broer, I., Hillemann, D., Straunch, E., and Pühler, A., *Gene*, 70, 25, 1988. With permission.)

REFERENCES

1. **Lal, R. and Lal, S.**, *Crop Improvement Utilizing Biotechnology*, CRC Press, Boca Raton, Florida, 1990.
2. **Peschke, V. M., Phillips, R. C., and Gengenbach, B. G.**, Transposable element activity in progeny of regenerated maize plants, *Int. Congr. Plant Tissue Cell Cult.*, University of Minnesota, Minneapolis, 1986, 285.
3. **Pental, D., Cooper-Bland, S., Harding, K., Cocking, E. C., and Muller, A. J.**, Cultural studies on nitrate reductase deficient *Nicotiana tabacum* mutant protoplasts, *Z. Pflanzenphysiol.*, 105, 219, 1982.
4. **Chaleff, R. S.**, Variants and mutants, in *Genetics of Higher Plants. Application of Cell Culture*, Newth, D. R. and Torrey, J. G., Eds., Cambridge University Press, London, 1981, 41.
5. **Chaleff, R. S. and Ray, T. B.**, Herbicide-resistant mutants from tobacco cell cultures, *Science*, 223, 1148, 1984.
6. **Singer, S. S. and McDaniel, C. N.**, Selection of glyphosate- tolerant tobacco calli and the expression of this tolerance in regenerated plants, *Plant Physiol.*, 78, 411, 1985.
7. **Shaner, D. L. and Anderson, P. C.**, Mechanism of action of the imidazolinones and cell culture selection of tolerant maize, in *Biotechnology in Plant Science*, Zaitlin, M., Day, P. R., and Hollaender, A., Eds., Academic Press, Orlando, FL, 1985, 187.
8. **Shaner, D. L., Anderson, P. C., and Stidham, M. A.**, Imidazolinones (potent inhibitor of acetohydroxyacid synthase), *Plant Physiol.*, 76, 545, 1984.
9. **Gressel, J.**, Biotechnologically conferring herbicide resistance in crops: the present realities, in *Molecular Form and Function of the Plant Genome*, van Vloten-Doting, L., Ed., Plenum Press, New York, 1985, 489.
10. **Oxtoby, E. and Hughes, M. A.**, Engineering herbicide-tolerance into Crops, *TIBTECH*, 8, 61, 1990.
11. **Gillette, J. R.**, Metabolism of drugs and other foreign compounds through enzymatic mechanisms, *Fortschr. Arzeimittelforsch.*, 6, 11, 1963.

12. Mozer, T. J., Tiemaier, D. C., and Jaworski, E. G., Purification and characterization of corn glutathione-s-transferase, *Biochemistry*, 22, 1068, 1983.
13. Jaworski, E. G., Mozer, T. J., Rogers, S. G., and Tiemaier, D. C., Herbicide target sites, mode of action and detoxification: chloacetanilides and glyphosate, in *Biosynthesis of the Photosynthetic Apparatus: Molecular Biology, Development and Regulation*, Thornber, J.P., Staehelin, L.A., and Hallick, R.B., Eds., Alan R. Liss, New York, 1984, 335.
14. Shah, D. M., Hironaka, D. M., Wiegand, R. C., Hardding, E. I., Krivi, G. G., and Tiemaier, D. C., Structural analysis of a maize gene coding for glutathione-S-transferase involved in herbicide detoxification, *Plant Mol. Biol.*, 6, 203, 1986.
15. Frear, D. S., Mansagar, E. R., Swanson, H. R., and Tanaka, F. S., Metribuzin metabolism in tomato: isolation and identification of N-glucoside conjugates, *Pestic. Biochem. Physiol.*, 19, 270, 1983.
16. Frear, D. S., Swanson, H. R., and Masanger, E. R., Alternative pathways of metribuzin metabolism in soybean: formation of N-glucoside and homoglutathione conjugates, *Pestic. Biochem. Physiol.*, 23, 56, 1985.
17. Shimabukuro, R. M., Detoxification of herbicides, in *Weed Physiology*, Vol. 2, Duke, S. O., Ed., CRC Press, Boca Raton, FL, 1985, 215.
18. Shimabukuro, R. M., Swanson, H. R., and Walsh, W. C., Glutathione conjugation:atrazine detoxification mechanisms in corn, *Plant Physiol.*, 46, 103, 1970.
19. Shimabukuro, R. H., Walsh, W. C., and Hoerauf, R. A., Metabolism and selectivity of dichlofop-methyl in wild oat and wheat, *J. Agric. Food Chem.*, 27, 615, 1979.
20. Loos, M. A., Phenoxyalanoic acids, in *Herbicides: Chemistry, Degradation and Mode of Action*, Vol. 1, Kearney, P.C. and Kaufman, D.D., Eds., Marcel Dekker, New York, 1975.
21. Schulz, A., Wengenmayer, F., and Goodman, H. M., Genetic engineering of herbicide resistance in higher plants, *Plant Sci.*, 9, 1, 1990.
22. Sweetser, P. B., Schow, G. S., and Hutchinson, J. M., Metabolism of chlorosulfuron by plants: biological basis for selectivity of new herbicide for cereals, *Pestic. Biochem. Physiol.*, 17, 18, 1982.
23. Brown, M. A., Chin, T. Y., and Miller, P., Hydrolytic activation versus oxidative degradation of assert herbicide, an imidazolinone arylcarboxylate, in susceptible wild oat virus tolerant to corn and wheat, *Pestic. Biochem. Physiol.*, 27, 24, 1987.
24. Hatzios, K. K. and Penner, D., *Metabolism of Herbicides in Higher Plants*, Burgess Publishing, Minneapolis, 1982.
25. Mumma, R. O. and Hamilton, R. H., Amino acid conjugates, in *Bound and Conjugated Pesticide Residues*, Vol. 29, Kaufman, D. D., Still, G. G., Paulson, G. D., and Bandal, S. K., Eds., American Chemical Society, Washington, D.C., 1976, 68.
26. Fleecker, J. and Owen, W. J., Influence of monooxygenase inhibitors on the metabolism of the herbicides chlorotoluron and metolachlor in cell suspension cultures, *Plant Sci.*, 50, 13, 1987.
27. Shimabukuro, R. M., Frear, D. S., Swanson, H. R., and Walch, W. S., Gluthione conjugation: and enzymatic basis for atrazine resistance in corn, *Plant Physiol.*, 47, 10, 1971.
28. Rueppel, M. L., Brightwell, B. B., Schaefer, J., and Marvel, J. T., Metabolism and degradation of glyphosate in soil and water, *J. Agric. Food Chem.*, 25, 517, 1977.
29. Moore, J. H., Braymer, H. D., and Larson, A. D., Isolation of *Pseudomonas* sp. which utilizes the phosphonate herbicide glyphosate, *Appl. Environ. Microbiol.*, 46, 316, 1983.
30. Talbot, H. W., Johnson, L. M., and Munnecke, D. M., Glyphosate utilization by *Pseudomonas* sp. and *Alcaligenes* sp. isolated from environmental sources, *Curr. Microbiol.*, 10, 225, 1984.
31. Balthazor, T. M. and Mallas, L. E., Glyphosate degrading microorganisms from industrial activated sludge, *Appl. Environ. Microbiol.*, 51, 432, 1986.

32. **Jacob, G. S., Schefer, J., Stejskalm, E. O., and Mekay, R. A.,** Solid state NMR determination of glyphosate metabolism in a *Pseudomonas* sp. *J. Biol. Chem.*, 260, 5899, 1985.
33. **Pipke, R., Amrhein, N., Jacob, G. S., and Kishore, G. M.,** Metabolism of glyphosate in an *Arthrobacter* sp. GLP-1, *Eur. J. Biochem.*, 165, 267, 1987.
34. **Kishore, G. M. and Jacob, G. S.,** Degradation of glyphosate by *Pseudomonas* sp. GP 2982 via a sarcosine intermediate, *J. Biol. Chem.*, 262, 64, 1987.
35. **Comai, L. and Stalker, D.,** Mechanism of action of herbicides and their molecular manipulation, *Plant Mol. Cell Biol.*, 3, 166, 1986.
36. **Stalker, D. M., McBride, K. E., and Malyi, L. D.,** Herbicide resistance in transgenic plants expressing a bacterial detoxification gene, *Science*, 242, 419, 1988.
37. **Stalker, D. M., Hiatt, W. R., and Comai, L.,** A single amino acid substitution in the enzyme 5-enolpyruvylshikimate-3-phosphate synthase confers resistance to the herbicide glyphosate, *J. Biol. Chem.*, 260, 4724, 1985.
38. **Fedke, C.,** *Biochemistry and Physiology of Herbicide Action,* Springer-Verlag, New York, 1982.
39. **Larkin, P. J. and Scowcroft, W. R.,** Somaclonal variation: novel source of variability from cell cultures for plant improvement, *Theor. Appl. Genet.*, 60, 197, 1981.
40. **Trebst, A. and Draber, W.,** Structure activity correlations of recent herbicides in photosynthetic reactions, in *Advances in Pesticide Science*, Vol. 2, Geissbuhler, H., Ed., Pergamon, New York, 1978, 223.
41. **Pfister, K., Steinback, K. E., Gardner, G., and Arntzen, C. J.,** Photoaffinity labeling of an herbicide receptor protein in chloroplast membranes, *Proc. Natl. Acad. Sci. U.S.A.*, 78, 981, 1981.
42. **Mardner, J. B., Matto, A. K., Golonbioff, P., and Edelman, M.,** Structure and physiological control of the rapidly metabolized 32.000 Dalton chloroplast membrane protein, in *Biosynthesis of the Photosynthetic Apparatus: Molecular Biology, Development and Regulation*, Vol. 14, Thornber, J. P., Staehelin, L. A., and Hallik, R. G., Eds., Alan R. Liss, New York, 1984, 309.
43. **Zurawski, G., Bohnert, H. J., Whitfeld, P. R., and Bottomley, W.,** Nucleotide sequence of the gene for the Mr 32.000 thylakoid membrane protein from *Spinacia oleracea* and *Nicotiana debneyi* predicts a totally conserved primary translation product of Mr 38,950, *Proc. Natl. Acad. Sci. U.S.A.*, 79, 7699, 1982.
44. **Erickson, J. M., Rahire, M., and Rochaix, J. D.,** *Chlamydomonas reinhardii* gene for the 32.000 mol. wt. protein of photosystem II contains four large introns and is located entirely within the chloroplast inverted repeat, *EMBO J.*, 3, 2753, 1984.
45. **Erickson, J. M., Rahire, M., Rochaix, J. D., and Mets, L. J.,** Herbicide resistance and cross-resistance: changes at three distinct sites in the herbicide-binding protein, *Science*, 228, 204, 1985.
46. **Erickson, J. M., Rahire, M., Bennoun, P., Delepelaire, P., Diner, B., and Rochaix, J. D.,** Herbicide resistance in *Chlamydomonas reinhardii* results from a mutation in the chloroplast gene for the 32-kilodalton protein of photosystem II; *Proc. Natl. Acad. Sci. U.S.A.*, 81, 3617, 1984.
47. **Erickson, J. M. and Rochaix, J. D.,** Abstr. 1st Int. Congr. Plant Molecular Biology, Galan, G.A., Ed., University of Georgia Center for Education, for the International Society for Plant Molecular Biology, Athens, GA, 1985, 54.
48. **Curtis, S. and Haselkoln, R.,** Isolation, sequence and expression of two members of the 32 kD thylakoid membrane protein gene family from cyanobacterium, *Proc. Natl. Acad. Sci. U.S.A.*, 81, 2693, 1984.
49. **Mulligan, B., Schultes, N., Chen, L., and Bogorad, L.,** Nucleotide sequence of a multiple-copy gene for the B protein of photosystem II of cyanobacterium, *Proc. Natl. Acad. Sci. U.S.A.*, 81, 2693, 1984.
50. **Hirschberg, J. and McIntosh, L.,** Molecular basis of herbicide resistance in *Amaranthus hybridus, Science*, 222, 1346, 1983.

51. **Hirschberg, J., Yehuda, A. B., Pecker, I., and Ohad, N.**, Mutations resistant to photosystem II herbicides, *Plant Mol. Biol.*, 140, 357, 1987.
52. **Mazur, J. B. and Falco S. C.**, The development of herbicide resistant crops, *Proc. Rev. Plant Physiol, Plant Mol. Biol.*, 40, 441, 1989.
53. **Michel, H., Epp, O., and Deisenhofer, J.**, Pigment-protein interactions in the photosynthetic reaction center from *Rhodopseudomonas viridis, EMBO J.*, 5, 2445, 1986.
54. **Michel, H.**, Three dimensional crystals of a membrane protein complex (the photosynthetic reaction center from *Rhodopseudomonas viridis*), *J. Mol. Biol.*, 158, 567, 1982.
55. **Deisenhofer, I., Epp, O., Miki, K., Huber, R., and Michel, H.**, Structure of the protein subunits in the photosynthetic reaction center of *Rhodopseudomonas viridis* at 3 Å resolution, *Nature*, 318, 618, 1985.
56. **Muhitch, M. J., Shaner, D. L., and Stidham, M. A.**, Imidazolinones and acetohydroxyacid synthase from higher plants, *Plant Physiol.*, 83, 451, 1987.
57. **Nafziger, E. D., Widholm, J. M., Steinrucken, H. C., and Killmer, J. L.**, Selection and characterization of a carrot cell-line tolerant to glyphosate, *Plant Physiol.*, 76, 571, 1984.
58. **Nomura, N. S. and Hilton, H. W.**, The adsorption and degradation of glyphosate in five Hawaiian sugarcane soils, *Weed Res.*, 17, 113, 1977.
59. **Cheung, A. Y., Bogorad, L., VanMontague, M., and Schell, J.**, Relocating a gene for herbicide tolerance: a chloroplast gene is converted into a nuclear gene, *Proc. Natl. Acad. Sci. U.S.A.*, 85, 391, 1988.
60. **Shah, D. M., Horsch, R. B., Klee, H. J., Kishore, G. M., Winter, J. A., Tumer, N. E., Hironaka, C. M., Sanders, P. R., Gasser, C. S., Aykent, S., Siegel, N. R., Rogers, S. R., and Fraley, R. T.**, Engineering herbicide tolerance in transgenic plants, *Science*, 233, 478, 1986.
61. **Helmer, G.**, Genetic engineering of atrazine tolerance in plants, paper presented at Int. Symp. Pestic. Biotechnol., Michigan State University, East Lansing, August 17th to 19th, 1986.
62. **Chaleff, R. S. and Mauvais, C. J.**, Acetolactate synthase is the site of action of two sulfonylurea herbicides in higher plants, *Science*, 224, 1143, 1984.
63. **Sebastian, S. A. and Chaleff, R. S.**, Soybean mutants with increased tolerance for sulfonylurea herbicides, *Crop Sci.*, 27, 948, 1987.
64. **Ray, T. B.**, The mode of action of chlorsulfuron: a new herbicide for cereals, *Pestic. Biochem. Physiol.*, 17, 10, 1982.
65. **Ray, T. B.**, The mode of action of chlorsulfuron: the lack of direct inhibition of plant DNA synthesis, *Pestic. Biochem. Physiol.*, 18, 262, 1982.
66. **LaRossa, R. A. and Smulski, D. R.**, ilv-B-encoded acetolactate synthase is resistant to the herbicide sulfometuron methyl, *J. Bacteriol.*, 160, 391, 1984.
67. **LaRossa, R. A., Van Dyk, T. K., and Smulski, D. R.**, Toxic accumulation of 2-ketobutyrate caused by inhibition of the branched-chain amino acid biosynthetic enzyme acetolactate synthase in *Salmonella typhimurium, J. Bacteriol.*, 169, 1372, 1987.
68. **Falco, S. C. and Dumas, K. S.**, Genetic analysis of mutants of *Saccharomyces cerevisiae* resistant to the herbicide sulfometuron methyl, *Genetics*, 109, 21, 1985.
69. **Falco, S. C., Dumas, K. S., and Livak, K. J.**, Nucleotide sequence of the yeast ILV2 gene which encodes acetolactate synthase, *Nucleic Acids Res.*, 13, 4011, 1985.
70. **Harnett, M. E., Newcomb, J. R., and Hodson, R. C.**, Mutations in *Chlamydomonas reinhardii* conferring resistance to the herbicide sulfometuron methyl, *Plant Physiol.*, 85, 898, 1987.
71. **Haughn, G. W., Smith, J., Mazur, B., and Somerville, C.**, Transformation with a mutant *Arabidopsis* acetolactate synthase gene renders tobacco resistant to sulfonylurea herbicides, *Mol. Gen. Genet.*, 211, 266, 1988.
72. **Haughn, G. W. and Somerville, C. R.**, Sulfonylurea-resistant mutants of *Arabidopsis thaliana, Mol. Gen. Genet.*, 204, 430, 1986.

73. **Kleswick, W. A., Eher, R. J., Gerwick, B. C., Monte, W. T., and Pearson, N. R.,** New 2-Aryloamino-sulphonyl-1,2,4-triazolo-(1,5-a)-pyrimidine(s) Useful as Selective Herbicides and to Supress Nitrification in Soil, Eur. Patent Appl., 0142152, 1984.
74. **Eoyang, L. and Silverman, P. M.,** Purification and subunit composition of acetohydroxyacid synthase I from *Escherichia coli* K-12, *J. Bacteriol.*, 157, 184, 1984.
75. **Eoyang, L. and Silverman, P. M.,** Role of small subunit (IIvN polypeptide) of acetohydroxyacid synthase I from *Escherichia coli* K-12 in sensitivity of the enzyme of valine inhibition, *J. Bacteriol.*, 166, 901, 1986.
76. **Friden, P., Donegan, J., Mullen, J., Tsui, P., Freundlich, M. et al.,** The *ilv*B locus of *Escherichia coli* K-12 is an operon encoding both subunits of acetohydroxyacid synthase I., *Nucleic Acids Res.*, 13, 3979, 1985.
77. **Wek, R. C., Hauser, C. A., and Hatfield, G. W.,** The nucleotide sequence of the *ilv*BN operon of *Escherichia coli*: sequence homologies of the acetohydroxyacid synthase isozymes, *Nucleic Acids Res.*, 13, 3935, 1985.
78. **Lawther, R. P., Calhoun, D. H., Adams, C. W., Hauser, C. A., Gray, J., and Hatfield, G. W.,** Molecular basis of valine resistance in *Escherichia coli* K-12, *Proc. Natl. Acad. Sci. U.S.A.*, 78, 922, 1981.
79. **Schloss, J. V., Van Dyk, D. E., Vasta, J. F., and Kutny, R. M.,** Purification and properties of *Salmonella typhimurium* acetolactate synthase isozyme II from *E. coli* HB 101/pDU9, *Biochemistry*, 24, 4952, 1985.
80. **DeFelice, M., Guardiola, J., Esposite, B., and Iaccarino, M.,** Structural genes for newly recognized acetolactate synthase in *Escherichia coli* K-12., *J. Bacteriol.*, 120, 1068, 1974.
81. **Squires, C. H., DeFelice, M., Devereux, J., and Calvo, J. M.,** Molecular structure of the *ilv*H in *Escherichia coli* K-12. *Nucleic Acids Res.*, 11, 5299, 1983.
82. **Squires, C. H., Defelice, M., Wessler, S. R., and Calvo, J. M.,** Physical characterization of the *ilvh* operon of *Escherichia coli* K-12, *J. Bacteriol.*, 147, 797, 1981.
83. **Ryan, G. F.,** Resistance of common groundsel to simazine and atrazine, *Weed Sci.*, 18, 614, 1970.
84. **Ryan, E. D. and Kohlhaw, G. B.,** Subcellular localization of isoleucine valine biosynthetic enzymes in yeast. *J. Bacteriol.*, 120, 631, 1974.
85. **Yadav, N., McDevitt, R. E., Bernard, S., and Falco, S. C.,** Single amino acid substitutions in the enzyme acetolactate synthase confer resistance to the herbicide sulfometuron methyl, *Proc. Natl. Acad. Sci. U.S.A.*, 83, 4418, 1986.
86. **Falco, S. C., Knowlton, S., LaRossa, R. A., Smith, J. K., and Mazur, B. J.,** Herbicides that inhibit amino acid biosynthesis: the sulfonylurea — a case study, *1987 Brit. Crop Protection Cong.-Weeds*, BCPC Publications, Surrey, U.K., 1987, 149.
87. **Bedbrook, J., Chaleff, R. S., Falco, S. C., Mazur, B. J., and Yadav, N.,** Nucleic Acid Fragment Encoding Herbicide Resistant Plant Acetolactate Synthase, Eur. Patent Appl., 0257993, 1988.
88. **Mazur, B. J., Chui, C.-F., and Smith, J. K.,** Isolation and characterization of plant genes coding for acetolactate synthase, the target enzyme for two classes of herbicides, *Plant Physiol.*, 85, 1110, 1987.
89. **Mazur, B. J., Falco, S. C., Knowlton, S., and Smith, J. K.,** Acetolactate synthase, the target enzyme of sulfonylurea herbicides, *Plant Mol. Biol.*, 140, 339, 1987.
90. **Jones, A. V., Yound, R. M., and Leto, K. J.,** Subcellular localization and properties of acetolactate synthase, target site of the sulfonylurea herbicides, *Plant Physiol.*, 77, 293, 1985.
91. **Lee, K. Y., Townsend, J., Tepperman, J., Black, M., Chui, C.-F. et al.,** The molecular basis of sulfonylurea herbicide resistance in higher plants, *EMBO J.*, 7, 1241, 1988.
92. **Comai, L., Sen, L. C., and Stalker, D. M.,** An altered *aro*A gene product confers resistance to the herbicide glyphosate, *Science*, 221, 370, 1983.

93. **Kishore, G. M. and Shah, D.**, Amino acid biosynthesis inhibitors as herbicides, *Annu. Rev. Biochem.*, 57, 627, 1988.
94. **Jawoski, E. G.**, Mode of action of N-Phospho-methylglycine: inhibition of aromatic amino acid biosynthesis, *J. Agric. Food. Chem.*, 20, 1195, 1972.
95. **Amrhein, N., Deus, D., Gehrke, P., and Steinrucken, H. C.**, The site of inhibition of the shikimate pathway by glyphosate, *Plant Physiol.*, 66, 830, 1980.
96. **Amerhein, N., Johanning, D., Schab, J., and Schulz, A.**, Biochemical basis for glyphosate tolerance in a bacterium and a plant tissue culture, *FEBS Lett.*, 157, 191, 1983.
97. **Hollander, H. and Amerhein, N.**, The site of inhibition of the shikimate pathway by glyphosate, *Plant Physiol.*, 66, 823, 1980.
98. **Comai, L., Faccioti, D., Hiatt, W. R., Thompson, G., Rose, R. E., and Stalker, D. M.**, Expression in plants of a mutant *aro*A gene from *Salmonella typhimurium* confers tolerance to glyphosate, *Nature*, 371, 741, 1985.
99. **Rogers, S. G., Brand, L. A., Holder, S. B., Sharps, E. S., and Brackin, M. J.**, Amplification of the *aro*A gene from *E. coli* results in tolerance to the herbicide glyphosate, *Appl. Environ. Microbiol.*, 46, 37, 1983.
100. **Della-Cioppa, G., Bauer, S. C., Taylor, M. L., Rochester, D. E., and Klien, B. K.**, Targeting a herbicide-resistant enzyme from *Escherichia coli* to chloroplasts of higher plants, *Bio/Technology*, 5, 579, 1987.
101. **Fillatti, J. J., Kiser, J., Roseem, R., and Comai, L.**, Efficient transfer of a glyphosate tolerance gene into tomato using a binary *Agrobacterium tumefaciens* vector, *Bio/Technology*, 5, 726, 1987.
102. **Steinrucken, H. C., Scholz, A., Amerhein, N., Porter, C. A., and Fraley, R. T.**, Overproduction of a 5-enolpyruvylshikimate- 3-phosphate synthase by a glyphosate-tolerant *Petunia hybrida* cell line, *Arch. Biochem. Biophys.*, 244, 169, 1986.
103. **Gasser, C. S., Winter, J. A., Hironaka, C. M., and Shah, D. M.**, Structure, expression and evolution of the genes encoding 5-enolpyruvylshikimate-3-phosphate synthase of petunia and tomato, *J. Biol. Chem.*, 263, 4280, 1988.
104. **Mollenhauer, C., Smart, C., and Amrhein, N.**, Glyphosate toxicity in the shoot apical region of the tomato plant, *Pestic. Biochem. Physiol.*, 29, 55, 1987.
105. **Kent-Moore, J., Bramern, H. D., and Larson, A. D.**, Isolation of *Pseudomonas* sp. which utilizes the phosphonate herbicide glyphosate, *Appl. Environ. Microbiol.*, 46, 316, 1983.
106. **Sprankle, P., Meggit, W. F., and Penner, D.**, Adsorption, mobility, and microbial degradation of glyphosate in the soil, *Weed Sci.*, 23, 229, 1975.
107. **Bandeen, J. D., Stephenson, G. R., and Cowett, E. R.**, Discovery and distribution of herbicide-resistant weeds in North America, in *Herbicide Resistance in Plants*, Le Baron, H. M. and Gressel, J., Eds., John Wiley & Sons, New York, 1982, 1.
108. **Romesser, J. A. and O'Keefe, D. P.**, Induction of cytochrome P-450 dependent sulfonylurea metabolism in *Streptomyces grisseolus*, *Biochem. Biophys. Res. Commun.*, 140, 650, 1986.
109. **Anderson, P. C. and Hibberd, K. A.**, Evidence for the interaction of an imidazolinone herbicide with leucine, valine, and isoleucine metabolism, *Weed Sci.*, 33, 479, 1987.
110. **Donn, G., Tischer, E., Smith, J. A., and Goodman, H. M.**, Herbicide-resistant alfalfa cells: an example of gene amplification in plants, *J. Mol. Appl. Genet.*, 2, 621, 1984.
111. **De Block, M., Botterman, T., Vandewiele, M., Dockx, T., Thoen, C., Gossle, V., Movva, N. R., Thompson, C., van Montagu, M., and Leemans, J.**, Engineering herbicide resistance in plants by expression of detoxifying enzyme, *EMBO J.*, 6, 2513, 1987.
112. **Tachibana, K., Watanabe, T., Sekizawa, T., and Takematsu, T.**, Action mechanism of bialphos. II. Accumulation of ammonia in plants treated with bialphos, *J. Pestic. Sci.*, 11, 33, 1986.

113. **Murakami, T., Anzai, H., Imai, S., Satoh, A., Nagaoka, K., and Thompson, C. J.**, The bialaphos biosynthetic genes of *Streptomyces hygroscopicus*: molecular cloning and characterization of the gene cluster, *Mol. Gen. Genet.*, 205, 42, 1986.
114. **Seto, H., Imai, S., Tsuruoka, T., Satoh, A., Kojima, M., Inouye, S., Sasaki, T., and Otake, N.**, Studies on the biosynthesis of bialaphos (SF-1293). I. Incorporation of ^{13}C and ^{3}H labeled precursors into bialphos, *J. Antibiot.*, 35, 1719, 1982.
115. **Anzai, H., Murakami, T., Imai, S., Satoh, A., Nagaoka, K., and Tompson, C. J.**, Transcriptional regulation of bialaphos biosynthesis in *Streptomyces hygroscopicus*, *J. Bacteriol.*, 169, 3482, 1987.
116. **Strauch, E., Wohlleben, W., and Pühler, A.**, Cloning of a phosphinothricin N-acetyltransferase gene from *Streptomyces viridochromogenes* Tü94 and its expression in *Streptomyces lividans* and *Escherichia coli, Gene*, 63, 65, 1988.
117. **Wohlleben, W., Arnold, W., Broer, I., Hillemann, D., Straunch, E., and Pühler, A.**, Nucleotide sequence of phosphinothricin N-acetyltransferase gene from *Streptomyces viridochromogenes* Tü94 and its expression in *Nicotiana tabacum, Gene*, 70, 25, 1988.
118. **De Greef, W., Delon, R., De Block, M., Leemans, J., and Botterman, J.**, Evaluation of herbicide resistance transgenic crops under field conditions, *Bio/Technology*, 7, 61, 1989.
119. **Harnett, M. E., Chui, C. F., Mauvais, C. J., McDevitt, R. E., and Knowlton, S.**, Herbicide resistant plants carrying mutated acetolactate synthase genes, in *ACS Symp. Fundamental and Practical Approaches to Combatting Resistance*, LeBaron, H., Moberg, W., and Green, M., Eds., ACS Books, Washington, D.C., 1989.
120. **Barsby, T. L., Kemble, R. J., and Yarrow, S. A.**, *Brassica* cybrids and their utility in plant breeding, *Plant Mol. Biol.*, 140, 223, 1987.
121. **Beversdorf, W. D., Weiss-Leman, J., Erickson, L. R., and Souza Machado, V.**, Transfer of cytoplasmically-inherited triazine resistance from bird's rape to cultivated *Brassica campestris* and *Brassica napus, Can. J. Genet. Cytol.*, 22, 167, 1980.
122. **Callahan, F. E., Edelman, M., Matoo, A. K., and Autar, K.**, Posttranslational acetylation and intra-thylakoid translocation of specific chloroplast proteins, in *Progress in Photosynthesis Research*, Biggens, J., Ed., Martinus Nijhoff, Dordrecht, 1987, 799.
123. **Calson, P.**, Methionine sulfoximine-resistant mutants of tobacco, *Science*, 180, 1366, 1973.
124. **Cole, D. J. and Owen, W. J.**, Influence of monooxygenase inhibitors on the metabolism of the herbicides chlororluron and metalachlor in cell suspension cultures, *Plant Sci.*, 50, 13, 1987.
125. **Darmency, H. M. and Pernes, J.**, Use of wild *Setaria viridis* Beauv. to improve triazine resistance in cultivated *S. italica* by hybridization, *Weed Res.*, 25, 175, 1985.
126. **Ellis, R. J.**, Chloroplast proteins: synthesis, transport and assembly, *Annu. Rev. Plant Physiol.*, 32, 111, 1981.
127. **Fowke, C. C.**, Plant protoplasts, in *Plant Protoplasts*, Constable, F., Ed., CRC Press, Boca Raton, FL, 1985.
128. **Fujimura, T., Sakurai, M., Akagi, H., Negishi, T., and Hirose, A.**, Regeneration of rice plants from protoplasts, *Plant Tissue Cult. Lett.*, 2, 74, 1985.
129. **Fedtke, C.**, *Biochemistry and Physiology of Herbicide Action*, Springer-Verlag, New York, 1982.
130. **Fleecker, J. and Steen, R.**, Hydroxylation of 2,4-D in several weed species, *Weed Sci.*, 19, 507, 1971.
131. **Galloway, R. E. and Mets, L. J.**, Atrazine, bromacil and diuron resistance in *Chlamydomonas*. A single non-mendelian genetic locus controls the structure of the thylakoid binding site, *Plant Physiol.*, 74, 469, 1984.
132. **Guddewar, M. B. and Dautermann, W. C.**, Purification and properties of glutathione-S-transferase from corn which conjugates S-triazine herbicides, *Phytochemistry*, 18, 735, 1979.

133. **Grebanier, A. E., Coen, D. M., Rich, A., and Bogorad, L.**, Membrane proteins synthesized but not processed by isolated maize chloroplasts, *J. Cell Biol.*, 78, 734, 1978.
134. **Golden, S. and Haselkorn, R.**, Mutation to herbicide resistance maps within the *psb*A gene of *Anacystis nidulans* R2, *Science*, 229, 1104, 1985.
135. **Hathway, D. F.**, Herbicide selectivity, *Biol. Rev.*, 61, 435, 1986.
136. **Hirsch, J., Bleecker, A., Kyle, D. J., and McIntosh, L.**, The molecular basis of triazine herbicide resistance, *Z. Naturforsch.*, 39, 412, 1984.
137. **Jensen, K. I. N.**, The role of uptake, translocation and metabolism in the differential intraspecific responses to herbicides, in *Herbicide Resistance in Plants*, Le Baron, H. and Gressel, J., Eds., John Wiley & Sons, New York, 1982, 133.
138. **Kartha, K. K., Michayluk, M. R., Nao, K. N., Gamborg, O. L., and Constable, F.**, Callus formation and regeneration from mesophyll protoplasts of rape plants, *Brassica napus* cultivar Zephyr, *Plant Sci. Lett.*, 3, 265, 1974.
139. **Trebst, A.**, The three dimensional structure of the herbicide binding niche on the reaction center polypeptides of PS II, *Z. Naturforsch.*, 42, 742, 1987.
140. **Mathis, P.**, Primary reaction of photosynthesis: discussion of current issues, in *Progress in Photosynthesis Research, Proc. Int. Congr. Photosynth., 7th*, Biggens, H., Ed., Martinus Nijhoff, Dordrecht, 1987, 1, 151.
141. **Mattoo, A. K., Hoffman-Falk, H., Marder, J. B., and Edelman, M.**, Regulation of protein metabolism: coupling of photosynthetic electron transport to *in vivo* degradation of the rapidly metabolized 32-kilodalton protein of the chloroplast membranes. *Proc. Natl. Acad. Sci. U.S.A.*, 81, 1380, 1984.
142. **Mets, L., Galloway, R. E., and Erickson, J. M.**, Prospects for genetic modification of plants for resistance to triazine herbicides, in *Biotechnology in Plant Science*, Academic Press, New York, 1985, 301.
143. **Miflin, B. J.**, The location of nitrate reductase and other enzymes related to amino acid biosynthesis in the plastids of root and leaves, *Plant Physiol.*, 54, 550, 1974.
144. **Mousdale, D. M. and Coggins, J. R.**, Subcellular localization of the common shikimate pathway enzymes in the *Pisum sativum. L.*, *Planta*, 163, 241, 1985.
145. **Smith, J. K., Mauvais, C. J., Knowlton, S., and Mazur, B. J.**, Molecular biology of resistance to sulfonylurea herbicides, in *ACS Symp. Biotechnology of Crop Protection*, 379, 25, 1988.
146. **Souza-Machado, V.**, Inheritance and breeding potential of triazine tolerance and resistance in plants, in *Herbicide Resistance in Plants*, Le Baron, H. M. and Gressel, J., Eds., Wiley, New York, 1982, 257.
147. **Strauch, E., Wohlleben, W., and Pühler, A.**, Cloning of phosphinothricin N-acetyltransferase gene from *Streptomyces viridochromogenes* Tü94 and its expression in *Streptomyces lividans* and *Escherichia coli*, *Gene*, 63, 65, 1988.
148. **Szabados, L., Hadlaczky, G., and Dudits, D.**, Uptake of isolated plant chromosomes by plant protoplasts, *Planta*, 151, 141, 1981.
149. **Thompson, C. T., Movva, N. R., Tizard, R., Crameri, R., Daives, J. E., Lauwereys, M., and Botterman, J.**, Characterization of the herbicide-resistant gene *bar* from *Streptomyces hygroscopicus*, *EMBO J.*, 6, 2519, 1987.
150. **Walbot, W. and Cullis, C. A.**, Rapid genomic change in higher plants, *Annu. Rev. Plant Physiol.*, 36, 367, 1985.
151. **Wiegand, R. C., Shah, D. M., Mozer, T. J., Harding, E. I., Diaz-Collier, J., Saunders, C., Jaworski, E. G., and Tiemaier, D. C.**, Messenger RNA encoding a gluthione-S-transferase responsible for herbicide tolerance in maize is induced in response to safener treatment, *Plant Mol. Biol.*, 7, 235, 1986.
152. **Zama, P. and Hatzios, K. K.**, Interactions between the metalachlor and the safner CGA-92194 at the levels of uptake and macromolecular synthesis in sorghum leaf protoplasts, *Pestic. Biochem. Physiol.*, 27, 86, 1987.

153. **Zama, P. and Hatzios, K. K.**, Effects of CGA-92194 on the chemical reactivity of metalachlor with glutathione and metabolism of metalachlor in grain sorghum-bicolor cultivar G-623, *Weed Sci.*, 34, 834, 1986.
154. **Zapata, F. J., Evans, P. K., Power, J. B., and Cocking, E. C.**, The effect of temperature on the division of protoplasts of *Lycopersicon esculentum* and *Lycopersicon peruvianum*, *Plant Sci. Lett.*, 8, 119, 1977.
155. **Bayer, E., Gugel, K. H., Hagele, K., Hagenmaier, H., Jessipow, S., Koing, W. A., and Zahner, H.**, Stoffwechselprodukte von Mikrooganismen. Phosphinothricin and Phosphinothricyl-alanyl-alanin, *Helv Chim. Acta*, 55, 224, 1972.
156. **Chater, K. F. and Bruton, C. F.**, Resistance, regulatory and production gene for the antibiotic methylnomycin are clustered, *EMBO J.*, 4, 1893, 1985.
157. **Kondo, Y., Shomura, T., Ogawa, Y., Tsuruoka, T., Watanabe, H., Totukawa, K., Suzuki, T., Moriyama, C., Yoshida, J., Inouye, S., and Niida, T.**, Studies on a new antibiotic SF-1293, I., Isolation and physio-chemical and biological characterization of SF-1993 substances, *Sci. Rep., Meiji Seika*, 13, 34, 1973.
158. **Stanzak, R., Matsushima, P., Baltz, R. H., and Rao, R. N.**, Cloning and expression in *Streptomyces lividans* of clustered erythromycin biosynthesis genes from *Streptomyces erythreus*, *Bio/Technology*, 4, 229, 1986.
159. **Vara, J., Malpartida, F., Hopwood, D. A., and Jimenez, A.**, Cloning and expression of purimycin N-acetyl transferase gene from *Streptomyces albiniger* in *Streptomyces lividans* and *Escherichia coli*, *Gene*, 33, 197, 1985.
160. **Murakami, T., Anzai, H., Imai, S., Satoh, A., Nagaoka, K., and Thompson, C. J.**, The bialaphos biosynthetic genes of *Streptomyces hydroscopicus*, molecular cloning and characterization of gene cluster, *Mol. Gen. Genet.*, 211, 424, 1988.
161. **Gressel, J. and Ben-Sinai, G.**, Low intra-specific competitive fitness in a trazine resistant, nearly isogenic line of *Brassica napus*, *Plant Sci.*, 38, 29, 1985.
162. **Jacob, G. S., Schaefer, J., Stejskal, E. O., and McKay, R. A.**, Solid-state NMR determination of glyphosate metabolism in *Pseudomonas* sp., *J. Biol. Schem.*, 260, 5899, 1985.
163. **Arntzen, C. J., Ditto, C. L., and Brewer, P. W.**, Chloroplast membrane alterations in triazine-resistant *Amaranthus retroflexus* biotypes, *Proc. Natl. Acad. Sci. U.S.A.*, 76, 278, 1979.
164. **Arntzen, C. J., Pfister, K., and Steinback, K. E.**, The mechanism of chloroplast triazine resistance: alterations in the site of herbicide action, in *Herbicide Resistance in Plants*, Le Baron, H. M. and Gressel, J., Eds. John Wiley & Sons, New York, 1985, 301.

Chapter 4

ENGINEERED RESISTANCE AGAINST PLANT VIRUS DISEASES

I. INTRODUCTION

Viral diseases of crop plants constitute a major economic problem through reduction in product yield.[1] At least seven hundred plant viruses are recognized which cause varied diseases and consequently significant crop losses.[2] The genetic material of viruses may be either DNA or RNA, and may be single or double stranded. Approximately 77% of the characterized plant viruses possess a single plus (messenger)-sense strand of RNA. Infection of plant tissue requires damage to the cell wall and/or plasma membrane, which, for insect-borne viruses, is achieved by the penetration by the insect stylet during feeding. Once inside the cell, the virus particle is uncoated to release its nucleic acid, and for at least some plus-stranded RNA viruses, such as tobacco mosaic virus (TMV), uncoating is achieved by cytoplasmic ribosomes which also translate the RNA. Plant virus nucleic acids are not integrated into the host genome. Common translational products among most, if not all, viruses, include (1) coat protein, (2) one or more proteins involved in the replication process, and (3) factors involved in the systemic transmission of the virus away from the site of infection.

A variety of strategies have evolved by which viruses produce the several proteins required. For example, cowpea mosaic virus, which has two RNAs, translates completely to produce two separate translational fusion polyproteins, which are subsequently cleaved to produce the mature polypeptides. Probably during a transcription process, tobacco mosaic virus separates its genomic RNA into monocistronic fragments, each of which is translated separately. Turnip yellow mosaic virus employs a combination of these strategies, and variations exist for other viruses.[1]

II. STRATEGIES FOR ENGINEERING GENES FOR PLANT PROTECTION

Although viruses can be removed from individual plants by meristem culture, the traditional methods to prevent the spread of viral disease generally rely on the prevention of infection. This can be achieved by the use of resistant cultivars, but such an approach is limited by the size of the partial available gene pool. As is the case with fungal and bacterial diseases, the mechanisms of resistance, and particularly the recognition systems involved, are not fully understood. An alternative approach to limiting virus infection has involved the deliberate infection of susceptible crop plants with a mild strain of a particular virus (which does not induce severe symptoms), for it has been shown that, on subsequent infection of the same plant material with a strongly virulent strain of the same or a closely related virus, disease symptoms may fail to develop. Such a phenomenon, superficially analogous to the vaccination of animals, was initially termed "cross-protection".[3] Thereafter, a variety of terms have been used to describe the resistance conferred to transgenic plants. These include coat protein-mediated resistance, genetically engineered virus resistance, virus protection, genetically engineered cross-protection, etc. The term "genetically engineered cross-protection" has been frequently used because, in many cases, the phenotype of resistance mimics that of cross-protection. However, cross-protection is a complex response caused by replication and gene expression from the entire viral genome, only one aspect of which is similar to the resistance conferred by the expression of a virus coat protein gene.[4-7] Thus, the term was thought to be less accurate than more narrowly defined terms.[3] In addition, "coat protein-mediated resistance" is used to refer to the resistance caused by the expression of a virus coat protein (CP) gene in transgenic plants. Accumulation of the CP confers resistance to infection and/or disease development by the virus from which the CP gene was derived and by related viruses. Recent advances in the field of molecular biology of several aspects of virus function have led to the proposal of three general strategies for plant protection against viruses using genetic engineering techniques: (1) modified cross-protection; (2) the use of satellite nucleic acids; and (3) the use of antisense RNA.

A. COAT PROTEIN EXPRESSION AND RESISTANCE IN TRANSGENIC PLANTS

Coat protein genes isolated from different single-stranded (+) sense RNA plant viruses have been cloned into plant expression vectors (Figure 1)[8] transferred into various crop plants (tobacco, tomato, and potato), using the *Agrobacterium tumefaciens*-mediated plant transformation system.[9] Analysis of RNA from transgenic plants is carried out by Northern blot hybridization (Figure 2)[8] and CP protein detected by Western blot analysis (Figure 3). In this way, transgenic plants are produced and this aspect has been discussed in several review articles.[10-13] The 35S or 19S promoters derived from cauliflower mosaic virus have been used most frequently to drive the expression of CP genes in transgenic plants.[14] A more detailed explanation of the construction of chimeric CP genes and their expression in transgenic plants has been reported.[15-19] It has also been demonstrated that transgenic tobacco plants expressing the CP gene of tobacco mosaic virus (TMV) show a significant delay in disease symptom development after TMV infection. This phenomenon was subsequently reported to be dependent upon the level of CP in the transgenic plants and the TMV inoculum concentration.[8,20] Resistance to viral infection effected by CP expression (designated CP-mediated protection)[16] was later demonstrated for other plant viruses from different virus groups

Chapter 4

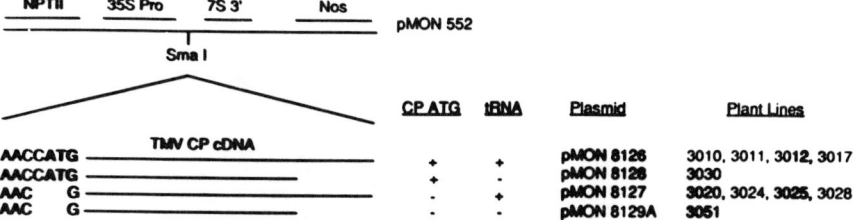

FIGURE 1. Construction of the chimeric genes used to transform tobacco plants. Plasmids pMON 8126 and pMON8128 contain the ATG initiation codon for the CP (CP ATG +), whereas pMON 8127 and pMON 8129A lack an ATG initiation codon (CP ATG −). Plasmids pMON 8128 and pMON 8129A lack sequences for the tRNA-like structure in the 3' nontranslated region of TMV RNA (tRNA −), whereas pMON 8126 and pMON 8127 contain the entire 3' nontranslated region in addition to the CP coding sequence (tRNA +). (From Powell, P. A., Sanders, P. R., Tumer, N., Fraley, R. T., and Beachy, R. N., *Virology*, 175, 124, 1990. With permission.)

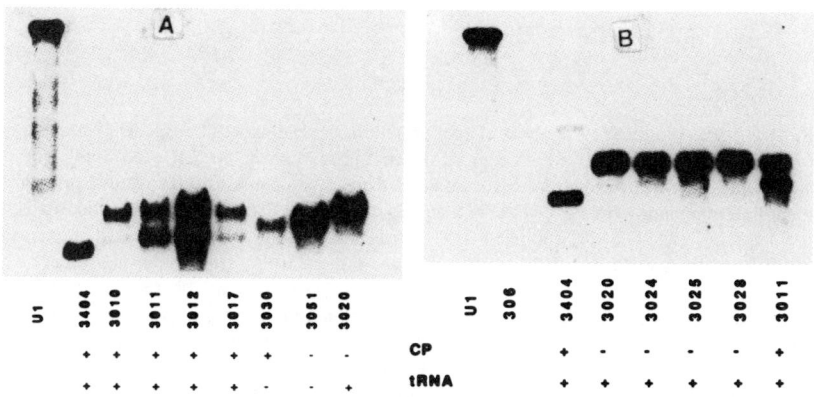

FIGURE 2. Analysis of RNA from transgenic plants. Autoradiograph of an RNA blot hybridized with a ^{32}P-labeled RNA transcript containing RNA complementary to the coat protein mRNA. (A) Lane 1 (numbering from the left) contains 2.45 ng of U_1 TMV RNA. Lane 2 Contains 25 μg of total RNA from plant line 3404 which expressed the plasmid of pTM 319. Lanes 3 through 9 contain 25 μg of total RNA from plants expressing the different constructs shown in Figure 1. Lanes 3 through 6 contain RNA from four different plants lines harboring pMON 8126; lane 7, pMON 8128; lane 8, pMON 8129A; lane 9, pMON 8127. (B) Lane 1 contains 2.45 ng of U_1 TMV RNA. The remaining lanes contain 25 μg of total RNA. Lane 2 contains RNA from a transformed plant harboring pMON 316, an intermediate plasmid lacking TMV cDNA sequences. Lane 3 contains RNA from line 3404 and the remaining lanes contain RNA from plants expressing the different constructs shown in Figure 1. Lanes 4 through 7 contain RNA from four different plant lines harboring pMON 8127. In lane 8 plant line 3011, expressing pMON 8126, was included on this blot for comparison of RNA size and amount from plant lines containing different plamids. (From Powell, P. A. et al., *Virology*, 175, 124, 1990. With permission.)

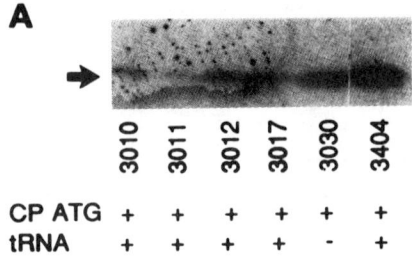

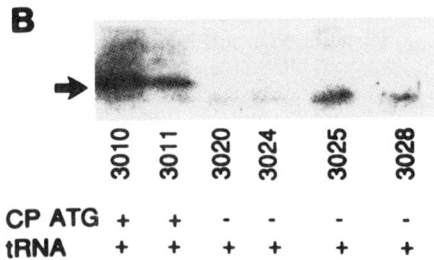

FIGURE 3. Western blot analysis of coat protein from seedlings of transgenic plants. Samples from seedlings of each line which express the coat protein or the NPT II gene were combined, using equal amounts of total protein from each. Each lane contains 50 μg of total protein. (A) Plant lines containing the CP cDNA with an intact ATG translational initiation codon (plasmids pMON 8126, pMON 8128, and pTM 319). Lanes 1 through 4 contain protein from expressors of different lines harboring 8126; lane 5 contains protein from a line harboring pMON 8128. Lane 6 contains protein from plant line 3404 which expresses pTM 319. (B) Plants containing the CP cDNA without the ATG translational initiation codon for CP (plasmid pMON 8127), lanes 3 to 6, compared to two lines containing the intact CP (plasmid pMON 8126), lanes 1 and 2. The arrows indicate the position of TMV CP. (From Powell, P. A. et al., *Virology*, 175, 124, 1990. With permission.)

(Table 1).[21] It is also evident from this table that in most of the studies transgenic tobacco plants are used, and in a few cases tomato and potato plants have been utilized. Two types of tobacco plants were used, those in which the virus is able to move from the initially inoculated leaf to the upper leaves (a systemic host); and those carrying the N gene (cvs Xanthi "ne" and Samsun NN), which localize the virus to areas on the inoculated leaf in necrotic "local" lesions.[21] In these studies, resistance to virus infection is recorded as a delay in or an escape from disease symptom development as compared to controls (a decrease in the severity of disease symptoms and virus accumulation in the plants, or as a reduction in the numbers of local lesions produced on the inoculated leaves).

The rationale of modified cross-protection technique is to identify the viral genes and gene products that are responsible for the observed protection phenomenon, and to separate them from genes encoding the proteins respon-

TABLE 1
Summary of Examples of Coat Protein-Mediated Resistance

Chimeric CP gene	Transgenic plant	Virus resistance	Ref.
P35S:[a]A1MV:NOS[b]	Tobacco	AlMV	44, 133
	Tomato	AlMV	44
P19S:A1MV:CaMV	Tobacco	AlMV	22
P35S:A1MV:T-DNA ORF25	Tobacco	AlMV	80
	Alfalfa	AlMV	80
P35S:CMVrbcS[c]-E9	Tobacco	CMV	25
P35S:TSV:NOS	Tobacco	TSV	54
P35S:PVX:rbcS-E9	Tobacco	PVX	18
eP35S:[d]PVX:rbcS-E9	Potato	PVX	43
P35S:PVX:NOS	Potato	PVX	55
eP35S:PVY:rbcS-E9	Potato	PVY	43
P35S:SMV:NOS	Tobacco	TEV, PVY	34
P35S:TRV:NOS	Tobacco	TRV, PEBV	67
P35S:TMV:NOS	Tobacco	TMV	20
		ToMV	42
	Tobacco	ORSV, PMMV, TMGMV	47
	Tomato	TMV, ToMV	45
eP35S:ToMV:rbcS-E9	Tomato	ToMV	38

[a] P35S: 35S promoter derived from cauliflower mosaic virus.
[b] NOS: nopaline synthase.
[c] rbc: ribulose biphosphate carboxylase.
[d] P35S:enhanced P35S.

sible for symptom formation. Transgenic plants can then be generated which possess the protective genes and would therefore be expected to be resistant to virulent strains without the dangers associated with introducing complete virus particles.

This technique has been used with some success in several plants. For example, Powell et al.[20] have isolated the TMV coat protein gene and introduced it into tobacco plants under the control of the CaMV 35S promoter. In those plants in which the gene was expressed, symptoms developed more slowly than in wild-type plants following inoculation with TMV, and 10 to 60% of the plants producing the gene product developed no disease symptoms at all. Loesch-Fries et al.[22] have demonstrated that tobacco plants transformed with the coat protein gene of alfalfa mosaic virus (AMV) are similarly protected against the full virus disease, and it was found that protection was specifically against AMV virus particles. Both AMV and TMV RNAs were as infectious in plants expressing the coat protein as in those plants which did not express it.

B. USE OF SATELLITE RNAs

A second approach for protection against viruses makes use of the observation that, in some plant RNA viruses, certain RNA sequences may ameliorate disease symptoms in infected plants. These small sequences, which are replicated and packaged normally, but bear no sequence homology to the main genomic RNA and are not required for virus replication and spread, are termed "satellite RNAs". Their replication and transmission are dependent on factors encoded by the viral genome, and they are presumably of strong selective advantage to the virus since they appear not to be lost either during virus proliferation or through natural selection. An alternative view might be to think of satellite RNAs as being parasitic on the virus. The strategy for genetically engineered protection is to introduce DNA copies of satellite RNA into susceptible plants, which, on transcription, might be expected to inhibit symptom formation.

Harrison et al.[23] have transformed tobacco plants with a DNA copy of a satellite RNA of cucumber mosaic virus (CMV) under the control of the CaMV 35S RNA promoter, using *Agrobacterium* as a vector. The accumulation of satellite RNA was correlated with a reduction in symptom formation. On infection with CMV, both control (untransformed) and transformed plants developed chlorotic lesions on the inoculated leaves. However, whereas control plants developed mosaic symptoms in all systemically invaded leaves and were stunted in their growth habit, transformed plants showed mosaic symptoms only in the first two or three leaves to be infected systemically, and leaves which were produced subsequently showed no symptoms.

The replication of CMV in the transformed plants was assayed by three methods:[23] (1) Northern blot analysis, using both viral genomic RNA and satellite RNA as probes to determine the effect of satellite RNA on replication of the genome; (2) immunoassay of the abundance of the viral coat protein; and (3) testing the infectivity of cell extracts on *Chenopodium amaranticolor*, a host of CMV which forms local lesions in proportion to the titer of biologically active virus particles. It was found that the reduction in symptoms was correlated with the decreased replication of the virus, as indicated by all three tests, and these effects of satellite RNA were apparent only in systemically invaded leaves. To determine the viral specificity of these effects, transformed tobacco plants were inoculated with a range of taxonomically diverse viruses. Only tomato aspermy virus (TAV), which is very closely related to CMV, induced the synthesis of CMV satellite RNA, and disease symptoms were correspondingly reduced. Interestingly, however, the symptom attenuation was not correlated with either reduced TMV genomic RNA replication or with reduced infectivity of tissue extracts, in contrast with the effects on CMV proliferation.

Using a similar approach, Gerlach et al.[24] investigated the effects of tobacco ringspot virus (TobRV) satellite RNA on symptoms produced by that virus in transgenic tobacco plants. In this system, it was also observed that transcription of the introduced cDNA equivalent of the satellite RNA was induced dramatically by infection with virus, provided that the cDNA comprised a trimer (i.e., three copies in tandem) of the monomeric satellite. If only a monomeric cDNA was integrated, the transgenic plants exhibited no large increase in satellite RNA, indicating an effect of cDNA copy number. This observation was also confirmed by Harrison et al.[23] The trimeric TobRV cDNA produced only monomer-sized transcripts, suggesting that cleavage of trimeric transcripts may have occurred in the transgenic plants. Furthermore, a reduction of ringspot symptoms was obtained in the trimer-, but not in the monomer-, satellite-transformed plants, and protection was correlated with a decreased level of replication of infectious virus particles.

The observation that TAV symptoms may be reduced without necessarily completely inhibiting viral replication is somewhat difficult to explain, but Harrison et al.[23] tentatively suggested the possibility that the CMV satellite RNA may have some effect on the early stages of TAV RNA replication, and symptom formation was dependent not so much on the final amount of viral RNA accumulated, but perhaps on the timing of its accumulation in relation to cell development — a retardation in replication — may account for reduced symptoms. It may also be possible that the satellite RNA or one or more of the proteins encoded by it directly interacts with (in an as yet uncharacterized way) the genomic RNA or the symptom-producing process.

An interesting feature of the experiments described by Harrison et al.[23] is that, even though the cDNAs corresponding to the satellite RNAs were under the control of the CaMV 35S RNA promoter, which is considered to be constitutive, transcription was amplified to a large extent only when the transgenic plants were inoculated with infective virus; the satellite was dependent upon the virus for replication. The protection was therefore "induced" to meet the needs of the plant, and this contrasts with the coat protein protection, which can be saturated by excess invading virus. A problem with the satellite RNA approach is that the sequences that are protective in one species can be virulent in a different species, or can mutate to a form that is virulent in the originally protected species.

C. USE OF ANTISENSE RNA

A third general approach to inhibit viral infection of plants involves the possibility of using antisense RNA. Antisense RNA (minus-strand RNA) binds to sense (plus-strand or messenger) RNA to prevent its translation, and virus replication, packaging, and/or systemic transmission could conceivably be inhibited in transgenic plants which encode an antisense strand to a specific viral sequence. Although there is evidence that antisense RNA may inhibit, for example, the synthesis of specific heat-shock proteins of the plants, there is as yet no decisive data on its use to reduce viral symptoms.

Antisense RNAs of several plant viruses have been incorporated and expressed in plants; some have provided weak protection against virus infection[5,19,25-27] and no protection in others.[15,26,27] The incorporation and expression of virus capsid protein genes, however, has provided the strongest virus resistance yet developed. However, the length and amount of antisense

RNA have not been fully investigated. Transgenic tobacco plants engineered to express either the potato virus X (PVX) coat protein (+) or the antisense coat protein transcript (CP-antisense) were protected from infection by PVX[18] as indicated by reduced lesion numbers on inoculated leaves, delay or absence of development of systemic symptoms, and reduction in virus accumulation.

It has been demonstrated that some types of RNA, including the satellite RNA of tobacco ringspot virus, are capable of spontaneous and specific cleavage. It appears that this autolytic reaction is related to particular short sequences in the RNA, which confer a characteristic "T-shape" or "hammerhead" secondary structure.[24,28] Artificial "ribozyme" molecules have been created which can cleave target mRNAs (e.g., that of the marker gene chloramphenicol acetyltransferase) and so prevent full expression of the encoding gene. The ribozyme molecule comprises two 8-nucleotide sequences which are homologous and so allow hybridization to regions of the target RNA which flank the sensitive autolytic site-cleavage at 5'-GUX-3', where X is an unpaired nucleotide. Like the use of antisense RNA, the potential for such a technique appears to be enormous, both in virology and in the study of other aspects of metabolic and developmental gene regulation.

III. CONSTRUCTION AND DESIGN OF CHIMERIC GENES ENCODING COAT PROTEINS AND THEIR INTRODUCTION INTO PLANTS

A. ISOLATION OF THE CP GENE: LIMITATIONS

The first step in the construction of the CP gene is isolation of cloned cDNA that represents the CP open reading frame (ORF). This is relatively straightforward for viruses with CPs that are encoded by subgenomic RNAs. However, for viruses such as those in the potyvirus group, where the CP is part of a larger polyprotein,[29,30] isolation of cDNAs of the CP gene is somewhat more difficult. In such cases site-specific mutagenesis may be required to ensure that the cDNA clone encodes, as nearly as possible, the precise amino acid sequence of the CP.

After isolating a CP cDNA clone it may be important to eliminate or alter nucleotide sequences that (may) affect the stability or translation of the transcript. For example, it may be important to place the AUG (initiator) codon within the context of a consensus sequence to assure efficient translational initiation (Figure 1).[8] The sequence ATXXAUGXC[31,32] is often used for the construction of plant genes. This sequence increases the expression of a gene encoding the CP of soybean mosaic virus (SMV).[33,34] The use of a context sequence may be less important when constructing a CP gene derived from a subgenomic mRNA that contains a favored initiation codon.

Some plant viral CP genes have large nontranslated regions at both the 5' and 3' ends of the coding region that are often cloned along the CP sequences. It is important to consider the possibility that intramolecular base pairing within these sequences will affect translation or stability of the mRNA. To reduce the possibility of base pairing it is advisable to keep these sequences short and generally low in potential G:C base pairs. A TMV CP gene that lacked the 3' nontranslated sequences produced more CP per mRNA molecule than a gene that included the nontranslated sequences.[8] In contrast, deletion of the nontranslated region beyond the 3' end of the soybean mosaic virus (SMV) CP gene had little or no effect on the accumulation of either mRNA or CP.

B. SELECTION OF APPROPRIATE PROMOTER

The second step in construction of a chimeric gene is selection of an appropriate transcriptional promoter, since plant RNA viruses do not contain promoters capable of transcription in plant chromosomes. The promoters that have been used to control CP genes in transgenic plants are listed in Table 1.[3] The most effective promoter, P35S, is derived from CaMV and directs the synthesis of the 35S-RNA transcript during infection.[35] Hanley-Bowdoin et al.[36] have constructed and introduced a CP gene of geminivirus tomato golden mosaic virus (TGMV) in *Nicotiana benthamiana* plants in which the AL1 open reading frame was transcribed under the control of the CaMV 35S promoter. The transgenic plants expressed functional AL1 protein and showed the viral double-stranded replication of TGMV DNA in the presence of host proteins. In addition, that AL1 protein did not prove to be a determinant of disease or pathogenesis. While P35S is an especially strong promoter,[37,38] it cannot compensate for a less than optimal gene construction. For example, Bevan and co-workers[39,40] introduced a gene containing the TMV CP cDNA clone under control of P35S into tobacco plants. Relatively low amounts of CP accumulated in plants and did not confer resistance to TMV infection.[39,40] The low expression may result from the 5' nontranslated sequence including an AUG sequence 5' of the CP ORF which interfered with efficient translation of the CP mRNA.[16]

Thus weaker promoters, such as the CaMV promoter P19S, can be useful when several copies of the chimeric gene are incorporated during transformation. A chimeric alfalfa mosaic virus (AlMV) CP gene directed by P19S conferred resistance to AlMV infection when two CP gene loci were expressed in plants, but not when only one was expressed.[22] In addition, tobacco plants that contained the TMV CP ORF under the control of P19S accumulated little CP and were not resistant to TMV infection.[41] Therefore, the choice of transcriptional promoter is especially important to achieve CP-mediated resistance.

C. IMPORTANCE OF THE 3' END SEQUENCE

The third component in the gene construct is the 3' end sequence that confers termination and polyadenylation (poly A) to the transcripts. Sequences from CaMV, storage protein genes, light-regulated genes, and the T-DNA region of the *Agrobacterium tumefaciens* Ti plasmid have all been used in the construction of chimeric genes. There is limited evidence to indicate that one 3' end is preferred over another. Powell et al.[20] suggested that readthrough of the poly A regulatory sequences from the napoline synthase (NOS) gene (from T-DNA) accumulated multiple sizes of TMV CP mRNA transcripts in tobacco plants. However, during expression of the same CP gene in tomato plants (produced by the poly A sequence of a seed storage-protein gene) the transcript was processed to the predicted size.[8]

D. SELECTION OF TRANSFORMED PLANTS

The most challenging aspect of producing virus-resistant plants is often the selection and regeneration of modified plants. For example, most of the resistant plants are produced via *Agrobacterium*-mediated gene transfer, largely because these plants were responsive to this transformation system (Table 1). Transgenic plants are generally selected on the basis of the presence of nopaline. For instance, a cDNA fragment encoding the cytoplasmic inclusion of tobacco vein mottling virus was inserted into the plant expression cassette of a Ti plasmid-based binary vector. The vector was transferred to *Agrobacterium tumefaciens*, which was then used to produce transgenic plants. Further confirmation of the transgenic plants is carried out by biochemical tests. For instance, analysis of poly(A) super (+) RNA from transgenic plants revealed a novel RNA of approximately 2100 nucleotides possessing tobacco vein mottling virus sequences. Also, immunoprecipitation of protein extracts of (super 35S) methionine-labeled transformed callus using cytoplasmic inclusion protein antiserum revealed a corresponding polypeptide. However, in certain situations it is difficult to transform a plant, especially in cereals where the target plant is difficult to transform and/or regenerate. A model system such as tobacco may be useful for such studies.

IV. DETECTION OF CP-MEDIATED PROTECTION AND ITS CHARACTERISTICS

An understanding of the molecular biology of viruses has greatly improved the opportunities for the early detection of the viral infection of plants. Antibodies raised to the coat protein provide an immunological method for the quantitative or semiquantitative determination of virus particles in purified or even crude plant extracts. Most commonly, antibodies are linked to fluorescent markers, such as fluorescein isothiocyanate (FITC), or to enzymes, such as alkaline phosphatase (enzyme-linked immunosorbent assays, ELISA). Viroids, however, are different from viruses in that, although they may cause virus-like diseases, they are simply RNA molecules, usually 200 to 400 nucleotides in size, and lack a coat protein; they themselves, therefore, cannot be detected by immunological techniques, and it is also believed that their RNA is not translated, so it is not possible to trace them by looking for antigenic gene products. Instead, it has been possible to use nucleic acid hybridization techniques, such as RNA dot blots, to detect the viroid "genome" in plant extracts using viroid RNA as radioactive or nonradioactive probe. The technique is both highly specific and sensitive, and nitrocellulose or nylon filters on which the extracted nucleic acid is bound can be reused, after washing, to test for the presence of a number of different viroids or viruses.

Assessing disease resistance involves inoculating plants that express the CP gene [CP(+)] and those that do not [CP(−)] with virus and comparing the numbers of infection sites and/or the development of disease symptoms on the two types of plants.

Both primary transformants and their R_1 progeny are tested for resistance. Because it is preferable to use populations of plants that are identical in age, growth rate, and size, the R_1 or successive generations of plants are generally used. Segregation of the introduced gene in the progeny can be followed immunologically using antibodies to coat proteins or by following the expression of a "reporter" gene.

A. DEVELOPMENT OF DISEASE SYMPTOMS ON CP(−) AND CP(+) PLANTS

In several examples of coat protein-mediated resistance, the resistance is manifested by several features. First, there are fewer sites where infection occurs on inoculated leaves. Nelson et al.[42] reported 95 to 98% fewer necrotic local lesions caused by TMV infection on CP(+) than on CP(−) Xanthi nc tobacco plants. Loesch-Fries et al.[22] and Hemenway et al.[18] likewise reported fewer lesions on AlMV CP(+) and on PVX CP(+) plants following inoculation with AlMV or potato virus $\propto$ (PVX), respectively, than on the corresponding CP(−) plants. These experiments indicate that the expression of the CP gene caused a reduction in the number of sites where infection should have occurred upon inoculation.

A second manifestation of resistance is a reduced rate of systemic disease development throughout the CP(+) plants. In each of the examples of CP-mediated resistance (Table 1), CP(+) plants are less likely to develop systemic disease symptoms than were CP(−) plants. Thus, if inoculation results in infection on the inoculated leaves, the likelihood that the infection will become systemic is considerably lower in CP(+) plants than in CP(−) plants.

A third manifestation of resistance is a lesser accumulation of virus in infected CP(+) plants compared to infected CP(−) plants. Virus accumulation in inoculated leaves and other plant parts have been quantified by ELISA or immunoblots. Nelson et al.[42] reported that the inoculated leaves of CP(+) plants contained less than 30% of the TMV as compared to the inoculated leaves of CP(−) plants and that virus accumulation in the young leaves of CP(+) plants was greatly inhibited. Transgenic plants that express the CP of CMV, PVX, and potato virus Y (PVY) accumulated much less virus, or none at all, following inoculation with the respective virus.[18,25,43] Lack of visual symptoms was generally correlated with a lack of virus accumulation. In those cases where no virus accumulated following inoculation, the plant can be considered as resistant to infection under the conditions of the test.

B. FACTORS AFFECTING VIRAL RESISTANCE IN TRANSGENIC PLANTS

1. Viral Inoculum Concentrations

All these manifestations of resistance can usually, but not always, be overcome by inoculating with increasing concentrations of virus. Powell et al.[20] reported that there was a lag of about 10 days in symptom expression in plants expressing the TMV CP gene compared to CP(−) controls when inoculated with 0.4 μg/ml TMV. CP(−) plants developed symptoms when inoculated with 0.01 μg/ml TMV. However, there was only a 1- to 2-d delay when the plants were inoculated with 2.0 μg/ml TMV. Also, as the inoculum concentration increased, there was a decrease in the number of CP(+) plants that escaped infection. In contrast, CP-mediated resistance provided near immunity to AlMV,[44] PVX,[18,43] PVY,[43] and tobacco etch virus (TEV),[34] even with inoculum concentrations as high as 50 μg/ml.

2. Nature of the Host Plant

A limited number of plant species have been developed with CP-mediated virus protection, although, as noted in Table 1, tobacco plants were used in most cases. To date the plants used have been those that are susceptible to infection by *A. tumefaciens* and amenable to subsequent plant regeneration.

Since plant species differ in their relative susceptibility to viruses, it is difficult to extrapolate the results obtained for one plant species to another, with respect to the efficiency of CP-mediated protection. No comparative study has been conducted to specifically address host-dependent effects, but the results reported from several different studies indicate that this might be an important factor. For example, young seedlings of transgenic tobacco plants that express the TMV CP gene are protected against inoculation with 1 μg TMV ml^{-1} for 3 to 5 days;[16] however, in tomato plants there was a 20-d or more delay after inoculation with 20 μg/ml TMV.[42,45] Nejidat and Beachy[46] reported that transgenic tomato and tobacco plants responded differently to elevated temperatures. However, the expression of an AlMV CP gene in transgenic tobacco or tomato plants[44] conferred comparable levels of resistance against AlMV in both plant types.

There may be differences in the degree of protection conferred by the same CP gene in two different plant cultivars. The expression of the TMV CP gene in cv. *Xanthi* "nc" tobacco provides significant protection against RMV (based on local lesion numbers); on the other hand, in the cultivar Xanthi (a systemic host) no protection was observed.[47] These results indicated that the local lesion assay may not necessarily be indicative of the resistance in systemic hosts. It is possible that in cases where resistance is incomplete, the establishment of an infection site results in a localized high inoculum concentration, which is able to overcome the resistance to systemic spread of the infection.

3. Growth Conditions and Plant Resistance

Plant growth conditions (light, temperature, humidity, water supply, nutrition) before and during or after infection can have a profound effect on plant susceptibility and the course of disease development.[2] Many of the CP-mediated protection assays to date have been performed under greenhouse or growth-chamber conditions. However, since transgenic plants in the field are subjected to changing environmental conditions, it is important to evaluate the effects of different environmental factors, separately and/or in combination, on the efficiency of CP-mediated protection.

Nelson et al.[45] demonstrated that tomato plants expressing the TMV CP gene retain their resistance to TMV infection under field conditions, although the level of CP was decreased slightly under these conditions compared to greenhouse conditions. Nejidat and Beachy[46] reported that resistance of tobacco plants, but not tomato plants, expressing the TMV CP gene is temperature dependent. Under high-temperature conditions (30 to 35°C) there was a 90% reduction in CP levels and less protection against TMV infection in tobacco. In contrast, TMV CP levels in transgenic tomato plants also dropped at elevated temperatures but retained high resistance to TMV. Transgenic plants which were moved from 35 to 22°C accumulated normal CP within several hours. Transgenic tobacco plants inoculated and held at 35/25°C day/night cycles retained resistance to TMV infection. Nejidat and Beachy[46] thus suggested that the low level of CP under elevated temperature is due to instability of the TMV CP. These results are consistent with the results reported by Nelson et al.[45] that decreased levels of CP in transgenic tomato plants under field conditions, where summer temperatures reached 36°C, had no effect on plant resistance. In contrast, there was no change in the level of CP in soybean mosaic virus (SMV) in transgenic tobacco plants grown under high temperature conditions. It was also suggested that CP-mediated protection should be tested throughout the life cycle of the plant to achieve a more complete understanding of resistance.[46] Additional studies to evaluate the effect of growth conditions (including extreme conditions) on the efficiency of the CP-mediated protection are, however, needed. Moreover, since the physiological conditions of the plant continually change during different stages in the growth cycle, plant susceptibility to viral infection may also change.

4. Specificity of the CP-Mediated Protection

The expression of a virus CP gene from one virus group does not confer effective protection against viruses from other groups. For instance, expression of an alfalfa mosaic virus CP gene in transgenic plants[22] was reported to confer resistance against AlMV (homologous protection), but not against the unrelated virus TMV. Anderson et al.[48] reported that TMV CP confers a very limited type of protection against the unrelated viruses PVX, PVY, and CMV.

There are indications that protection against viruses other than that from which the CP gene was isolated may occur between related strains and related viruses (from the same group). In addition, CP amino acid sequence homology may determine, in part, the degree of intra-virus group protection obtained by expression of single CP gene. Nelson et al.[42] demonstrated that the TMV (U1 or vulgare strain) CP protects the plants against infection by a severe yellow strain of TMV (PV230). van Dun and Bol[49] showed that the CP of tobacco rattle virus, strain TCM, protects the plants against infection by an isolate of pea early browning virus, which shares a high degree of homology in its genome with that of TCM, while conferring no protection against the PLB strain of TMV (which has only 39% amino acid sequence homology in its CP with TCM CP). Anderson et al.[48] showed that the TMV CP confers only weak protection against SHMV (C-TMV), which shares 39% homology in CP amino acid sequence with TMV CP. Stark and Beachy[34] reported that the expression of SMV CP in tobacco plants confers significant protection against infection by two other potyviruses, TEV and PVY, even though SMV is a nonpathogen on tobacco. TEV shares 62% and PVY shares 58% CP amino acid sequence homology with SMV (Figure 4).

This aspect of plant protection was analyzed in greater detail by Nejidat and Beachy.[47] They tested protection on tobacco plants expressing the TMV CP gene against tobamoviruses (ToMV) sharing different degrees of CP amino acid sequence homology with TMV CP. Their results indicated that the TMV CP confers high resistance against four ToMV; pepper mild mottle virus, PMMV; tobacco mild green mosaic virus TMGMV; and odontoglossum ringspot virus, ORSV, which share 82 to 60% homology in CP amino acid sequence with TMV. In addition, a weak protection (in the local lesion host only) against ribgrass mosaic virus (RMV), which shares only 45% homology in the CP amino acid sequence with TMV, was also observed. Thus heterologous CP-mediated protection within the same virus group is possible, provided that there is a minimal degree of homology (perhaps 60%) in the CP amino acid sequences. This characteristic may be of great value when CP-mediated protection is considered for applied agricultural use, since the expression of one CP gene may provide protection against multiple viruses from the same group. It is not clear whether the absolute homology in the amino acid sequence between the CPs is important to obtain broad protection, or whether conservation of particular structural dominants is important.

It is worth noting that interference between members of the tobamovirus group was previously studied, and the results are in agreement with the results obtained via CP-mediated protection. Siegel[50] demonstrated that TMV (strain U1), which systemically infects *Nicotiana sylvestris,* inhibits the production

Chapter 4

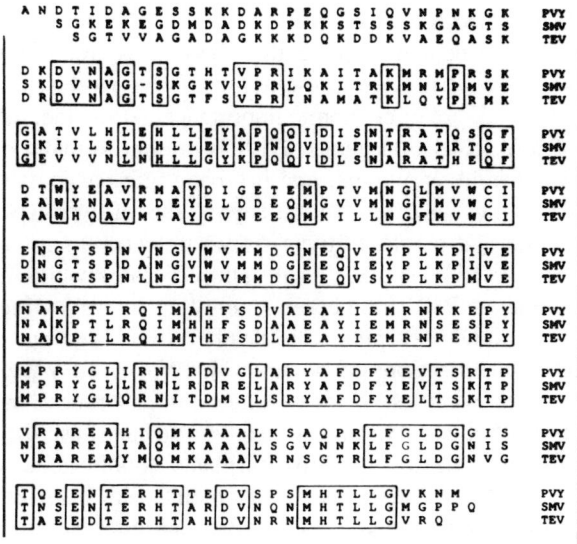

FIGURE 4. A comparison of the CP amino acid sequence of PVY, SMV, and TEV (D. Stark, unpublished results). Boxed regions indicate amino acid homology in all three viruses. The CP gene from the strain of TEV used in this study was cloned and sequenced as described. This strain of TEV shares 98 to 99% nucleic and amino acid sequence homology with two other strains of TEV. This supports the observation that strains of the same potyvirus exhibit extremely high coat protein sequence homology. (From Stark, D. M. and Beachy, R. N., *Bio/Technology*, 7, 1257, 1989. With permission of *Bio/Technology Magazine*.)

of necrotic local lesions by the U2 strain of TMV (TMGMV) in mixed or separate infection. Likewise, U2 inhibits necrotic local lesion production by U1 in *Phaseolus vulgaris*.[51] There was no cross-protection between SHMV and TMV.[52]

V. EXAMPLES OF COAT PROTEIN-MEDIATED RESISTANCE

In the last few years, "coat protein-mediated resistance," has been described for plant viruses in several different virus groups. Since 1986, there have been a number of reports of coat protein-mediated resistance in plants of the Solanaceae and Leguminosae (Table 1). Resistance has been developed against viruses in several different groups. Genetically engineered resistance to virus infection by expression of viral CP in transgenic plants has been demonstrated for TMV,[20,53] AlMV,[44,49,54] CMV,[25] PVX,[18,55] tobacco streak virus (TSV),[54] and tobacco rattle virus (TRV).[54] Each virus group is considered in the following sections, which summarize results and attempts to identify the common features and the differences among the examples.

A. POTATO VIRUSES X AND Y

PVX and PVY, included among the potyviruses, represent the largest group of plant viruses, causing disease in most of the worlds' important crop plants. PVX and PVY are of economic importance worldwide, affecting yield and vitality of infected potato plants. PVX is present in many potato fields, generally causing mild symptoms, and has been referred to as the "healthy potato virus".[56] It can yield production depressions of over 10%, depending on the virus strain and potato cultivar.[57] PVY infection is a major problem in potatoes because PVY spreads easily and depresses yields up to 80%.[56] Primary symptoms of PVY are necrosis, mottling, or yellowing of leaflets, leaf dropping, or sometimes premature death.[58] In addition to losses caused by each virus, dual infections of potato with PVX and PVY often show a synergistic increase in disease severity referred to as "rugose mosaic". Synergistic effects of PVX and PVY infection have also been described in tobacco. Rochow et al.[59] have shown that tobacco plants doubly infected with PVX and PVY may produce severe disease symptoms with a direct relationship between severity of symptoms and PVX concentration. Potyvirus genomes are (+) sense RNAs approximately 10 kb in length and are translated to produce a polyprotein that is cleaved by one or more proteases encoded within the polyprotein. The CP is proximal to the carboxyl terminus of the polyprotein.[29,30] Shukla and Ward[60] compared the amino acid sequences of more than 20 different potyvirus CPs and noted that, on the whole, potyvirus CPs are 50 to 60% homologous within the group, with blocks of identical amino acids throughout the protein.

Potato virus $\propto$ (PVX), the type member of the potexvirus group, is a flexous rod-shaped virus with a genome of (+) sense RNA of approximately 6.4 kb.[61] The PVX CP is approximately 26 kDa in mass and is encoded by a subgenomic mRNA that is coterminal with the 3' end of PVX RNA.

Hemenway et al.[18] have demonstrated that expression of PVX coat protein in tobacco results in protection against PVX. Protected plants show lower numbers of lesions on their inoculated leaves, delay or absence of systemic symptom development, and reduction in virus accumulation in both inoculated and systemic leaves. The extent of protection in the CP+ plants correlates with the level of expression of the CP.

Several groups have reported the expression of CP genes from transgenic plants. Hemenway et al.[18] and Lawson et al.[43] reconstructed the PVX CP cDNA clone by adding a systemic DNA fragment to a partial cDNA. Eighteen nucleotides of noncoding sequence were found on 840 nucleotide cDNA, the 3' nontranslated region of the viral RNA, and a poly(A) stretch. The cDNA clone was ligated with the P35S and the 3' sequence from the E9 gene.[62] The chimeric gene constructed by Hoekema et al.[55] included the P35S promoter, the 3' end from the NOS gene, and a cDNA from PVX X3. The CP cDNA included 8 nucleotides beyond (3' end) the termination (UAA) codon.

Chimeric genes were introduced with *A. tumefaciens* into tobacco[18] or potato,[43,55] and modified plants were regenerated. Regenerated potato lines included the commercially important cultivars Bintje and Escort[55] and Russet Burbank.[3] Regenerated potato and tobacco plants accumulated PVX CP at 0.05 to 0.3% of total plant protein.

Resistance to PVX in CP(+) potato plants was tested on rooted cuttings. The first symptoms (slight mottling) appeared after 6 to 7 days postinoculation. Following inoculation by PVX, CP(+) Bintje or Escort lines accumulated 20- to 50-fold less virus than CP(−) plants and CP(+) developed symptoms more slowly than CP(−) plants.[55] Furthermore, CP(+) lines with higher concentrations of CP were more resistant than lines with lower concentrations of CP. Lawson et al.[43] showed that CP(+) Russet Burbank lines did not accumulate any PVX at three different inoculum concentrations tested, while 100% of the controls were infected at each inoculum concentration.

CP(+) tobacco plants that were inoculated with 0.05 μg/ml PVX accumulated much less virus in inoculated leaves and systemically infected leaves than CP(−) plant lines.[18] Furthermore, the number of starch lesions on CP(+) plants was 10 to 15% of the number on CP(−) plants. Only 10 to 30% of the CP(+) plants developed symptoms by day 18 while all of the CP(−) plants developed symptoms by day 11. Furthermore, for tobacco plants there was correlation between the level of CP in modified plants and the degree of resistance. As described in other sections of this review, CP-mediated resistance against TMV and AlMV was largely overcome by inoculation with viral RNA. In contrast, the relative number of starch lesions produced by PVX RNA on CP(+) plants was essentially equivalent to that produced by PVX.[18] These results may indicate that a mechanism of CP-mediated resistance against PVX is different from resistance against other viruses.

Potato virus Y genome consists of a single positive sense RNA approximately 10,000 nucleotides in length.[63] The 5' terminus has a small protein covalently attached (Vpg) and the 3' terminus is polyadenylated. Subgenomic RNAs are not found in infected tissues; rather, the genome of potyviruses is expressed as a polyprotein precursor that is generated by cleavages from the polyprotein by a virus-encoded protease.[29] Unlike PVX, which is only mechanically transmissible, PVX is transmitted both mechanically and by aphids in a nonpersistent manner. As mentioned earlier, Lawson et al.[43] transformed Russet Burbank with both PVX and PVY coat protein genes and the expression of these two different coat proteins was examined in transgenic plants. The transgenic plants were then analyzed for protection against PVX and PVY inoculated separately or in combined inoculum, to determine whether the synergistic reaction between PVX and PVY would interfere with expression of CP-mediated protection. In addition, they also investigated whether CP-mediated protection is effective when PVY is transmitted by aphids. Transgenic plants that expressed both CP genes were resistant to infection by PVX and PVY by mechanical inoculation.[43] One line was also resistant when PVY was inoculated with viruliferous green peach aphids. These experiments demonstrate that CP protection is effective against mixed infection by two different viruses and against mechanical and aphid transmission of PVY. The results further demonstrate that it is possible to engineer multiple traits such as resistance to two different viruses into a commercial potato cultivar. This is the first example of achieving dual virus resistance. By evaluating several different transgenic lines expressing PVX and PVY CP, Lawson et al.[43] have also identified a transgenic potato line, 303, that was essentially immune to infection by PVX and PVY. Potatoes expressing PVX CP have previously been evaluated in the field of agronomic traits, and it has been shown that intrinsic properties of the commercial potato cultivars are preserved after transformation with the PVX CP gene.[55]

B. POTATO VIRUS S

Potato virus S (PVS) is a member of the carlavirus group and is transmitted by aphids in a nonpersistent manner to members of the Solanaceae and Chenopodiaceae. Because of its near symptomless morphology in potato, PVS has been a difficult virus to control and can occur in seed lots at levels of 70% or more.[64] In combination with potato virus X (PVX), it can cause significant reduction in tuber yields.[65] Elite seed stocks in Canada are produced from virus-free material by heat therapy and meristem tip culture,[66] but such stocks become rapidly reinfected with PVS once grown in the field.[67] Contributing to this is the fact that the many potato viruses, including PVS, overwinter in tubers missed in the previous harvest, and volunteer plants emerging from such tubers are a primary source of inoculum and play a significant role in virus epidemiology.[68-72]

The PVS RNA is encapsidated in an approximately 33-kDa coat protein to make slightly flexous filamentous particles of 650 × 12 nm.[68] The viral genome of PVS consists of one single-stranded, positive sense, RNA molecule with an estimated M_r of 2.39×10^6 which contains a 3'-terminal polyadenylated region.[69] Translation of PVS RNA *in vitro* using a rabbit reticulocyte lysate has been reported to yield primarily four products with M_r values of 124K, 112K, 98K, and 36K.[69]

Mackenzie et al.[72] studied the sequence of 3553 nucleotides from the 3' terminus of the Andean (Peruvian) strain of PVS RNA and examined the amino acid sequence homologies between the coat protein gene and five other predicted open reading frames (ORFs), with similar regions from the potato virus X genome. The sequence obtained contained six open reading frames (ORFs) encoding proteins of M_r 7222, M_r 11802, M_r 25092, and at least M_r 41052. The sequence of the 33K ORF has been confirmed to be that of the viral coat protein gene. The nucleotide sequence of this ORF was obtained from plasmids which were isolated by colony hybridization with a specific monoclonal antibody to PVS, and the expression of coat protein fusion products was verified by Western blots of bacterial cell lysates. The deduced amino acid sequence of a 70-amino acid portion from the central region of the PVS coat protein was 59%, identical to the analogous region of the potato virus X. In addition, the 7K, 12K, and 25K ORFs displayed significant sequence homology with the similarly sized ORFs from a number of potexviruses. The partial 41K ORF product was homologous with the C-terminal portion of the viral replicase proteins of potato virus X and white clover mosaic virus.

Recently MacKenzie and Tremaine[70] reported the production of transgenic *Nicotiana debneyii* which expressed the coat protein gene from PVS and which were resistant to inoculation with PVS or purified viral RNA. In their experiment, the coat protein gene from PVS was introduced into *Nicotiana debneyii* by leaf disc transformation using *Agrobacterium tumefaciens*. Transgenic plants expressing the viral coat protein were highly resistant to subsequent infection by the ME strain of PVS, as indicated by the absence of symptom development and a lack of accumulation of virus. As in reported experiments with plants expressing potato virus X coat protein, plants expressing PVS coat protein were also protected from inoculation with PVS RNA.[70]

These results provide further evidence that coat protein-mediated resistance for these two groups of viruses, which share similar genome organizations, may involve inhibition of some early event in infection other than or in addition to virus uncoating.[70-74] While the molecular mechanism of PVS disassembly and/or assembly in infected cells is unknown, it is believed that the viral coat protein is translated from an encapsidated subgenomic RNA species of approximately 1.3 kb.[74] The presence of encapsidated subgenomic RNAs encoding the viral coat protein has also been reported for at least two members of the potexvirus group, white clover mosaic virus and narcissus mosaic virus.

Recently MacKenzie et al.[72] described the transformation and regeneration of Russet Burbank potato plants expressing PVS coat protein that were resistant to inoculation with PVS or purified PVS RNA and showed a measure of resistance to infection by potato virus M (PVM). They introduced the coat protein gene (Figure 5) from PVS into potato cultivar Russet Burbank by leaf disk transformation using *Agrobacterium tumefaciens*. Transgenic plants expressing the viral coat protein were resistant to subsequent infection following mechanical inoculation with the Andean ME strain of PVS as indicated by a lack of accumulation of virus in the upper leaves (Figure 6). In agreement with previous reports for plants expressing potato virus X coat protein, plants expressing PVS coat protein were also protected from inoculation with PVS RNA, and in addition they showed a measure of resistance to inoculation with a related carlavirus, potato virus M. The coat protein-mediated protection afforded by these transgenic plants was sufficient to prevent the accumulation of virus in the tissues of nontransformed Russet Burbank shoots that had been grafted onto transgenic plants inoculated with PVS (Figure 7). In reciprocal

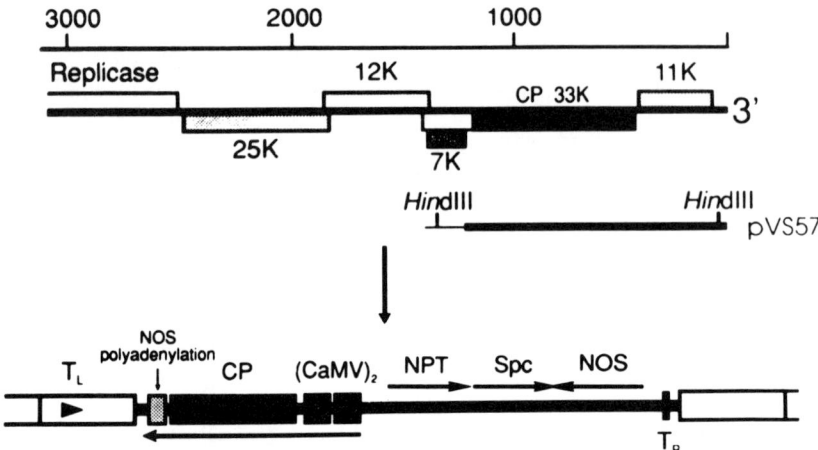

FIGURE 5. Genome organization of potato virus S (PVS) RNA and a schematic representation of the arrangement of the coat protein (CP) cistron following triparental mating and homologous recombination into the disarmed pTiB6S3SE plasmid carried by *Agrobacterium tumefaciens* GV3111SE. A *Hin*dIII fragment derived from pVS57 was cloned into the pBluescript KS+ vector and subsequently excised by double digestion with *Xho*I and *Eco*RI for insertion into pCDX1, which had been digested with *Xho*I and *Eco*RI. The left and right T-DNA border sequences are represented by T_L and T_R, respectively, and genes for nopaline synthase (NOS), spectinomycin resistance (Spc), and neomycin phosphotransferase (NPT) are indicated. The arrow below indicates the direction of transcription of the PVS coat protein gene under the control of the duplicated cauliflower mosaic virus (CaMV) 35S promoter. (From MacKenzie, D. J., Tremaine, J. H., and McPherson, J., *Mol. Plant-Microbe Interact.*, 4, 95, 1991. With permission.)

grafts, shoots from transgenic plants contained significantly lower virus concentrations following grafting into plants systematically infected with PVS.

C. ALFALFA MOSAIC VIRUS (AlMV)

Alfalfa mosaic virus (AlMV) strains comprise the monotypic alfalfa mosaic virus group. The three AlMV RNAs plus RNA 4 are separately encapsidated into bacilliform particles. The RNA 4 is derived from RNA 3 and is the mRNA for the 24-kDa CP. AlMV is most closely related to viruses in the ilarvirus group in that the three genomic RNAs are not infectious alone or in combinations, but require CP or its mRNA for virus replication.[73-76] The CP binds to the RNA at sites near the 3' end and may serve as a recognition signal for the viral replicase.[77,78] The CPs of some members of the ilarvirus group and AlMV are interchangeable for the initiation of replication even though the CP sequences are not homologous.[75,79]

Several workers have been successful in introducing CP genes of AlMV into plants. Tumer et al.[44] constructed a chimeric gene using the CaMV 35S promoter, whose cDNA was derived from AlMV RNA 4, and the NOS 3' end. Transformation mediated by *A. tumefaciens* was used to produce both transgenic tobacco and tomato plants. The amounts of CP in the transgenic

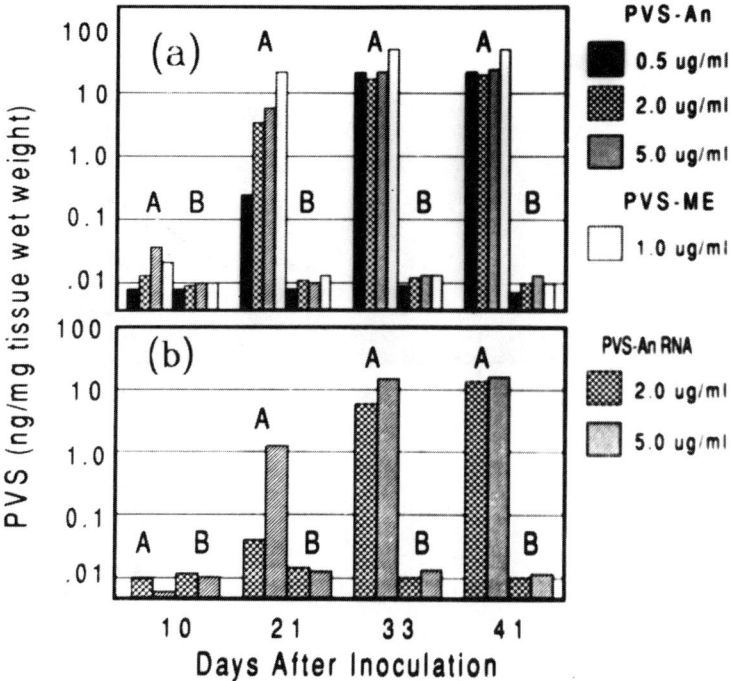

FIGURE 6. Accumulation of potato virus S (PVS) coat protein antigen in the upper leaves of nontransformed (A) and RB58 transgenic Russet Burbank potato (B) plants after inoculation with either intact PVS particles (panel a) or PVS RNA (panel b) at various days after inoculation. Plants were mechanically inoculated with preparations of the Andean (An) strain of PVS (PVS-An; 0.5, 2.0, and 5.0 µg/ml), the ME strain of PVS (PVS-ME; 1.0 µg/ml), or PVS-An RNA (2.0 and 5.0 µg/ml). Viral coat protein concentrations were determined by double-antibody sandwich ELISA and are expressed as the $\log_{10}$ of nanograms of virus coat protein per milligram wet weight of tissue. (From MacKenzie, D. J., Tremaine, J. H., and McPherson, J., *Mol. Plant-Microbe Interact.*, 4, 95, 1991. With permission.)

plant lines, estimated by immunoblot analysis, accounted for 0.1 to 0.4% of the total extractable leaf protein in tobacco and 0.1 to 0.8% in tomato. Several different transgenic tobacco plant lines were inoculated with AlMV at about 5 and 50 µg/ml to determine their susceptibility to infection. The CP(+) plants did not show symptoms 2 weeks after inoculation, whereas the control plants developed lesions and systemic mosaic. Furthermore, the inoculated leaves of CP(−) plants contained as much as 400 times more virus than the leaves of CP(+) plants.

Transgenic tobacco plants and alfalfa plants were developed by Loesch-Fries and colleagues.[22,80] Two chimeric genes were constructed; one contained a gene comprising P19S:AlMV CP:CaMV 3′ end, the other contained P35S:AlMV CP:T-DNA ORF 25 3′ end. In the 33 tobacco lines containing the P19S:CP construct, CP accumulated to 0.05% of the total leaf protein. In the 34 tobacco lines containing the P35S:CP construct, CP accumulated

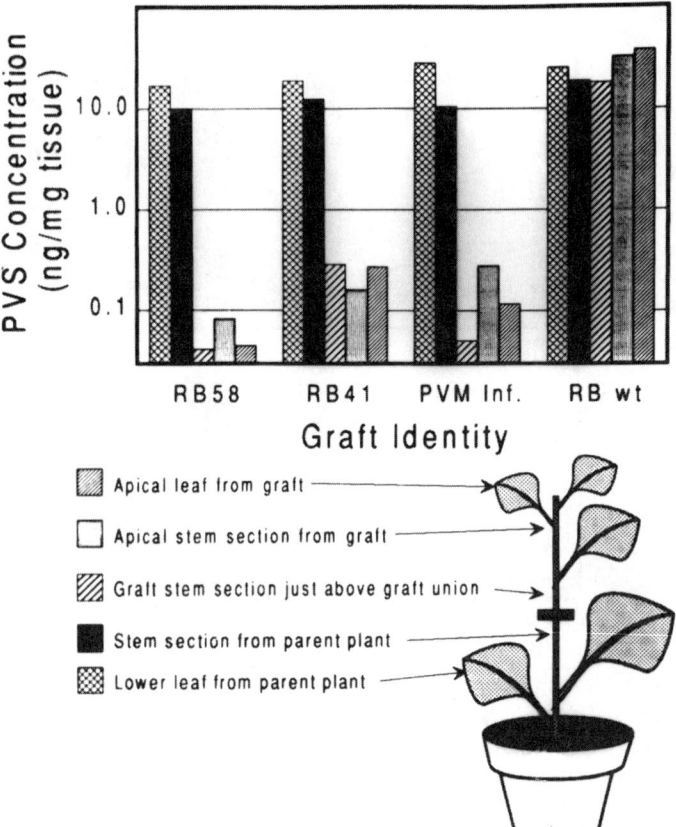

FIGURE 7. Concentration of potato virus S (PVS) coat protein antigen in the tissues of systemically infected plants containing shoot grafts from transgenic plants (RB41 and RB58), plants infected with potato virus M (PVM), or nontransformed plants. Tissue samples from the lower infected leaves or stem segments of the parent plant, as well as stem segments from just above the graft union, and apical stem segments or apical leaves from the graft were removed 42 days after grafting and assayed for the presence of PVS by double-antibody sandwich ELISA. Viral coat protein concentrations are expressed as the $\log_{10}$ of nanograms of virus coat protein per milligram wet weight of tissue. (From MacKenzie, D. J., Tremaine, J. H., and McPherson, J., *Mol. Plant-Microbe Interact.*, 4, 95, 1991. With permission.)

to about 0.13% of the total leaf protein. Plants containing the P19S:CP gene were analyzed for susceptibility to AlMV infection.[22] Progeny of plants containing the largest amounts of CP showed fewer primary infections and developed systemic infection more slowly than CP(−) plants. Plants containing 0.006% CP were not resistant. Resistance was effective against infection by two strains of AlMV but not against infection by AlMV RNAs nor against TMV. Anderson et al.[48] confirmed that plants containing AlMV CP were susceptible to TMV, but reported that they exhibited delayed symptom development after inoculation with low concentrations, such as 0.01 to 0.5 μg/ml, of PVX or CMV.

The P35S:AlMV CP construct was also introduced into alfalfa plants.[47] Of the 240 plants regenerated from *Agrobacterium*-mediated transformation, 40% expressed the CP gene, with accumulation to as much as 0.05% of total protein. Protoplasts isolated from strong expressors were resistant to infection by AlMV as were alfalfa plants.[80] None of the transgenic plants developed infections up to 5 months after inoculation.

A chimeric gene consisting of P35S:AlMV CP:NOS 3' end was inserted into tobacco by van Dun et al.[49] Eleven of the 15 regenerants expressed CP, accumulating up to 0.005% of extractable leaf protein. Transgenic plants were resistant to infection by AlMV strain YSMV, but were not resistant to AlMV RNAs nor to infection by the ilavirus tobacco streak virus (TSV). A modified gene containing a frame-shift mutation accumulated CP mRNA but not CP. These plants were not resistant to AlMV,[49,54] indicating that the CP rather than the transcript is responsible for resistance.

D. CUCUMBER MOSAIC VIRUS (CMV)

A single report has described CP-mediated resistance against cucumber mosaic virus (CMV), the type of the cucumovirus group. Viruses in this group have a tripartite genome, and a coat protein of 24 kDa encoded by a subgenomic RNA derived from the smallest segment of the viral genome. RNAs 1 and 2 are separately encapsidated in the same particle.[30] Cuozzo et al.[25] used a cDNA clone of RNA 4 from CMV strain D to construct a gene that included the P35S:CMV-CP:E9 3' end. The E9 3' end was derived from a gene encoding the small subunit of ribulose biphosphate carboxylase.[62] The cDNA included an ATG that is out of frame with respect to the CP initiation codon. The cDNA clone also included 300 nucleotides of 3' nontranslated sequences.

Regenerated plants accumulated CMV CP to approximately 0.002% of cellular protein. Two CP(+) plant lines were resistant to infection by CMV strain C.[25] After inoculation with CMV, both CP(+) lines accumulated less CMV in inoculated leaves and in upper (systemic) leaves than did CP(−) lines. In these experiments, inoculum concentrations of 5 μg CMV/ml led to disease in each CP(−) plant, while 50 μg CMV/ml caused little or no infection of CP(+) plants.

E. TOBACCO STREAK VIRUS (TSV)

Ilarviruses contain three molecules of (+) sense RNA that are separately encapsidated in icosahedral particles. CP molecules of TSV are approximately 25 kDa in mass. Like AlMV, TSV coat protein production is dependent upon synthesis of a subgenomic RNA (RNA 4) derived from RNA 3. Also like AlMV, replication of TSV requires CP or the subgenomic RNA 4.[75,76] Although TSV CP can activate its own replication and that of AlMV, the sequences of the CP of these viruses are not closely related. van Dun et al.[49,54] constructed a CP gene from the WC strain of TSV[81] comprised of the P35S:TSV-CP:NOS 3' end. The chimeric gene was introduced into Samsun NN and Xanthi nc tobacco and two lines were kept for study. Each produced chimeric

FIGURE 8. Diagrammatic representation of the chimeric gene constructed to express the SMV CP cDNA. A cDNA clone containing the SMV CP gene was isolated as described. An oligonucleotide was designed both to engineer an AUG codon in the context of an *Nco*I restriction site at the beginning of the SMV CP gene and to create a *Bgl* II restriction site upstream of the *Nco*I site (changes indicated). The modified SMV CP cDNA was then ligated as a *Bgl* II-Eco RI fragment into the integrative intermediate plasmid pMON316, creating a chimeric gene under the control of the cauliflower mosaic virus 35S promoter and the nopaline synthase 3' flanking region (open boxed regions). This modified plasmid was then used for plant transformation as described. The hatched box represents the coding region, and the closed box represents the 3' noncoding region of SMV. (From Stark, D. M. and Beachy, R. N., *Bio/Technology*, 7, 1257, 1989. With permission of *Bio/Technology Magazine*.)

CP mRNA of the expected size: the level of TSV CP in the transgenic plant lines was not quantitated.

Resistance was assessed by inoculating seedling progeny of one of the lines, TSV-1, with TSV (10 μg/ml); lesions were observed on CP(−) plants but not on CP(+) plants.[49,54] Likewise, systemic disease symptoms developed on CP(−) plants within 3 weeks, whereas no symptoms developed on CP(+) plants, and TSV could not be recovered from CP(+) plants 3 weeks after inoculation. TSV CP(+) plants were not resistant to infection by AlMV, although TSV CP(+) could potentiate the replication of AlMV in TSV CP(+) plants. Likewise, tobacco plants that express the AlMV gene were not resistant to TSV.[49,54]

F. SOYBEAN MOSAIC VIRUS (SMV)

Soybean mosaic virus (SMV) is also included among potyviruses. SMV has a host range limited almost entirely to the Leguminosae and it does not infect *Nicotiana tabacum* cv. Xanthi, nor does it replicate in inoculated tobacco mesophyll protoplasts. Stark and Beachy[34] introduced a gene (P35S:SMV CP:NOS 3') encoding a CP gene (Figure 8) from soybean mosaic virus (SMV) strain N (15) into Xanthi tobacco. The CP cDNA was altered to place the AUG initiation codon in a consensus sequence favorable for efficient translation. Plant lines harboring the SMV CP gene accumulated levels of SMV CP ranging from 0.001% to 0.23% of extractable protein.[34] Because SMV is not a pathogen of tobacco, these lines were inoculated with tobacco etch virus (TEV) and PVY, which are antigenically distinguished from each virus (TEV and PVY), while one line, 1052, was highly resistant to inoculum levels up to 50 μg/ml. Resistance was greater in plants inoculated 37 d after planting than in plants inoculated 31 d after planting. Plants that expressed the SMV

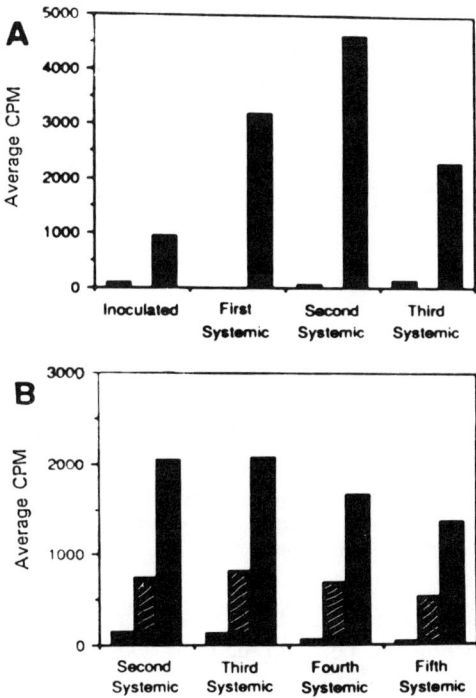

FIGURE 9. Accumulation of TEV in CP(+) and CP(−) plants. 1052 CP(+) and 306 CP(−) plants (8 each) were inoculated with 5 μg/ml TEV. (A) 5 days after inoculation (DAI) the entire lower inoculated leaf was harvested, while three leaf punches were taken from each of the three leaves above the upper inoculated leaf. The solid bars represent the average amount of TEV in CP(+) plants, as determined by the amount of ^{125}I-labeled antibody (cpm) bound to dot-blots. The lightly hatched bars represent the average amount of TEV in CP(−) plants. (B) At 10 d after inoculation three leaf punches were taken from the second through the fifth leaves above the upper inoculated leaf. The solid bars represent virus in CP(+) plants that did not show disease symptoms (2/8). The dark hatched bars represent virus accumulation in CP(+) plants that exhibit disease symptoms (6/8), and the lightly hatched bars, CP(−) plants. (From Stark, D. M. and Beachy, R. N., *Bio/Technology*, 7, 1257, 1989. With permission of *Bio/Technology Magazine*.)

CP gene showed reduced accumulation of TEV in both inoculated and systemic leaves (Figure 9). Stark and Beachy[34] used a subjective disease rating system to quantitate symptoms on sytemically infected plants. Although many CP(+) plants escaped infection, others became infected, yet developed very mild symptoms on one or several leaves by 12 d postinoculation. Low disease ratings were correlated with low levels of virus replication.

CP-mediated resistance against potyviruses has several important features. First, it is possible to produce plant lines that are highly resistant to infection by mechanical or aphid inoculation. Second, a CP gene from a potyvirus that is not a pathogen can protect transgenic tobacco plants against pathogenic potyviruses (PVY and TEV). This is referred to as "protection against heterologous viruses". Third, while CP accumulation leads to high levels of

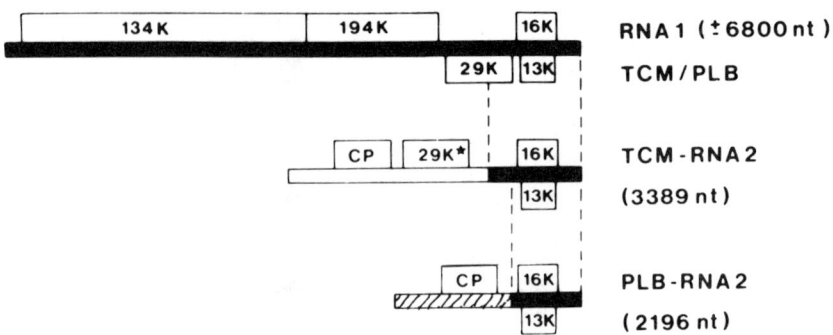

FIGURE 10. Schematic representation of the genome structure of TRV strains TCM and PLB. RNA 1-specific sequences are represented by a solid bar; sequences that are homologous in RNAs 1 and 2 are connected by dotted lines. Sequences that are unique to TCM RNA 2 and PLB RNA 2 are indicated by open and hatched boxes, respectively. The locations of cistrons encoding proteins with molecular weights of 134K, 194K, 29K, 16K, 13K, 29K*, and coat protein (CP) are indicated. (From Angenent, G. C., van Den Owweland, J. M. W., and Bol, J. F., *Virology,* 175, 191, 1990. With permission.)

resistance in some lines, other CP(+) lines express little or no resistance to infection. The differences between the most resistant and least resistant lines have not yet been determined.

G. TOBACCO RATTLE VIRUS (TRV)

Tobacco rattle virus (TRV), the type member of the tobraviruses, possesses a bipartite genome of (+) sense RNA. RNA 1 carries functions required for replication and systemic spread, and RNA 2, which varies in length depending upon the virus isolate, encodes the CP and at least one additional protein (Figure 10).[82] TRV RNAs are encapsidated by a single type of capsid protein of approximately 22 kDa to produce rod-like particles. DNA sequencing of different TRVs has shown that the CP gene is located on RNA.[83]

A TRV CP gene from the TCM strain was constructed by combining the P35S promoter and the NOS 3′ end.[49,84] The cDNA contained a short 5′ nontranslated region and a 3′ nontranslated region greater than 400 nucleotides in length. The chimeric gene was introduced into Samsun NN tobacco. In CP(−) plants, the TCM strain of TRV produced necrotic spots on inoculated leaves and severe necrosis on systemically infected stems and leaves. In contrast, CP(+) plants STRV-3, which contained 0.05% TRV CP, developed mild necrosis and accumulated little or no TRV RNA after inoculation with 1 μg/ml TRV strain TCM.[49] Transgenic tobacco plants expressing the CP gene of TRV strain TCM were found to be transient to infection with homologous virus but not to infection with the PLB strain of TRV.[49] The amino acid sequence identity between the CP of TRV strains TCM and PLB was 39%, and the two CP genes did not cross-hybridize.

Plants of line STRV-3 were also inoculated with TRV strain PLB or pea early browning virus (PEBV). Strains PLB and TCM of TRV have approximately 39% amino acid sequence homology in their CPs, and each CP can encapsidate the other RNA. The PEBV isolate has considerably more homology with the TCM strain.[84] CP(+) plants were not resistant when inoculated with TRV PLB, although they were resistant to PEBV. van Dun et al.[54] concluded that resistance to PEBV reflected the high degree of relatedness of the PEBV and TRV TCM CP molecules that interfere in some way with infection by PEBV as in TRV TCM. It is unlikely that interference is at the level of CP:RNA interactions since CP molecules of each strain are able to encapsidate the RNAs of the other.

H. TOBACCO MOSAIC VIRUS (TMV)

Tobacco mosaic virus is the most thoroughly characterized plant virus, and contains (+) sense RNA of approximately 6400 nucleotides. The 126- and 183-kDa proteins that are required for virus replication[85] are translated directly from genomic RNA from the same initiation codon.[86-88] The 30-kDa movement proteins (MP) and 17.5-kDa coat protein are translated from separate 3' coterminal subgenomic mRNAs.[89] The MP has been shown to function in cell-to-cell movement of the virus.[90,91] The CP encapsidates the viral RNA and is required for long distance movement in the plant.[92-96] CP monomers encapsidate TMV virion RNA during infection, and a full-length 1360 nm long virion is formed (Figure 11).[97]

Transgenic tobacco plants expressing chimeric TMV CP or MP genes have been produced.[20,21,90,97] A chimeric gene (P35S:CP:NOS 3') was used to produce modified tobacco plants (cv Xanthi and Xanthi nc) and tomato plants (cv VF36). Lines that accumulated CP levels from 0.001 to −0.2% of total plant protein were recovered. Following inoculation with TMV, CP(+) lines either escaped infection or developed disease symptoms significantly later than CP(−) lines.[20] The degree of resistance in CP(+) plants was measured by inoculating them with low (0.001 μg/ml) to high (10 to 100 μg/ml) concentrations of TMV and comparing disease development in CP(+) and CP(−) plants.

CP(+) plants required at least 10^4 times higher concentrations of virus to overcome CP-mediated resistance than required for disease in CP(−) plants.[16] Powell et al.[8] confirmed the requirement of TMV CP for resistance, rather than CP RNA sequences per se, by specific mutagenesis to delete the AUG initiation codon from the gene. Plant lines that accumulated gene transcripts but not CP were not resistant to infection, and plant lines that accumulated low amounts of CP (i.e., 0.01% or less) were, in general, less resistant than lines that accumulated high amounts of CP (0.01 to 0.2%).

CP-mediated resistance against TMV was also demonstrated in tomato, cv VF36.[98] The CP(+) plants were resistant to TMV at inoculum concentrations of 20 μg/ml, and to several strains of related tobamovirus, which has

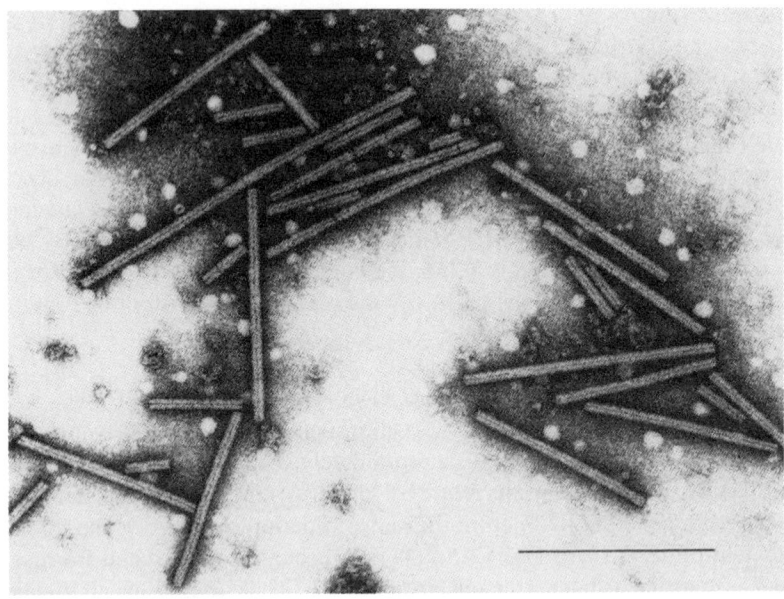

FIGURE 11. Electron micrograph of virions purified from Xanthi/CP(+) leaves 7 days after inoculation with transcripts from CP(−) TMV clone U3/12CPfs. Virions were stained with 2% aqueous phosphotungstate and viewed by transmission electron microscopy. The horizontal bar at the lower right represents 300 nm. Magnification approximately 67,500 ×. (From Holt, C. A. and Beachy, R. N., *Virology*, 181, 109, 1991. With permission.)

a CP 80 to 85% homologous in amino acid sequence to TMV CP.[99] While the CP(+) lines were resistant to each strain of ToMV tested, resistance was not as effective against ToMV as against TMV.[98] Protection is largely overcome when TMV RNA is applied to inoculum which brings the interaction between endogenous CP and challenge TMV (Figure 12).[100] Plants that expressed the CP gene [(CP+)] and those that did not [(CP−)] accumulated equivalent amounts of virus in the inoculated leaves after inoculation with TMV RNA, but the [(CP+)] plants showed a delay in the development of systemic symptoms and reduced virus accumulation in the upper leaves. The two transgenic tomato plants that expressed the coat protein (CP) of the common strain of tobacco mosaic virus (TMV) were produced from cultivar VF36 using gene transfer[45] techniques. Yields from one CP-expressing line were equal to that of the uninoculated VF36 plants, suggesting that expression of the CP gene does not intrinsically cause a decrease in yield.

As discussed earlier, resistance to viral infection is not only expressed to the same virus in CP(+) plants, but also to other viruses representing different degrees of relatedness.[47,48] Based upon a comparison of amino acid sequence and/or composition of virus CPs,[99] ToMV is most closely related to TMV followed by, in order of relatedness: pepper mild mottle virus (PMMV);

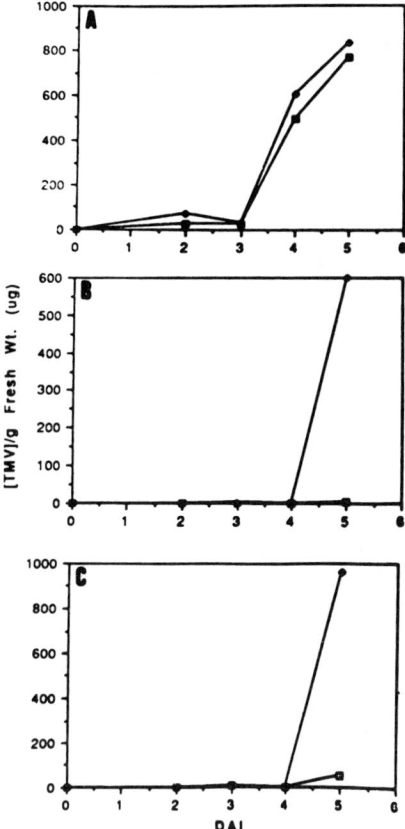

FIGURE 12. ELISA assays of TMV accumulation in CP(+) and CP(−) plants over time. (A) TMV accumulation in the inoculated leaves of CP(+) and CP(−) plants. CP(+) (□) and CP(−) (◆) plants were inoculated with TMV RNA and accumulation was monitored by ELISA. (B) TMV accumulation in the stem between the inoculated leaf and systemic leaves of CP(+) and CP(−) plants. CP(+) and CP(−) plants were inoculated with TMV RNA and accumulation was monitored by ELISA. (C) TMV accumulation in the systemically infected leaves of CP(+) and CP(−) plants. CP(+) and CP(−) plants were inoculated with TMV RNA and accumulation was monitored by ELISA. (From Wisniewski, L. A., Powell, P. A., Nelson, R. S., and Beachy, R. N., *Plant Cell*, 2, 559, 1990. With permission of the American Society of Plant Physiologists.)

tobacco mild mosaic virus (TMGMV;U_2 strain of TMV); ondontoglossum ringspot virus (ORSV); ribgrass mosaic virus (RMV) and sunhemp mosaic virus (SHMV). The CP homology of these viruses to TMV CP ranges from 85% homology for ToMV to 40% homology for SHMV. A CP(+) Xanthi nc plant line (748) was tested for resistance to each of the viruses by comparing the numbers of local lesions produced on CP(+) and CP(−) plants.[47] Infection of TMV was inhibited by 95 to 98%, ToMV by 95 to 98% PMMV by 95 to 98%, TMGMV by 95 to 98%, ORSV by 80 to 95%, and RMV (Table 2)

TABLE 2
Number of Necrotic Lesions on Control (CP−) and Transgenic (CP+) Tobacco Plants (Line 748) after Inoculation with Different Tobamoviruses of Their RNAs

Inoculum	Concentration (µg/ml)	Necrotic lesions per plant[a] CP(−)	CP(+)	%[b]
ToMV	1.0	771 ± 66	1	0.13
ToMV RNA	10.0	191 ± 31	148 ± 17	77.00
PMMV	1.0	308 ± 12	1	0.30
PMMV RNA	10.0	380 ± 49	272 ± 25	71.00
TMGMV	1.0	906 ± 103	1	0.10
TMGMV RNA	10.0	264 ± 42	107 ± 19	40.00
RMV	0.01	13 ± 2	6 ± 1	46.00
	0.10	143 ± 19	62 ± 7	43.00

[a] For each treatment, the number of necrotic lesions was scored on two leaves of each of five plants at 4 days after inoculation. The numbers are mean ± SE. The results are representative of four repeats at different concentrations with similar results.

[b] The numbers are the percentage of the necrotic lesions produced on the CP(+) plants compared to those produced on the CP(−) control plants.

From Nejidat, A. and Beachy, R. N., *Mol. Plant-Microbe Interactions,* 3, 247, 1990. With permission.

and SHMV (least related to TMV) by 40 to 60%.[47] In contrast to CP(+) tomato plants, CP(+) Xanthi tobacco plants showed similar resistance to TMV and ToMV. Furthermore, if a systemic infection developed in CP(+) Xanthi tobacco, it was significantly delayed in CP(+) lines following inoculation with ToMV and TMGMV, but not RMV. Thus, viruses with CPs of 60% or greater homology to TMV CP are less able to infect the resistant tobacco lines than are more distantly related tobamoviruses.

Anderson et al.[48] inoculated TMV CP(+) plant lines with members of different virus groups. There was no resistance against infection by CMV, AlMV, PVX, or PVY on inoculated leaves of CP(+) plants; however, rates of systemic spread of CMV, PVX, and PVY were reduced in CP(+) compared with CP(−) plants. There was also no effect on infection or systemic spread of AlMV in TMV CP(+) plants.

Transgenic tobacco (*Nicotiana tabacum* 'Xanthi') plants that express the coat protein (CP) gene from the U_1 strain of tobacco mosaic virus (TMV) are resistant to infection by TMV48. These plants are also protected against other viruses; they were inoculated with low concentrations of PVX, PVY, CMV, AlMV, and CMV and the cowpea strain of TMV. Although the accumulation of virus in inoculated leaves was equivalent in plants that express the CP gene (CP +) and plants that do not express the CP gene (CP −), there was a delay of 1 to 3 d in the development of systemic disease symptoms in CP(+) plants infected with PVX, PVY, CMV, and A1MV as compared with CP(−) plants. The magnitude of protection, however, was significantly lower than against TMV. Protection against CP(−) TMV, assayed on a CP(+) local lesion host, was lower than against TMV U_1.

VI. MECHANISMS OF CP-MEDIATED RESISTANCE

The mechanisms of CP-mediated protection are not precisely defined at present. While it is clear that there are differences between the examples of CP-mediated resistance, the unifying observation is that CP molecules are responsible for the resistance. The question of how CP confers resistance remains unanswered; furthermore, the nature of virus infection and disease development make it likely that the resistance is effected at several stages.

Upon introduction of virus into its host, either by mechanical wounds or by vector transmission, infection starts with the release of the nucleic acid for translation by host ribosomes. It has been proposed that for many, if not all, viruses, swelling of the particle precedes the release of the nucleic acid.[101] This is followed by binding of ribosomes and disassembly of the virus concurrent with translation and initiation of the infection process. It is believed that CP interferes in some way with this process. It is this interference which has been exploited for cross-protection.

A. ROLE OF CP GENE CROSS-PROTECTION

As mentioned earlier, the mechanisms of CP-mediated protection are not precisely defined at present. In many respects, protection resembles infected plants which are less susceptible to superinfection by a related strain of the virus.[42] There is a delay in the development of systemic symptoms in plants that express a CP(+) gene after inoculation with TMV, compared with control CP(−) plants, and some CP(+) plants escape disease development altogether.[20] Characteristics of the CP-dependent system include the following:

(1) Increasing the concentration of the inoculum decreases the delay in symptoms and the proportion of plants that escape infection[20]
(2) The number of chlorotic and necrotic lesions is reduced by 70 to 95% on inoculated leaves of CP(+) plants compared with CP(−) plants.[42] This decrease in the number of primary infection sites accounts in part for the delay in systemic symptom development
(3) There is a delay in movement of the virus from the inoculated leaves to the upper leaves of an infected CP(+) plant when an equal number of infection sites are present on the inoculated leaves of transgenic and control plants
(4) Inoculation with viral RNA produces more lesions on CP(+) plants than virion inoculum but only about 50% of the number produced on

CP($-$) plants.[42] This indicates that at least part of the protection is dependent on the presence of the CP of the challenger and that some event related to the establishment of infection is inhibited in CP($+$) plant lines.

Similar characteristics to (1), (2), and (4) have been reported for classical cross-protection.[4]

Plants which are systemically infected with CaMV exhibit a characteristic mosaic patterning of the leaves, the cells of which accumulate viral inclusion bodies in the cytoplasm. The major protein of these inclusion bodies, P_{66}, is encoded by gene VI of CaMV, translated from the 19S tRNA, of the two transcripts produced by this double-stranded DNA virus. Gene VI has been implicated in host range control, and in symptom production Baughman et al.[102] have produced further evidence to support this view. A segment of the viral genome carrying gene VI was introduced into tobacco plants by *Agrobacterium*-mediated gene transfer, and the transgenic plants were found to display viral-like symptoms. Deletions and frame-shift mutations of gene VI prevented the formation of disease symptoms in transgenic plants. There was also a direct correlation between the appearance of symptoms and of the amount of P_{66} accumulated by the plant, as detected by immunoblotting. In order to study the possible functions of the inclusion matrix protein (IBMP) encoded by gene VI of CaMV,[94] the *Xba*I fragment containing the gene VI of a Japanese strain of CaMV (CaMV S-Japan) was transferred to tobacco plants. Eight out of 18 kanamycin-resistant plants (40%) expressed detectable levels of IBMP. Those transgenic plants expressing IBMP produced leaves with a light green color, and their growth was suppressed as compared with control plants. Symptom-like necrotic spots also appeared on the leaves and stems of the mature transgenic plants. Furthermore, in these transgenic plants, pathogenesis-related proteins 1a, 1b, and 1c were highly expressed.

Transgenic tobacco plants which expressed a chimeric gene encoding the tobacco mosaic virus CP and the TMV 3' untranslated region are protected against infection by TMV. Genes that encode the sequences representing the TMV CP subgenomic RNA, but do not produce protein (because of removal of the initiation codon), and RNA that lacks the tRNA-like sequence of the TMV 3' end of TMV RNA, were expressed on transgenic plants. Only plants that accumulated CP regardless of the presence or absence of 3' end of TMV RNA were protected against infection by TMV.[8] The results indicate that the CP per se, rather than TMV RNA, is responsible for the resistance to infection by TMV. Furthermore, the degree of protection is dependent upon the level of accumulated CP.

It has been suggested that cross-protection may confer resistance by preventing the uncoating of other invading proteins. This would account for the susceptibility of cross-protected plants to naked viral RNA, which obviously is already "uncoated". Perhaps the coat protein produced in transgenic plants competes for factors required by invading viruses for the uncoating process. Other proteins, such as the 30 kDa "movement" protein of TMV, do not confer cross-protection when expressed in transgenic plants. It was observed by Powell et al.[20] that protection against symptoms was not absolute, and this may be related to the relatively low levels of coat proteins produced by the transgenic plants (less than about 1% of the coat protein found in virus-infected plants). This suggests that the concentration of coat protein may become a limiting factor which may determine, to some extent, the protection obtainable. The use of stronger gene promoters than the CaMV 19S (used in the AMV work) might produce higher levels of resistance.

Transgenic plant lines have been reported to express variable levels of the CP gene product. This variation may be due to somaclonal variation[103] or may be a result of the position in the genome into which the gene was integrated. CP levels of 0.15[33,34] and 0.001%[25] of the total soluble leaf proteins have been reported. The level of protection conferred against infection by PVX or TMV in transgenic tobacco plants expressing the CP genes of PVX and TMV, respectively, were demonstrated to be dependent upon the level of the CP expressed, i.e., higher levels of the CP conferred better protection against virus infection. On the other hand, no such correlation was observed for PVY.[43] An explanation for this result may be deduced from a report by Clark et al.[104] In this study the transgenic tobacco plants that expressed the TMV CP gene under the control of a promoter from a ribulose bisphosphate carboxylase small subunit (rbcS) gene were compared to plant lines that expressed the CP gene under the control of the 35S promoter. A plant line in which the CP gene was driven by the 35S promoter was more resistant to TMV infection than a plant line in which the rbcS promoter was used, although they accumulated the same level (on a w/w basis) of CP in leaf extracts. Since the 35S and the rbcS promoters are differentially expressed in plants, it is possible that not only the absolute level of the CP is important to obtain protection, but that tissue, or subcellular localization of the CP may determine the efficiency of CP-mediated protection. On the other hand, the possibility that different mechanisms of protection are involved in the case of PVY protection cannot be ruled out. Thus, while the amounts of CP required to produce resistance are different in various host-virus systems, there is convincing evidence that the presence of the CP in transgenic plants is essential. No protection was observed when translationally defective CP gene transcripts of AlMV[54] or TMV[8] were expressed in transgenic tobacco plants.

Evidence is accumulating that CP-gene expression interferes with an early event in infection. First, resistance can be largely overcome, in some but not all examples, by inoculation with viral RNA other than virions. This is the case for resistance against TMV,[45] AlMV,[22] and TSV.[54] On the other hand, inoculation with PVX RNA did not overcome CP-mediated resistance against PVX.[18] Studies with protoplasts have also yielded important information about early events leading to CP-mediated resistance. Loesch-Fries et al.[22] reported

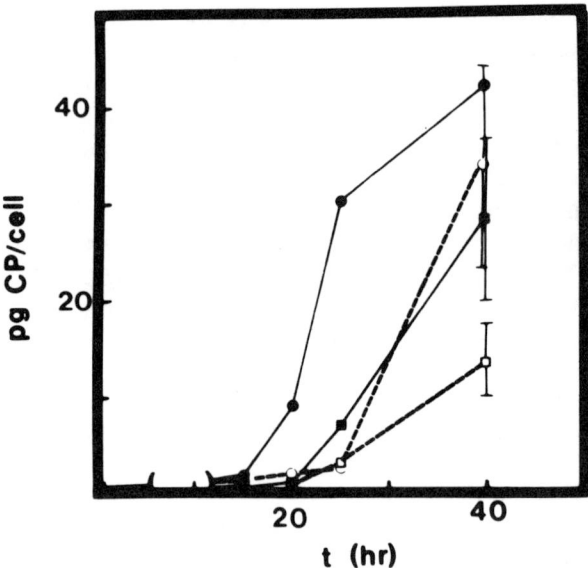

FIGURE 13. TMV CP aggregation state affects transient protection against TMV in tobacco protoplasts. Tobacco protoplasts were inoculated with 1.5 µg of TMV/ml alone (●) or in the presence of the 615 µg of TMV CP/ml, preincubated at pH 5.7 (□), pH 7.0 (■), or pH 8.0 (○). These 5 results are the averages of the experiments. Brackets on 40-h time points are explained in the legend to Figure 1. (From Register, J. C., III, and Beachy, R. N., *Virology*, 173, 656, 1989. With permission.)

resistance against AlMV in CP(+) protoplasts but not against AlMV RNA. Likewise, Register and Beachy[105] reported less resistance against TMV RNA and TMV that was briefly treated at pH 8, to swell the virions against TMV CP(+) plants and protoplasts. Protoplasts from transgenic tobacco plants that expressed the TMV CP gene (CP +) were used to study the initial events of infection,[105] and these experiments showed that blockage of a very early event of infection, probably an initial stage of virus uncoating, is responsible for this protection. Although it is not known how uncoating of TMV is blocked in CP(+) plants, it is likely that CP is responsible for protection. This prompted Beachy and co-workers to present a model for CP-mediated protection.[105,106]

Register and Beachy[107] have shown that purified viral CP can confer resistance in tobacco protoplasts that do not express viral CP gene (CP−) when the CP is introduced into the protoplasts either concomitant with or shortly after incubation with the challenge virus (Figure 13). One limitation in the use of this assay, they pointed out, is obtaining sufficient quantities of virus and CP. This difficulty can be circumvented by transiently expressing a viral CP gene of interest from a plasmid introduced into protoplasts.

B. THE ROLE OF NONSTRUCTURAL PROTEINS IN CROSS-PROTECTION

The release of RNA from virus particles is an early step in infection; the results of several experiments support the hypothesis that CP interferes with an early event in infection that releases the encapsidated RNA. This does not necessarily eliminate the role for viral CP in binding to viral RNA blocking either uncoating or gene expression. However, there are exceptions to this, and the example of CP-mediated protection with TRV also argues against a protein/RNA role in resistance. For instance, van Dun and Bol[49] reported that, although the CPs encoded by TCM and PBL strains of TRV could encapsidate the RNA molecules of the other strain, the CP of the TCM strain did not protect tobacco plants from infection by the PCB strain. This result, and the lack of resistance against both viral RNA and preswelled virus particles (in the case of TMV), argues against reencapsidation of viral RNA as a primary mechanism of resistance.

Correlation of the temporal and special pattern of induction of the pathogenesis-related (PR) genes PR1a, PR1b, and PR1c with viral infections in certain tobacco cultivars has implicated PR proteins in viral resistance.[108] To test whether the PR1 proteins of tobacco are involved in viral resistance, transgenic *Nicotiana tabacum* plants were constructed which constitutively express the PR1b gene. This protein was secreted from cells of transgenic plants and accumulated in the extracellular space at levels equivalent to those found in nontransgenic plants in association with disease resistance. Transgenic plants derived from the cultivar Xanthi exhibited no delayed onset or reduction in the severity of systemic symptoms after TMV infection. These data indicate that the PR1b protein of tobacco is not sufficient for TMV resistance, and imply that the PR1 proteins may not function as unique antiviral factors. Similar observations were made by Carr et al.[109] They found that the transgenic tobacco plants constitutively expressing the coat protein (CP) of tobacco mosaic virus (TMV) exhibit enhanced resistance to TMV. To determine if this enhanced resistance might be mediated through the PR1 family of pathogenesis-related (PR) proteins, their synthesis was examined. In transgenic plants derived from *NN* genotypic tobacco, a high proportion (similar to 80%) of those producing CP also expressed the PR1 genes at low levels. However, this correlation between TMV CP and PR1 gene expression was not observed in the similarly transformed *nn* genotypic tobacco plants. Therefore, it appears unlikely that PR1 proteins play a critical role in genetically engineered resistance in transgenic plants producing TMV CP.

To obtain further insight into the mechanism, Hamilton et al.[110] transformed tobacco with structural and nonstructural genes of TRV. As mentioned earlier, RNA 1 of TRV encodes the putative viral replicase proteins with molecular weights of 134,000 and 194,000 (134K/194K proteins), a 29K protein with a putative function in cell to cell movement and a 16K protein with an unknown function.[110] The 16K gene completely overlaps the 13K

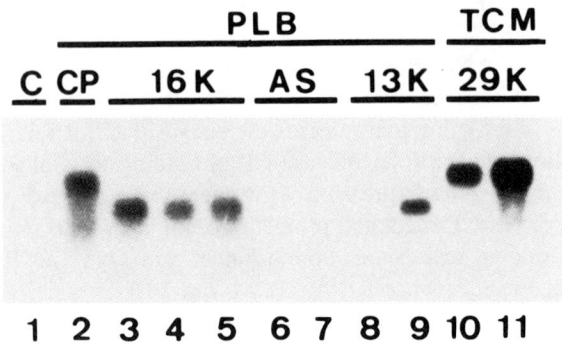

FIGURE 14. Accumulation of virus-specific transcripts in plants transformed with different TRV genes. For plants transformed with nonstructural genes, the analysis of several independent transformants is shown. A Northern blot was loaded with polyadenylated RNAs from a vector-transformed control plant (lane 1), PLB-CP plant 3 (lane 2), PLB-16K plant 0 (lane 3), PLB-16K plant 4 (lane 4), PLB-16K plant 7 (lane 5), PLB-AS plant 1 (lane 6), PLB-AS plant 2 (lane 7), PLB-13K plant 0 (lane 8), PLB-13K plant 1 (lane 9), TCM2-29K plant 2 (lane 10), and TCM2-29K plant 3 (lane 11). The blots were hybridized to labeled cDNA corresponding to nucleotides 502 to 2087 of TRV PLB RNA 2 (lanes 1-9) or to nucleotides 1437 to 1941 of TRV TCM RNA 2 (lanes 10 and 11). (From Angenent G. C., van Den Owweland, J. M. W., and Bol, J. F., *Virology*, 175, 191, 1990. With permission.)

gene.[111,112] The 16K is expressed in TRV-infected protoplasts.[111,112] Expression of 13K gene and other nonstructural genes has not yet been detected. The amino acid sequence similarity between the CPs of strains PLB and TCM is 39%, while that of 16K/13K proteins of two strains is over 90%. Angenent et al.[82] transformed tobacco with the CP gene, the 16K and 13K of PLB, and the 29K gene of TCM. Acccumulation of RNA transcripts from the integrated viral genes was detectable in all types of transformants (Figure 14).[82] Plants expressing CP were resistant to infection with virions of the homologous strain (PLB), but susceptible to infection with RNA of the homologous strain or nucleoprotein of the heterologous strain. When protoplasts from plants expressing TCM CP were inoculated with TCM virions, there was a normal production of genomic RNAs and CP, but the synthesis of mRNA and protein corresponding to the 16K gene was selectively defective (Figure 15). However, the defect was not observed when protoplasts from plants expressing PLB CP were inoculated with PLB virions,[82] suggesting that an interaction between CP and plus-stranded viral RNA in the parental virus particles results in the synthesis of minus-strand RNAs that do not serve as templates for the synthesis of RNA (16K/13K protein). Endogenous CP may further prevent uncoating of incoming virus particles and CP molecules that remain bound to the parental RNA which could induce mistakes in the transcription of minus strand RNA (RNA-1b) thus producing defects in 16K/13K proteins.

The consistent but low-level resistance against infection by TMV RNAs in CP(+) plants[42] and protoplasts[105] implies another mechanism of resistance

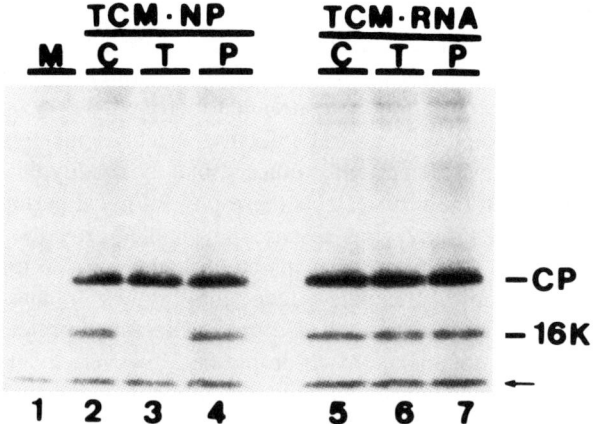

FIGURE 15. Autoradiograph of a SDS-polyacrylamide gel showing the accumulation of CP and 16K protein in protoplasts from transgenic plants inoculated with nucleoprotein (lanes 2, 3, and 4) or RNA (lanes 5, 6, and 7) of TRV TCM. The protoplasts were isolated from nontransformed tobacco plants (lane M), vector-transformed control plants (lanes C), TCM CP plants (lanes T), and PLB CP plants (lanes P). The sample shown in lane 1 is from mock-inoculated protoplasts. 35S-labeled proteins extracted from the protoplasts were subjected to immunoprecipitation with a mixture of antisera to TCM CP and PLB 16K protein. The positions of the CP and 16K proteins are indicated in the right margin; the arrow indicated a host protein that was precipitated by the antisera. (From Angenent, G. C., van Den Owweland, J. M. W., and Bol, J. F., *Virology*, 175, 191, 1990. With permission.)

that acts after the uncoating event(s). Osbourn et al.[113] inoculated protoplasts from CP(+) plants with reconstructed virions. When RNA of TMV U_1 was encapsidated by the CP of SHMV, there was a low but consistent level of resistance, implying that while the uncoating of the virus was not affected, replication of TMV RNA was reduced. By contrast, virus particles produced by encapsidation of SHMV RNA in TMV CP were not able to infect CP(+) protoplasts. These authors suggest that TMV CP may play a role in regulating the replication of TMV RNA in this system. CPs of several RNA-containing bacteriophages can affect replication and expression of their RNAs.[114]

C. VIRAL GENOME REPLICATION AND DISSEMINATION OF THE VIRUS THROUGH THE PLANT

Since eukaryotic cells synthesize RNA from DNA and genomic replication is restricted to the S phase of the cell cycle, RNA viruses encode the protein comprising a RNA-dependent RNA polymerase (RdRp) complex for their own replication. The caulimoviruses, as typified by the cauliflower mosaic virus, have double-stranded DNA as their nucleic acid, and replicate in a biphasic process. The viral DNA is transcribed by the plant-encoded DNA-dependent RNA polymerase II, and the RNA produced acts as a template for transcription back to DNA using virus-encoded reverse transcriptase. After replication of the nucleic acids, it is packaged into newly synthesized coat protein to produce daughter virus particles. The only other group of plant DNA viruses, the gemini viruses, contain single-stranded DNA which is thought to be replicated via a double-stranded intermediate.

Although the symptoms of viral infection, such as the hypersensitive response, chlorosis, and PR protein synthesis, may be readily detectable, the mechanisms by which the virus effects these physiological perturbations are largely undetermined. Many virus diseases are named after the phenotypic changes which occur in leaves of the infected plants (tobacco mosaic virus; cowpea chlorotic mottle virus; the sugar-beet "yellow"), and this effect appears to be due to the inhibition of chloroplast rRNA and protein synthesis, leading to a reduced abundance of chloroplasts. However, these effects are not necessarily correlated simply with virus replication, and presumably viral gene products interact with the plant cell in a very complex way.

Following replication, the successful establishment of the infection requires the dissemination of the virus through the plant. In incompatible plant-pathogen interactions, a hypersensitive response may be elicited, and the spread of the particles is prevented. PR proteins may also be synthesized by the host. If, however, the plant is susceptible, the systemic movement of the virus may occur. Electron microscopic evidence indicates that virus particles are transmitted from cell to cell simplistically, i.e., via the cytoplasmic connections (plasmodesmata) that exist between adjacent cells. It is assumed that these plants contain a single, continuous cytoplasm, which is, with perhaps the exception of the very tip of the shoot meristem, accessible to virus particles (although infection of seed and pollen is rare). There is also evidence that the transmissibility of the virus is genetically encoded. A strain of tobacco mosaic virus (LS-1) was isolated which, at elevated temperatures, was unable to move through the plant.[1] *In vitro* studies in protoplasts have shown that the replication *per se* of the virus is not temperature sensitive, and a specific 30-kDa protein has been implicated as determining transmissibility. Sequence analysis has shown that the gene for this protein from LS-1 differs from that of the temperature-resistant strain by a single base pair mutation, resulting in the protein possessing a serine instead of a proline. Transfer of the wild-type gene to transgenic plants showed that the LS-1 virus moves through those plants at the restrictive temperature, demonstrating the requirement of the virus for this protein.[1] Proteins of similar function have been identified in other viruses, although their mode of action is unknown.

In most studies of CP-mediated resistance, systemic infection was either prevented or delayed in CP(+) compared with CP(−) plant lines. This could result from interference with the spread of virus from cell to cell in inoculated tissue, ingress of virus from the inoculated leaf into the vascular tissue, movement through the vascular tissue, ingress into upper (or lower) non-inoculated leaves, and/or initiation of infection in other leaves.

To detect effects of the CP on local and systemic spread of virus, it is necessary to study the systemic infection of the host expressing the CP gene. Several groups reported a delay in development of systemic disease symptoms[22,42,47] in transgenic plants that expressed the TMV and AlMV CP genes after inoculation with TMV or AlMV, respectively. In each case the inoculated leaves of transgenic plants were less heavily infected than were the control plants. This low level of virus and delay in systemic symptoms could be due in part to the lower level of virus in the inoculated leaves. Dodds et al.[115] used a mild strain of CMV to protect against a more severe strain. The challenge strain replicated to spread into the upper leaves. Dodds et al.[115] postulated that the mechanism that prevents initial infection with virion challenge virus may also function to prevent movement of infection within the plant.

A recent study by Wisniewski et al.[100] compared the spread of TMV infection in CP(+) and CP(−) plants by inoculating leaves at a single point with TMV RNA and monitoring the spread of virus. The spread of virus to closely adjacent tissue (1 to 3 mm) was similar in CP(+) to CP(−) plants. However, spread to more distant tissues (5 to 10 m distant) and to other leaves was significantly reduced in CP(+) compared to CP(−) plants.

Tissue-printing techniques were used to demonstrate that TMV appeared in vascular tissues and in apical parts several days earlier in CP(−) plants than in CP(+) plants. These data were interpreted to indicate that entry of virus into the vascular tissue was impeded in CP(+) plants.[100]

Simple grafting experiments were undertaken to determine the effect of TMV CP on systemic spread of infection (Figure 16).[100] A CP(+) stem section was placed between the root stock and apical section of a CP(−) plant; the leaves of the CP(−) root stock were inoculated with TMV, and the rate of virus spread was monitored by ELISA. When the CP(+) stem section included a leaf, infection of the CP(−) apex was dramatically reduced compared to a CP(+) stem section without a leaf or a CP(−) section with or without a leaf. This suggests that systemic spread includes infection and replication in the intervening leaf, if present. Since infection of CP(+) leaves was very low, infection of the apical section was prevented. Absence of CP(+) leaves, on the other hand, resulted in the same rate of virus movement through both types of stem sections. In addition, these experiments indicate that protection against systemic spread in CP(+) plants is caused by one or more mechanisms that, in correlation with the protection against inital infection upon inoculation, result in a phenotype of resistance to TMV.

It is apparent, however, that CP-mediated protection against TMV follows in two stages. The first stage involves an early event in infection which proceeds by uncoating of the virus,[105] while the second stage involves movement of TMV from the inoculated leaf into the upper leaves as demonstrated by Wisniewski et al.[100] It is also apparent that mechanisms of CP protection are not the same in all viruses, because the characteristics of protection vary

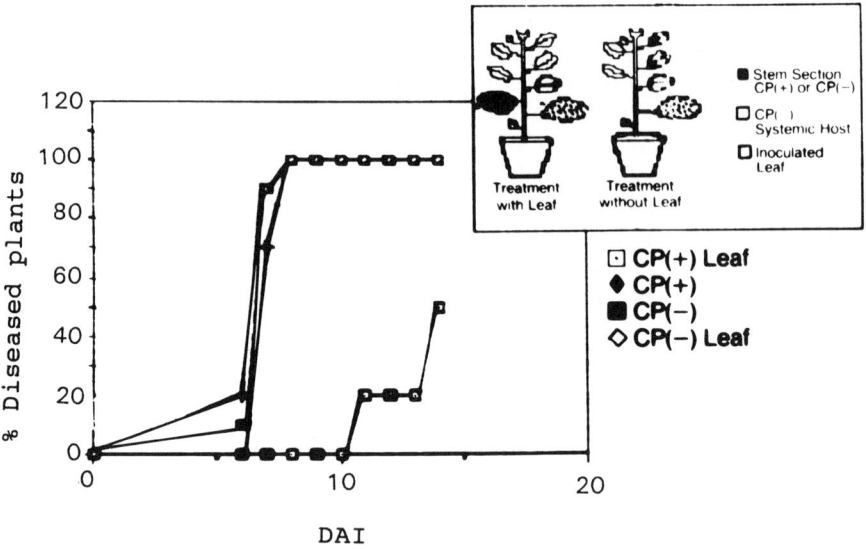

FIGURE 16. Percent TMV infection in plants containing CP(+) or CP(−) intersections. Percentage of grafted plants with stem sections that are either CP(+) or CP(−) and either contain or do not contain a leaf showing systemic symptom development on the systemic leaves over time after inoculation. The largest leaf on the rootstock was inoculated with 0.35 μg/ml TMV. Inset shows the grafted plants. (From Wisniewski, L. A., Powell, P. A., Nelson, R. S., and Beachy, R. N., *Plant Cell,* 2, 559, 1990. With permission of the American Society of Plant Physiologists.)

among different viruses. For example, protection against TMV is largely overcome by inoculation with TMV RNA.[42] In contrast, protection against PVX is equally effective against inoculation by PVX and PVX RNA.[18] Similarly, protection against CMV is effective against high concentrations of virus,[25] while TMV protection is overcome with high concentrations of virus.[20]

In the majority of the reports describing CP-mediated protection, the different virus CP genes were expressed from the CaMV 35S promoter, and the 35S promoter is nominally a constitutive promoter which expresses itself efficiently in vascular tissues. To avoid this bias, Clark et al.[104] chose to test whether expression of the TMV CP cistron from a promoter other than the 35S promotor in transgenic tobacco plants would affect virus protection. They chose the ribulose biphosphate carboxylase small unit (rbcS) promoter which has been shown to be primarily active in mesophyll tissue.[116] To accomplish this, Clark et al.[104] ligated the CP gene of U1 strain of TMV to petunia rbcS promoter (Figure 17) and transferred the construct to tobacco plants. Plant lines expressing comparable levels of CP from the rbcS and CaMV 35S promoters were compared for resistance to TMV. In whole plant assays, the 35S:CP construct gene gave higher resistance than rbcS:CP constructs (Figure 18). On the other hand, leaf mesophyll protoplasts isolated from both plant lines were equally resistant to infection by TMV (Figure 19). The differences

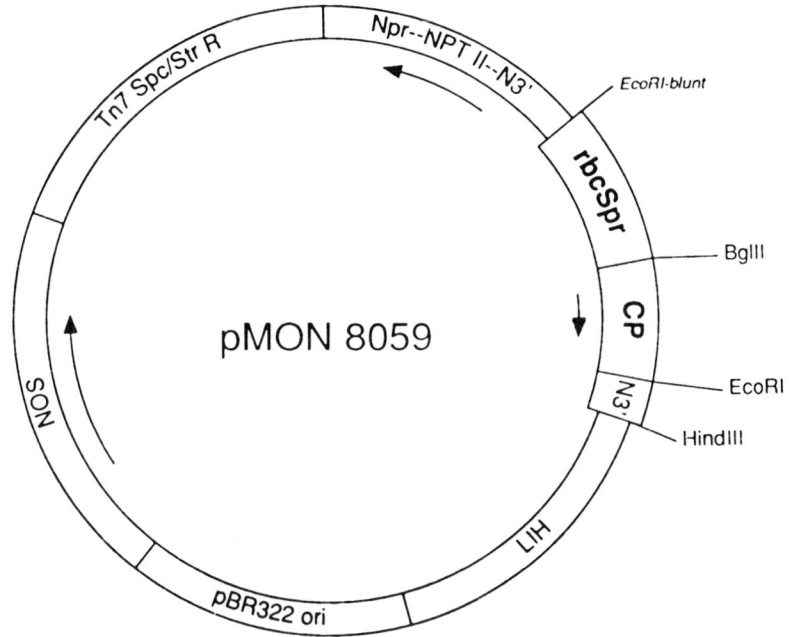

FIGURE 17. Plasmid pMON8059 was constructed by first inserting a rbcS promoter, isolated from the petunia 11A rbcs gene as a 900-bp *Eco* RI-*Bgl* II fragment, into pMON200. A multilinker and the nopaline synthase poly(A)-addition sequence were then added as a *Bgl* II-*Hin*d III fragment. Finally the TMV CP ORF was inserted into the multilinker region as a 700-bp *Bgl* II-*Eco* RI fragment. This plasmid was then triple-mated into *Agrobacterium tumefaciens*. (From Clark, W. G. et al., *Virology*, 179, 640, 1990. With permission.)

in the levels of virus protection between the two lines (rbcS:CP and 35S:CP) correlated with tissue specificity of gene expression. This prompted them to propose that the difference in virus resistance between lines 3012 (35S:CP) and 4109 (rbcS:CP) is due to differences in tissue-specific expression of the two promoters, and that this differential resistance phenomenon is a combination of enhanced protection at the point of virus entry, the leaf epidermal cell layer, and at the point of long distance spread of the virus.

Transgenic MP(+) plants have been reported to complement the temperature sensitive defect in movement of the Ls1 mutant of TMV, allowing cell-to-cell and long distance movement of the virus at the restrictive temperature.[90,117] Complementation of defective challenge virus in transgenic plants that express the appropriate viral gene has also been observed in other systems, including alfalfa mosaic virus[54] and tomato golden mosaic geminivirus.[36] The recent development of *in vitro* expression systems that allow production of infectious TMV RNAs from cloned full length genomes[91,118] has permitted the direct manipulation of the TMV genome at the DNA level. This permitted Holt and Beachy[97] to generate infectious full length cDNA

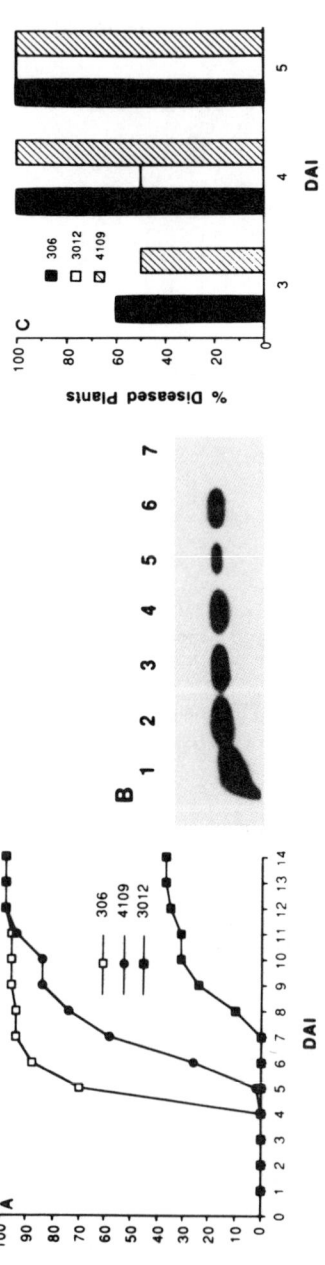

FIGURE 18. (A) Time course of disease development in transgenic plants accumulating comparable levels of TMV CP. The plants were inoculated with 0.2 μg/ml of U_1 TMV. The data for lines 4109 (rbcS:CP) and 3012 (35S:CP) were collected on CP(+) plants. Line 306 contains no CP gene and serves as the negative control. (B) Immunoblot analysis of epidermal protoplasts, mesophyll tissue, and root tissue from lines 3012 (35S:CP) and 4109 (rbcS:CP). Lane 1, 10 ng of TMV CP; lane 2, 3012 epidermal protoplasts; lane 3, 3012 mesophyll tissue; lane 4, 3012 root tissue, lane 5, 4109 epidermal protoplasts; lane 6, 4109 mesophyll tissue; and lane 7, 4109 root tissue. Lanes 2 through 7 were each loaded with 15 μg of soluble protein extracted from each tissue. (C) Time course of disease development in transgenic plants accumulating comparable levels of TMV CP when inoculated with TMV RNA. (From Clark, W. G. et al., *Virology*, 179, 640, 1990. With permission.)

EXPRESSION OF TMV COAT PROTEIN AFFECTS PROTECTION

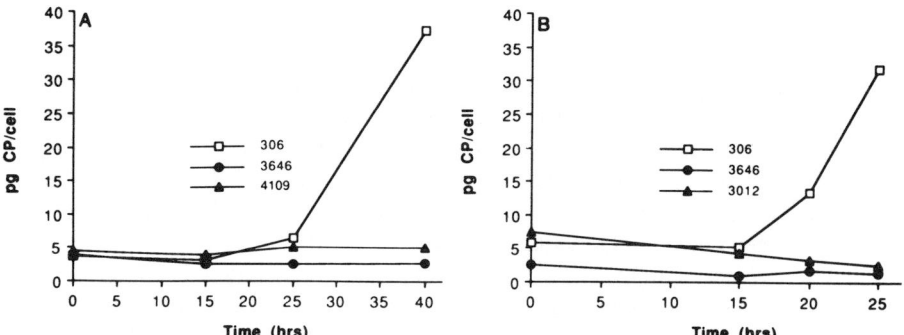

FIGURE 19. Relative virus protection in isolated mesophyll protoplasts from transgenic plants. Mesophyll protoplasts were isolated from the leaves of transgenic plants of different lines and inoculated with TMV at 1 mg/ml to compare the relative levels of virus protection afforded in each line. A and B represent independent trials. The level of TMV CP was measured by ELISA as a marker for virus replication and was plotted against time after inoculation. Lines: 3646 (35S:CP), 4109 (rbcS:CP), 3012 (35S:CP), and 306 as the negative (CP−)control. (From Clark, W. G. et al., *Virology*, 179, 640, 1990. With permission.)

clones of the U_1 strain of TMV and mutagenized cDNA by frame shift mutations in either the MP or CP gene. Although transcripts from mutant clones could replicate efficiently, cell-to-cell (local) and or long distance (systemic) movements of the viral progeny were abolished in nontransgenic tobacco plants. However, inoculation of transgenic tobacco plants that expressed a wild-type TMV MP gene resulted in both local and systemic viral infection. The CP- frameshift mutant, although unable to move systemically, in nontransformed tobacco, exhibited systemic movement in transgenic plants that expressed a wild-type TMV CP gene. Transgenic tobacco plants that expressed the appropriate wild-type TMV gene were thus able to complement, in trans, mutant virus lacking a functional MP or CP gene.

VII. FIELD TRIALS

Work on the production of CP(+) transgenic plants is either in progress or has been completed for several plants.[119-136] Field trials for such plants have already started. A detailed account of the field trials was given by Beachy et al.[3] In this regard, TMV CP gene has been very important, and transgenic plants with this gene are being tested in the fields. Tobacco plants expressing the AlMV CP ToMV CP and PVX and PVY CP genes have been tested in the field: each test received prior USDA-APHIS approval.[3] The first field test with TMV-resistant tomato plants was conducted in 1987 by researchers from Washington University and the Monsanto Company.[42,45] The R4 and R2 generations of CP(+) tomato lines produced from cultivar VF36 were grown in the field with CP(−) plants. There was no significant difference in the amounts of CP in field-grown plants compared to growth chamber or greenhouse-grown plants. CP(+) plants, mechanically inoculated with TMV, exhibited a delay in the development of disease symptoms or did not develop symptoms compared with the CP(−) plants. No more than 5% of the CP(+) plants developed symptoms by fruit harvest, whereas 99% of the VF36 plants developed symptoms. Lack of visual symptoms was associated with a lack of virus accumulation in the CP(+) plants. Fruit yields of the infected VF36 plants decreased 26 to 35% compared to healthy plants, whereas yields from CP(+) lines were equal to those of uninoculated VF36 plants, which indicated that the expression of the CP gene does not intrinsically cause yield depression.

In tomatoes, most of the yield loss is attributed to infection by strains of ToMV that can reduce yields in fresh market tomatoes by 10 to 50%. In addition to TMV, which causes marginal loss, the use of virus-resistant cultivars containing the locus Tm-1 or Tm-2 has provided the best control against ToMV infection.[131] To determine if the TMV CP gene conferred protection against infection by field isolates of ToMV, tests were conducted in 1988 in Florida and Illinois. Progeny that were homozygous for the TMV CP gene and VF36 CP(−) controls were challenged with a Florida isolate of ToMV, Naples C, in Florida. Eight weeks after inoculation, all the CP(−) plants were infected, while about 29% of the CP(+) plants were infected. The field test in Illinois was conducted to determine if expression of the TMV CP gene in tomato would protect against a number of different strains of ToMV. Two TMV strains, U_1 and PV-230, and four ToMV strains, L, Epcot, Naples C, and Aucubam, were used.[3] By 8 weeks after inoculation, all of the CP(−) control plants were infected by the TMV strains and 94 to 100% were infected by the ToMV strains. In contrast, only 3% of the CP(+) plants were infected with the TMV strains; however, 57 to 89% of the CP(+) plants were infected with the ToMV strains. These results indicate that the TMV CP gene conferred resistance against ToMV-Naples C infection under Florida field conditions and against two strains of TMV under Illinois field conditions. Only weak protection was conferred against infection by the ToMV strains under Illinois field conditions.

Plants that expressed a CP gene derived from ToMV-Naples C were produced in order to enhance protection against ToMV.[128] These plants were evaluated under field conditions in Illinois, along with the control tomato line UC82B, lines expressing the TMV CP gene, and lines expressing both the TMV and ToMV CP genes. At 4 weeks after inoculation with TMV or ToMV-Naples C, all of the control plants were infected by TMV, and 93% were infected by ToMV. The TMV CP(+) lines were resistant to TMV infection, as shown in the earlier field test; however, they were less resistant to infection by ToMV-Naples C. The tomato lines expressing the ToMV CP gene were highly resistant to infection by ToMV-Naples C; however, some plants from each line became infected with TMV. Plants that expressed both TMV and ToMV CP genes were equally well protected against TMV and ToMV.

Field tests of tobacco plants that expressed the CP gene of AlMV were carried out in Wisconsin in 1988 by researchers at the Agrigenetics Advanced Science Company. Symptom development and virus content of mechanically inoculated plants of the line that contained the P35S: AlMV CP:CaMV 3' end gene and CP(−) Xanthi nc tobacco followed throughout the growing season. CP(+) plants developed disease more slowly or not at all compared to CP(−) plants; symptom development was correlated with virus accumulation. At 85 days after inoculation only 9% of the CP(+) plants had developed a systemic infection, while 93% of the CP(−) plants had developed an AlMV accumulation as compared to CP(+) plants that were three to six times less infected.

Beachy and co-workers[3] conducted the experiments with transgenic potato plants in 1989 under permission granted by USDA-APHIS at the Monsanto Research Farm near Jerseyville, Illinois. In their study, they inoculated transgenic Russet Burbank clones expressing PVX and PVY CP genes simultaneously with PVX and PVY and transplanted them into the field. They examined disease development under field conditions and measured the effect of primary infection on the yield of transgenic and control Burbank plants. Their results showed that expression of PVX and PVY CP genes confers a very high level of resistance to PVX and PVY infection in clone 303, one of the four different transgenic potato clones tested. After inoculation with PVX and PVY, tuber yield of control Russet Burbank plants decreased, while the yield of clone 303 was unaffected. Four different uninoculated transgenic clones had tuber yields as high as control Russet Burbank. These results confirm that resistance to PVX and PVY is effective in the field and can prevent yield losses due to dual infection by these viruses.

In this field trial, expression of the CP genes prevented disease development in clone 303. After virus inoculation, yields of potato tubers from this clone were similar to those of uninoculated plants. These results show that CP-mediated protection has the potential for maintaining yields in the field. Uninoculated plants from line 303 exhibited growth characteristics and yield similar to the transformed Russet Burbank plants. Further testing will confirm whether clone 303 has retained all the intrinsic properties of Russet Burbank with the addition of PVX and PVY resistance.

VIII. CONCLUSIONS

The development of genetic engineering techniques has enabled the production of transgenic plants that are resistant to viral diseases. Expressing the coat protein gene of virus in transgenic plants confers resistance against the virus from which the gene was isolated and to other closely related strains and viruses. This approach has been demonstrated to be effective in conferring protection against viruses from different virus groups including alfalfa mosaic virus, cucumber virus, potexvirus, polyvirus, tobravirus, and tobamovirus.

The data available indicate that several factors may affect the efficiency of the protection obtained, including the level of the CP in the transgenic plants, the plant in which the CP gene is expressed, and environmental conditons.

It is clear that coat protein-mediated resistance can reduce virus infection and disease development for a number of different host-virus systems. Each of these examples reported between 1986 and 1992 may represent the greatest number of viral CP genes identified in such a short period of time. Although the scientific and technical advances have been rapid, additional studies must be carried out to better understand the molecular basis of resistance and before resistant plants can be used in agriculture.

It is also important that other virus groups and different hosts be studied to extend the applications of the technology, and develop a more complete understanding of the mechanisms of resistance. Each example of CP-mediated resistance has involved a virus with (+) sense RNA that is encapsidated by a single type of protein. Examples of important diseases caused by viruses with alternate composition include: the geminiviruses, which contain single-stranded DNA such as maize streak and wheat dwarf viruses, tomato yellow leaf curl virus, and a large variety of other whitefly-transmitted viruses; tomato spotted wilt virus, and enveloped virus with multiple structural proteins and a genome that includes nonmessage-sense RNA; viruses causing rice tingro, one containing ss-RNA and the other ds-DNA, each encapsidated by multiple capsid proteins; caulimoviruses, which contain double-stranded DNA and replicate by retrovirus-like mechanisms; rice dwarf virus, the genome of which contains multiple segments of ds-RNA encapsidated in several capsid proteins. Each of these viruses is quite different from the simple (+) sense RNA viruses, and it will be important to determine whether or not CP-mediated resistance will be effective for their control.

It is also important to understand fully how the CP provides resistance; through understanding these mechanisms, it may be possible to improve and extend resistance beyond its current limitations. To date, most mechanistic studies have involved TMV and tomato and tobacco plants. It is highly likely that other plant virus/host combinations in CP-mediated resistance will be different from the TMV systems. Until further studies are completed, it is not wise to speculate on general mechanisms of resistance, or to propose ways to alter or improve resistance.

The field trials completed to date have been extremely valuable for evaluating disease resistance and the agronomic characteristics of transgenic plants. These tests have been conducted with artificial rather than natural viral infections and have demonstrated that CP genes confer disease resistance under a limited set of field conditions. It is now important that tests be extended to other sites with a greater variety of soil and climatic conditions as well as sites with high incidences of natural infection.

Cross protection has been successfully employed for the control of viral diseases of, for example, tobacco, *Citrus* spp., *Papaya* spp., etc. There are, however, problems with this technique: the mild strain of virus may mature to a virulent form; it may act synergistically with other viruses present to adversely affect plant growth and development; a virus which is mild for one species may have virulent effects on a second crop, and so cross-contamination may be a problem; and mild strains of virus, while not inducing severe symptoms on a crop, may nevertheless reduce product yield to some extent.

REFERENCES

1. **Zaitlin, M. and Hull, R.**, Plant virus-host interactions, *Annu. Rev. Plant Physiol.*, 38, 291, 1987.
2. **Matthews, R. E. F.**, *Plant Virology*, 2nd ed., Academic Press, New York, 1981.
3. **Beachy, R. N., Loesch-Fries, S., and Tumer, N. E.**, Coat protein-mediated resistance against virus infection, *Annu. Rev. Phytopathol.*, 28, 451, 1990.
4. **Hamilton, R. I.**, Defenses triggered by previous invaders-virus, in *Plant Disease, An Advanced Treatise*, Vol. 5, Horsfall, J. G. and Cowling, G. R., Eds., Academic Press, New York, 1980, 279.
5. **Ponz, F. and Bruening, G.**, Mechanism of resistance to plant viruses, *Annu. Rev. Phytopathol.*, 24, 355, 1986.
6. **Sequeira, L.**, Cross protection and induced resistance. Their potential for plant disease control, *Trends Biotechnol.*, 2, 25, 1984.
7. **Sherwood, J. L.**, Mechanisms of cross-protection between plant virus strains, in *Plant Resistance to Viruses*, Evered, D. and Harnett, S., Eds., Wiley, Winchester, UK, 1987, 136.
8. **Powell, P. A., Sanders, P. R., Tumer, N., Fraley, R. T., and Beachy, R. N.**, Protection against tobacco mosaic virus infection in transgenic plants requires accumulation of coat protein rather than coat protein RNA sequences, *Virology*, 175, 124, 1990.
9. **Horsch, R. B., Fry, J. E., Hoffmann, N. L., Eichholtz, D., Rogers, S. G., and Fraley, R. T.**, A simple and general method for transferring genes into plants, *Science*, 227, 1229, 1985.
10. **Fraley, R. T., Rogers, S. G., and Horsch, R. B.**, Genetic transformation in higher plants, *Critical Rev. Plant Sci.*, 4, 1, 1986.
11. **Klee, H., Horsch, R., and Rogers, S.**, *Agrobacterium*-mediated plant transformation and its further applications to plant biology, *Annu. Rev. Plant Physiol.*, 38, 467, 1987.
12. **Schell, J.**, Transgenic plants as a tool to study the molecular organization of plant genes, *Science*, 237, 1176, 1987.
13. **Weising, K., Schell, J., and Kahl, G.**, Foreign genes in plants: transfer, structure, expression, and applications, *Annu. Rev. Genet.*, 22, 421, 1988.
14. **Pfeiffer, P. and Hohn, T.**, Cauliflower mosaic virus as a probe for studying gene expression in plants, *Physiol. Plant*, 77, 625, 1989.
15. **Beachy, R. N., Stark, D. M., Deom, C. M., Oliover, M. J., and Fraley, R. T.**, Expression of sequences of tobacco mosaic virus in transgenic plants and their role in disease resistance, in *Tailoring Genes for Crop Improvement*, Bruening, G., Harada, J., Kosuge, T., and Halender, J., Eds., New York, Plenum, 1987, 169.
16. **Beachy, R. N.**, Virus cross-protection in trangenic plants, in *Plant Gene Research. Temporal and Spatial Regulation of Plant Genes*, Verma, D. P. S. and Goldberg, R. B., Eds., Springer-Verlag, New York, 1988, 323.

17. **Beachy, R. N., Roger, S. G., and Fraley, R. T.**, Genetic transformation to confer resistance to plant virus disease, in *Genetic Engineering: Principles and Methods*, Vol. 9, Setlow, J. K. Ed., Plenum Press, New York, 1987, 229.
18. **Hemenway, C., Fang, R. X., Kaniewski, J. J., Chua, N. H., and Tumer, N. E.**, Analysis of the mechanism of protection in transgenic plants expressing the potato virus X coat protein or its antisense RNA, *EMBO J.*, 7, 1273, 1988.
19. **Hemenway, C., Tumer, N. E., Powell, P. A., and Beachy, R. N.**, Genetic engineering of plants for viral disease resistance, in *Cell Culture and Somatic Cell Genetics of Plants*, Vol. 6, Schell, T. and Vasil, I. K., Eds., Academic Press, New York, 1989, 405.
20. **Powell, P. A., Nelson, R. S., De, B., Hoffman, N., Rogers, S. G., Fraley, R. T., and Beachy, R. N.**, Delay of disease development in transgenic plants that express the tobacco mosaic virus coat protein gene, *Science*, 232, 738, 1986.
21. **Nejidat, A., Clark, W. G., and Beachy, R. N.**, Engineered resistance against plant virus diseases, *Physiol. Plant*, 80, 662, 1990.
22. **Loesch-Fries, L. S., Morlo, D., Zinnen, T., Burhop, L., Hill, K., Krahan, K., Jarivs, N., Nelson, S., and Halk, E.**, Expression of alfalfa mosaic virus RNA4 in transgenic plants confers virus resistance, *EMBO J.*, 6, 1845, 1987.
23. **Harrison, B. D., Mayo, M. A., and Baulcombe, D. C.**, Virus resistance in transgenic plants that express cucumber mosaic virus satellite RNA, *Nature*, 328, 799, 1987.
24. **Gerlach, W. L., Llewellyn, D., and Haseloff, J. P.**, Construction of a plant disease resistance gene from the satellite RNA of tobacco ringspot virus, *Nature*, 328, 802, 1987.
25. **Cuozzo, M., O'Connell, K. M., Kaniewski, W., Fang, R.-X., Chua, N. H., and Tumer, N. E.**, Viral protection in transgenic tobacco plants expressing the cucumber mosaic virus coat protein or its antisense RNA, *Bio/Technology*, 6, 549, 1988.
26. **Baulcombe, D. C., Hamilton, S. D. O., Mayo, M. A., and Harrison, B. D.**, Resistance to viral disease through expression of viral genetic material from the plant genome, in *Plant Resistance to Viruses*, Evered, D. and Harnett, S., Eds., Wiley, Winchester, U.K., 1987, 133.
27. **Rezaian, M. A., Skene, K. G. M., and Ellis, J. G.**, Anti-sense RNAs of cucumber mosaic virus in transgenic plants assessed for control of the virus, *Plant Mol. Biol.*, 11, 463, 1988.
28. **Gerlach, W. L., Llewellyn, D., and Haseloff, J. P.**, Plant virus resistance based on the satellite RNA of tobacco ringspot virus, *J. Cell Biochem. Suppl.*, 12C, 239, 1988.
29. **Dougherty, W. G. and Carrington, J. C.**, Expression and function of potyviral gene products, *Annu. Rev. Phytopathol.*, 26, 123, 1988.
30. **Dougherty, W. G. and Hiebert, E.**, Genome structure and gene expression of plant RNA viruses, in *Molecular Plant Virology, Replication and Gene Expression*, Davies, J. W., Ed., CRC Press, Boca Raton, FL, 1985, 23.
31. **Kozak, M.**, Point mutations close to the AUG initiator codon affect the efficiency of translation of rat preproinsulin *in vivo*, *Nature*, 308, 241, 1984.
32. **Lutcke, H. A., Chow, K. C., Mickel, F. S., Moss, K. A., Kern, H. F., and Scheele, G. A.**, Selection of AUG codons differs in plants and animals, *EMBO J.*, 43, 1987.
33. **Stark, D. M.**, Coat Protein-Mediated Protection against Potyviruses in Transgenic Plants, Ph.D. thesis, Washington University, St. Louis, MO, 1989.
34. **Stark, D. M. and Beachy, R. N.**, Protection against potyvirus infection in transgenic plants, evidence for broad spectrum resistance, *Bio/Technology*, 7, 1257, 1989.
35. **Guilley, H., Dudley, R. K., Jonard, G., Balazs, E., and Richards, K. E.**, Transcription of cauliflower mosaic virus DNA: detection of promoter sequences, and characterization, *Cell*, 30, 763, 1982.
36. **Hanley-Bowdoin, L., Ulmer, J. S., and Rogers, S. G.**, Expression of fuctional replication protein from tomato golden mosaic virus in transgenic plants, *Proc. Natl. Acad. Sci., U.S.A.*, 87, 1446, 1990.

37. **Lawton, M. A., Tierney, M. A., Nakamura, I., Anderson, E., Komeda, Y., et al.,** Expression of a soybean β-conglycinin gene under the control of the cauliflower mosaic virus 35S and 19S promoters in transformed tobacco tissue, in *Molecular Strategies for Crop Protection*, Arntzen, C. J. and Ryan, C. A., Eds., Alan R. Liss, New York, 1987, 221.
38. **Sanders, P. R., Winter, J. A., Barnason, A. R., Rogers, S. G., and Fraley, R. T.,** Comparison of cauliflower mosaic virus 35S and napoline synthase promotes in transgenic plants, *Nucleic Acids Res.*, 15, 1543, 1987.
39. **Bevan, M. W. and Harrison, B. D.,** Genetic engineering of plants for tobacco mosaic virus resistance using the mechanisms of cross-protection, in *Molecular Strategies for Crop Protection*, Arntzen, C. J. and Ryan, C. A., Eds., Alan R. Liss, New York, 1986, 215.
40. **Bevan, M. W., Mason, S. E., and Goelet, P.,** Expression of toabacco mosaic virus coat protein by a cauliflower mosaic virus promoter in plants transformed by *Agrobacterium*, *EMBO J.*, 4, 1921, 1985.
41. **Powell, P. A.,** Genetically Engineered Protection against Tobacco Mosaic Virus Infection by the Expression of Viral Sequences in Transgenic Plants, Ph.D. thesis, Washington University, St. Louis, MO, 1988.
42. **Nelson, R. S., Powell-Abel, P., and Beachy, R. N.,** Lesions and virus accumulation in inoculated transgenic tobacco plants expressing the coat protein gene of tobacco mosaic virus, *Virology*, 158, 126, 1987.
43. **Lawson, C., Kaniewski, W., Haley, L., Rozman, R., Newell, C., Sanders, C., and Tumer, N. E.,** Engineering resistance to mixed virus infection in a commercial potato cultivar: resistance to potato virus X and potato virus Y in transgenic Russet Burbank, *Bio/Technology*, 8, 127, 1990.
44. **Tumer, N. E., O'Connell, K. M., Nelson, R. S., Sanders, P. R., Beachy, R. N., Fraley, R. T., and Shah, D. M.,** Expression of alfalfa mosaic virus coat protein gene confers cross protection in transgenic tobacco and tomato plants, *EMBO J.*, 6, 1181, 1987.
45. **Nelson, R. S., McCormick, S. M., Delanny, X., Dube, P., Layton, J., Anderson, E. J., Kaniewska, M., Proksch, R. K., Horsch, R. B., Rogers, S. G., Fraley, R. J., and Beachy, R. N.,** Virus tolerance, plant growth, and field performance of transgenic tomato plants expressing coat protein from tobacco mosaic virus, *Bio/Technology*, 6, 403, 1988.
46. **Nejidat, A. and Beachy, R. N.,** Decreased levels of TMV coat protein in transgenic tobacco plants at elevated temperatures reduce resistance to TMV infection, *Virology*, 173, 531, 1989.
47. **Nejidat, A. and Beachy, R. N.,** Transgenic tobacco plants expressing a tobacco virus coat protein gene are resistant to some tobamoviruses, *Mol. Plant Microb. Interact.*, 3, 247, 1990.
48. **Anderson, E. J., Stark, D. M., Nelson, R. S., Tumer, N. E., and Beachy, R. N.,** Transgenic plants that express the coat protein gene of TMV or AlMV interfere with disease development of non-related viruses, *Phytopathology*, 12, 1284, 1989.
49. **van Dun, C. M. P. and Bol, J. F.,** Transgenic tobacco plants accumulating tobacco rattle virus coat protein resist infection with tobacco rattle virus and pea early browning virus, *Virology*, 167, 649, 1988.
50. **Siegel, A.,** Natural exclusion of strains of tobacco mosaic virus, *Virology*, 8, 470, 1959.
51. **Wu, J. H. and Rappaport, I.,** An analysis of interference between two strains of tobacco mosaic virus on *Phaseolus vulgaris* L., *Virology*, 14, 259, 1961.
52. **Zinnen, T. M. and Fulton, R. W.,** Cross-protection between sunn-hemp mosaic and tobacco mosaic viruses, *J. Gen. Virol.*, 67, 1679, 1986.
53. **Powell, P. A., Stark, D. M., Sanders, R. P., and Beachy, R. N.,** Protection against tobacco mosaic virus in transgenic plants that express tobacco mosaic virus antisense RNA, *Proc. Natl. Acad. Sci., U.S.A.*, 86, 6949, 1989.

54. **van Dun, C. M. P., Overduin, B., van Vloten-Doting, L., and Bol, J. F.**, Transgenic tobacco expressing tobacco streak virus or mutated alfalfa mosaic virus coat protein does not cross-protect against alfalfa mosaic virus infection, *Virology,* 164, 383, 1988.
55. **Hoekema, A., Huisman, M. J., Molendijk, L., van den Elzen, P. J. M., and Cornelissen, B. J. C.**, The genetic engineering of two commercial potato cultivars for resistance to potato virus X, *Bio/Technology,* 7, 273, 1989.
56. **Beemster, A. B. R. and deBokx, J. A.**, Survey of properties and symptoms, in *Viruses of Potatoes and Seed Potato Production,* deBokx, J. A. and Van der Want, J. P. H., Eds., Wageningen, Netherlands, 1987, 84.
57. **Munro, J.**, Potato Virus X, in *Compendium of Potato Diseases,* Hooker, W. J., Ed., American Phytopathological Society, St. Paul, 1986, 72.
58. **de Bokx, J. A.**, Potato virus Y, in *Compendium of Potato Diseases,* Hooker, W. J., Ed., American Phytopathological Society, St. Paul, 1986, 70.
59. **Rochow, W. F., Ross, A. F., and Siegel, B. M.**, Comparison of local-lesion and electron microscope particle count methods for assay of potato virus X from plants doubly infected by potato viruses X and Y, *Virology,* 1, 28, 1955.
60. **Shukla, D. D. and Ward, C. W.**, Structure of potyvirus coat proteins and its application in the taxonomy of the potyvirus group, *Adv. Virus Res.,* 36, 273, 1989.
61. **Milne, R. G.**, *The Plant Viruses,* Vol. 4, *The Filamentous Plant Viruses,* Plenum, New York, 1988.
62. **Coruzzi, G., Broglie, R., Edwards, C., and Chua, N.-H.**, Tissue-specific and light-regulated expression of a pea nuclear gene encoding the small subunit of ribulose-1,5-biphosphate carboxylase, *EMBO J.,* 3, 1671, 1984.
63. **Hiebert, E. and Dougherty, W. G.**, Organization and expression of the viral genome, in *The Plant Viruses,* Vol. 4, Milne, R. G., Ed., Plenum Press, New York, 1988, 159.
64. **MacKinnon, J. P. and Bagnall, R. H.**, Use of *Nicotiana debneyii* to detect viruses S, X, and Y in potato seed stocks and relative susceptibility of six common varieties to potato virus S, *Potato Res.,* 15, 81, 1972.
65. **Wright, N. S.**, The effect of separate infections by potato viruses X and S on Netted Gem potato, *Am. Potato J.,* 54, 147, 1987.
66. **Wright, N. S.**, Assembly, quality control and use of a potato cultivar collection rendered virus-free by heat therapy and tissue culture, *Am. Potato J.,* 65, 181, 1987.
67. **Hahan, T., Slack, S. A., and Stattery, R. J.**, Reinfection of potato seed stocks with potato virus and potato virus X in Wisconsin, *Am. Potato J.,* 58 117, 1981.
68. **Koenig, R.**, Carlavirus group. No. 259, in *Descriptions of Plant Viruses,* Commonwealth Mycolological Institute/Association for Applied Biology, Kew, Surrey, England, 1982.
69. **Monis, J. and de Zoeten, G. A.**, Characterization and translation studies of potato virus S RNA, *Phytopathology,* 80, 441, 1990.
70. **MacKenzie, D. J. and Tremaine, J. H.**, Transgenic *Nicotiana debneyii* expressing viral coat protein are resistant to potato virus S infection, *J. Gen. Virol.,* 71, 2167, 1990.
71. **Foster, G. D., Millar, A. W., Meehan, B. M., and Mills, P. R.**, Nucleotide sequence of the 3'-terminal region of Helenium virus S RNA, *J. Gen. Virol.,* 71, 1877, 1990.
72. **MacKenzie, D. J., Tremaine, J. H., and McPherson, J.**, Genetically engineered resistance to potato virus S in potato cultivar Russet Burbank, *Mol. Plant-Microbe Interact.,* 4, 95, 1991.
73. **Bol, J. F., van Vloten-Doting, L., and Jaspars, E. M. J.**, A functional equivalence of top component a RNA and coat protein in the initiation of infection by alfalfa mosaic virus, *Virology,* 46, 73, 1971.
74. **Foster, G. D. and Mills, P. R.**, Evidence for the role of subgenomic RNAs in the production of potato virus S coat protein during *in vitro* translation, *J. Gen. Virol.,* 71, 1247, 1990.
75. **van Vloten-Doting, L.**, Coat protein is required for infectivity of tobacco streak virus: biological equivalence of the coat proteins of tobacco streak and alfalfa mosaic virus, *Virology,* 65, 215, 1975.

76. **van Vloten-Doting, L. and Jaspars, E. M. J.**, Plant covirus systems: Three component systems, in *Comprehensive Virology*, Fraenkel-Conrat, H. and Wagner, R., Eds., Plenum, New York, 1977, 1.
77. **Zuidema, D., Bierhuizen, M. F. A., Cornelissen, B. J. C., Bol, J. F., and Jaspars, E. M. J.**, Coat protein binding sites on RNA 1 of alfalfa mosaic virus, *Virology*, 125, 361, 1983.
78. **Houwing, C. J. and Jasparsa, E. M. J.**, Protein binding sites in nucleation complexes of alfalfa mosaic virus RNA4, *Biochemistry*, 21, 3408, 1982.
79. **Gonsalves, D. and Fulton, R. W.**, Activation of Prunus necrotic ringspot virus and rose mosaic virus by RNA4 components of some Ilaviruses, *Virology*, 81, 398, 1977.
80. **Halk, E. L., Merlo, D. J., Liao, L. W., Jarvis, N. P., Nelson, S. E., et al.**, Resistance to alfalfa mosaic virus in transgenic tobacco and alfalfa, in *Molecular Biology of Plant-Pathogen Interactions*, Staskwicz, B., Ahlquist, P., and Yoder, O., Eds., Liss, New York, 1989, 101.
81. **Cornelissen, B. J. C., Janssen, H., Znidema, D., and Bol, J. F.**, Complete nucleotide sequence of tobacco streak virus RNA 3, *Nucleic Acids Res.*, 12, 2407, 1984.
82. **Angenent, G. C., van Den Owweland, J. M. W., and Bol, J. F.**, Susceptibility to virus infection of transgenic tobacco plants expressing structural and nonstructural genes of tobacco raffle virus, *Virology*, 175, 191, 1990.
83. **Pefersen, S. G., Lehmbeck, J., and Borkhardt, B.**, Analysis of RNA2 of pea early browning virus strain SP5, *Plant Mol. Biol.*, 13, 735, 1989.
84. **Angenent, G. C., Lindhorst, H. J. M., van Belkum, A. F., Cornelissen, B. J. C., and Bol, J. F.**, RNA 2 of tobacco rattle virus strain TCM encodes an unexpected gene, *Nucleic Acid Res.*, 14, 4673, 1986.
85. **Shikawa, M., Meshi, T., Motoyoshi, F., Takamatsu, N., and Okada, Y.**, In vitro mutagenesis of the putative replicase genes of tobacco mosaic virus, *Nucleic Acids Res.*, 14, 8291, 1986.
86. **Hunter, T. R., Hunt, T., Knowland, J., and Zimmern, D.**, Messenger RNA for coat protein of tobacco mosaic virus, *Nature (London)*, 260, 759, 1976.
87. **Pelham, J.**, Strain-genotype interaction of tobacco mosaic virus in tomato, *Ann. Appl. Biol.*, 71, 219, 1972.
88. **Pehlam, H. R. B.**, Leaky UAG termination codon tobacco mosaic virus RNA, *Nature (London)*, 272, 469, 1978.
89. **Beachy, R. N. and Zaitlin, M.**, Characterization and *in vitro* translation of the RNAs from less-than-full-length, virus related, nucleoprotein rods present in tobacco mosaic virus preparations, *Virology*, 81, 160, 1977.
90. **Deom, C. M., Oliver, M. J., and Beachy, R. N.**, The 30-kildalton gene product of tobacco mosaic virus potentiates virus movement, *Science*, 237, 389, 1987.
91. **Meshi, T., Ishikawa, M., Motoyoshi, F., Samba, K., and Okada, Y.**, In vitro transcription of infectious RNAs from full-length cDNAs of tobacco mosaic virus, *Proc. Natl. Acad. Sci. U.S.A.*, 83, 5013, 1986.
92. **Siegel, A., Zaitlin, M., and Sehgal, O. P.**, The isolation of defective tobacco mosaic virus strains, *Proc. Natl. Acad. Sci. U.S.A.*, 48, 1845, 1962.
93. **Atabekov, J. G. and Dorokhov, Y. L.**, Plant virus-specific transport function and resistance of plants to viruses, *Adv. Virus Res.*, 29, 313, 1984.
94. **Takahashi, H., Shimamoto, K., and Ehara, Y.**, Cauliflower mosaic virus gene VI causes growth suppression, development of necrotic spots and expression of defense-related genes in transgenic tobacco plants, *Mol. Gen. Genet.*, 216, 188, 1989.
95. **Takamatsu, N., Ishikawa, M., Meshi, T., and Okada, Y.**, Expression of bacterial chloramphenicol acetyltransferase gene in tobacco plants mediated by TMV-RNA, *EMBO J.*, 6, 307, 1987.
96. **Dawson, W. O., Bubrick, P., and Grantham, G. L.**, Modifications of the tobacco mosaic virus potentiate virus movement, *Science*, 237, 389, 1987.

97. **Holt, C. A. and Beachy, R. N.**, *In vitro* complementation of infectious transcripts from mutant tobacco mosaic virus cDNA in transgenic plants, *Virology*, 181, 109, 1991.
98. **Nelson, R. S., McCormick, S. M., Delannay, X., Dube, P., Layton, J., Anderson, E. J., Kaniewska, M., Proksch, R. K., Horsch, R. B., Rogers, S. G., Fraley, R. T., and Beachy, R. N.**, Virus tolerance, plant growth, and field performance of transgenic tomato plants expressing coat protein from tobacco mosaic virus, *Bio/Technology*, 6, 403, 1988.
99. **Gibbs, A.**, Tobamovirus classification, in *The Plant Viruses, The Rodshaped Plant Viruses*, van Regenmortel, M. H. V., van Regenmortel, H. V., and Fraenkel-Conrat, H., Plenum, New York, 1986, 168.
100. **Wisniewski, L. A., Powell, P. A., Nelson, R. S., and Beachy, R. N.**, Local and systemic movement of tobacco mosaic virus (TMV) in tobacco plants that express the TMV coat protein gene, *Plant Cell*, 2, 559, 1990.
101. **Wilson, T. M. A.**, Nucleocapsid disassembly and early gene expression by positive strand RNA viruses, *J. Gen. Virol.*, 66, 1201, 1985.
102. **Baughman, G. A., Jacobs, J. D., and Howell, S. H.**, Cauliflower mosaic virus gene VI produces a symptomatic phenotype in transgenic plants, *Proc. Natl. Acad. Sci. U.S.A.*, 85, 733, 1988.
103. **Larkin, P. L. and Scowcroft, W. R.**, Somaclonal variation a novel source of variability from cell cultures for plant improvement, *Theor. Appl. Genet.*, 60, 197, 1981.
104. **Clark, W. G., Register, J. C., III, Nejidat, A., Echholtz, D. A., Sanders, P. R., Fraley, R. T., and Beachy, R. N.**, Tissue sepcific expression of the TMV coat protein in transgenic tobacco plants affects the level of coat protein mediated protection, *Virology*, 179, 640, 1990.
105. **Register, J. C., III and Beachy, R. N.**, Resistance to TMV in transgenic plants results from interference with and early event in infection, *Virology*, 166, 524, 1988.
106. **Register, J. C., III, Powell, P. A., Nelson, R. S., and Beachy, R. N.**, Genetically engineered crop protection against TMV interferes with initial infection and long-distance spread of virus, in *Molecular Biology of Plant-Pathogen Interactions*, Staskawicz, B., Ahlquist, P., and Yoder, A. R., Eds., Alan R. Liss, New York, 1988, 269.
107. **Register, J. C., III and Beachy, R. N.**, Effect of protein aggregation state on coat protein mediated protection against tobacco mosaic virus using a transient protoplasts assay, *Virology*, 173, 656, 1989.
108. **Cutt, J. R., Harpster, M. H., Dixon, R. A., Carr, J. P., Dusmuir, P., and Klessing, D. F.**, Disease response to tobacco mosaic virus in transgenic tobacco plants that constitutively express the pathogenesis-related PR1b gene, *Virology*, 173, 89, 1989.
109. **Carr, J. P., Beachy, R. N., and Klessig, D. F.**, Are the proteins of tobacco involved in genetically engineered resistance to TMV?, *Virology*, 169, 470, 1989.
110. **Hamilton, W. D. O., Boccara, M., Robinson, D. J., and Baulcombe, D. C.**, The complete nucleotide sequence of tobacco rattle virus RNA-1, *J. Gen. Virol.*, 68, 2563, 1987.
111. **Angenent, G. C., Verbeek, H. B. M., and Bol, J. F.**, Expression of the 16K cistron of tobacco rattle virus in protoplasts, *Virology*, 169, 305, 1989.
112. **Angenent, G. C., Posthumus, E., Bredeforde, F. T., and Bol, J. F.**, Genome structure of tobacco rattle virus strain PLB: Further evidence on the occurrence of RNA recombination among tobraviruses, *Virology*, 171, 271, 1989.
113. **Osbourn, J. K., Watts, J. W., Beachy, R. N., and Wilson, M. A.**, Evidence that nucleocapsid disassembly and a later step in virus replication are inhibited in transgenic tobacco protoplasts expressing TMV coat protein, *Virology*, 172, 370, 1989.
114. **Eoyang, L. and August, J. T.**, Reproduction of RNA bacteriophages, in *Comprehensive Virology. Reproduction, Small and Intermediate RNA Viruses*, Vol. 2, Fraenkel-Conrat, H. and Wagner, R.R., Eds., Plenum, New York, 1974, 1.
115. **Dodds, J. A., Lee, S. Q., and Tiffany, M.**, Cross protection between strains of cucumber mosaic virus: effect of host and type of inoculum on accumulation of virions and double-stranded RNA of the challenge strain, *Virology*, 144, 301, 1985.

116. **Jefferson, R. A., Kavanagh, T. A., and Bevan, M. W.**, GUS fusions: β-glucuronidase as a sensitive and versatile gene fusion marker in higher plants, *EMBO J.*, 6, 3901, 1987.
117. **Deom, C. M., Oliver, M. J., and Beachy, R. N.**, The 30-kilodalton gene product of tobacco mosaic virus potentiates virus movement, *Science*, 237, 239, 1987.
118. **Short, M. N. and Davies, J. W.**, Narcissus mosaic virus: A potexvirus with an encapsidated subgenomic messenger RNA from coat protein, *BioSci. Rep.*, 3, 837, 1983.
119. **Arntzen, C. J. and Ryan, C. A.**, *Molecular Strategies for Crop Protection*, Alan R. Liss, New York, 1986.
120. **Stiekema, W. J., Heidekamp, F., Louwerse, J. D., Verhoeven, H. A., and Dijkhuis, P.**, Introduction of foreign genes into potato cultivars 'Bintje' and 'Desiree' using an *Agrobacterium tumefaciens* binary vector, *Plant Cell Rep.*, 7, 47, 1988.
121. **Cassab, G. and Varner, J. E.**, Immunocytolocalization of extensin in developing soybean seed coats by immunogold- silver staining and tissue printing in nitrocellulose paper, *J. Cell Biol.*, 105, 2581, 1987.
122. **Davison, J.**, Plant beneficial bacteria, *Bio/Technology*, 6, 282, 1988.
123. **Eggenberger, A. L., Stark, D. M., and Beachy, R. N.**, The nucleotide sequence of a soybean mosaic virus coat protein coding region and its expression in *Escherichia coli, Agrobacterium tumefaciens*, and tobacco callus, *J. Gen. Virol.*, 70, 1853, 1989.
124. **Fraser, R. S. S. and Loughlin, S. A. R.**, Effect of temperature on the Tm-1 gene for resistance to tobacco mosaic virus in tomato, *Physiol. Plant Pathol.*, 20, 109, 1982.
125. **Gasser, C. S. and Fraley, R. T.**, Genetically engineering plants for crop improvement, *Science*, 244, 1293, 1989.
126. **Huisman, M. J., Lindhorst, H. J. M., Bol, J. F., and Cornelissen, B. J. C.**, The complete nucleotide sequence of potato virus X reveals homologies at the amino acid level with various plus-stranded RNA viruses, *J. Gen. Virol.*, 69, 1789, 1988.
127. **Monis, J. and de Zoeten, G. A.**, Characterization and translation studies of potato virus S RNA, *Phytopathology*, 80, 441, 1990.
128. **Sanders, P. R., Tumer, N., Fraley, R. Tk., and Beachy, R. N.**, Protection against tobacco mosaic virus infection in transgenic plants requires accumulation of coat protein rather than coat protein RNA sequences, *Virology*, 175, 124, 1990.
129. **van Beynum, G. M. A., DeGraf, J. M., Castell, A., Kraal, B., and Bosch, L.**, Structural studies on the coat protein of alfalfa mosaic virus genes, *Eur. J. Biochem.*, 72, 63, 1977.
130. **van Loon, L. C.**, Polyacrylamide disk electrophoresis of the soluble leaf proteins from *Nicotiana tabacum* var. "Samsun" and "Samsun NN". III. influence of temperature and virus strain on changes induced by tobacco mosaic virus, *Physiol. Plant Pathol.*, 6, 289, 1975.
131. **Hall, T. J.**, Resistance at the Tm-2 locus in the tomato to tomato mosaic virus, *Euphytica*, 29, 189, 1980.
132. **Evered, D. and Harnett, S.**, *Plant Resistance to Viruses*, Wiley, Winchester, U.K., 1987.
133. **Goelet, P., Lomonosoff, G. P., Butler, P. J. G., Akram, M. E., Gaait, M. J., and Karn, J.**, Nucleotide sequence of tobacco mosaic virus RNA, *Proc. Natl. Acad. Sci. U.S.A.*, 79, 5818, 1982.
134. **MacKenzie, D. J., Tremaine, J. H., and Stace-Smith, R.**, Organization and interviral homologies of the 3'-terminal portion of potato virus S RNA, *J. Gen. Virol.*, 70, 1053, 1989.
135. **Smit, C. H., Roosein, J., van Vloten-Doting, L., and Jaspers, E. M. J.**, Evidence that alfalfa mosaic virus infection starts with three RNA-protein complexes, *Virology*, 112, 169, 1981.
136. **van Dun, C. M. P., Bol, J. F., and van Vloten-Doting, L.**, Expression of alfalfa mosaic virus and tobacco rattle virus protein genes in transgenic tobacco plants, *Virology*, 159, 299, 1987.

Chapter 5

IMPROVEMENT OF THE NUTRITIONAL QUALITY OF PLANTS BY MANIPULATION OF SEED STORAGE GENES

I. INTRODUCTION

Seeds are one of the richest sources of plant proteins. Among major crop plants, seeds contain 10 to 50% protein, and most of this is a storage protein. Storage proteins have no enzymatic activity, and simply provide a source of amino acids, nitrogen, and carbon skeletons for the developing seedling. Storage proteins are deposited in the seed in an insoluble form in protein bodies and survive desiccation for long periods of time.

Eight species of cereal contribute over 50% of total world food calories; seven species of grain legume make a small total contribution, but are important in certain geographical areas. In addition to carbohydrate, these seed crops supply a large proportion of the protein required by man and animals. The nutritional problem with cereals and legumes is that they contain limited amounts of certain amino acids which are essential to man and monogastric animals. Most cereals are deficient in lysine and to a lesser extent threonine, while legumes are deficient in sulfur amino acids. Some seed crops, notably rice, have low overall protein levels, but a somewhat better amino acid balance. Thus there is a need to add these deficient amino acids to storage proteins or to over produce some of the storage proteins to increase the nutritional quality of seed proteins.

During the past 5 years, these proteins have been characterized at the molecular level, and the structures of a number of storage proteins as well as the mechanisms by which they are synthesized and deposited in the seed have been elucidated. The details of the organization and structure of the genes encoding storage proteins and the nature of the DNA sequences regulating their expression have recently been worked out. This has made it possible to manipulate these genes in order to improve the nutritional value of seeds.

The four major categories of seed proteins are: (1), albumins (soluble in water); (2), globulins (soluble in salt solutions); (3), prolamins (soluble in aqueous alcohol) and (4), glutelins (soluble in dilute acid or alkali). The nomenclature and properties of the major storage proteins of some important seed crops are summarized in Table 1. Although the classification is fairly crude, it provides a useful basis of reference. At one time, there was often confusion between prolamins and glutelins due to insufficient separations, but current techniques now give better separations.

II. GENETIC MANIPULATIONS OF SEED STORAGE PROTEINS

There have been a variety of genetic approaches improving nutritional quality of seeds. Early attempts concentrated on screening of existing cultivars for unusual amino acid compositions (using dye-binding assays). The high-lysine barley 'Hiproly' was selected in this way, but failed to give rise to any commercial varieties. An extension of this approach has been to screen plants grown from mutagenized seed or young excised embryos for overproduction of certain free amino acids[1] or for unusual protein patterns. A mutant of barley, Ris 1508, produced in this way has a very high lysine content due to severe depletion of the storage protein, hordein. The mutation is pleiotropic and not linked to the storage protein loci. In addition to the inhibition of storage protein synthesis, there are also effects on endosperm structure, starch metabolism, and ribonuclease and amylase levels. The grains of Ris 1508 were shrunken and gave a poor yield. The variety has therefore never been grown commercially, nor has the character been incorporated into any agronomically viable line. Thus the contribution that conventional breeding can make towards achieving these aims is limited. Biologists have now turned to genetic engineering for a possible solution.

The rapid advances in the field of recombinant DNA has led to the use of Ti plasmid-derived vectors and characterization of a number of seed protein genes from both dicots and monocots. These genes have been transferred into tobacco and petunia, and their expression properties have been studied in the heterogeneous host.[2,3] In addition to the abundance and strictly regulated synthesis of storage proteins in seeds, the expression of seed storage protein genes has also been used as a model system for studying organ-specific gene expression in plants. The goal has been to localize on these genes cis-acting DNA sequences that are necessary for seed-specific expression, and eventually to isolate from seed nuclear extracts transcriptional factors which bind to these sequences. As a step toward accomplishing the first of these aims, different seed storage protein genes have been introduced into tobacco or petunia. Experiments involving the transfer of seed protein genes from bean and soybean into either tobacco or petunia have demonstrated that the transferred genes were expressed specifically in the embryos of seeds from transgenic plants.[4-7] These studies have also shown that both the temporal control and tissue specificity of these genes were preserved in transgenic plants. For several seed protein genes, deletion analysis of the upstream regions have been used to determine the minimum elements required for tissue specificity. So far, the results of such studies have revealed that approximately one hundred to a few hundred base pairs are sufficient for tissue-specific regulation and temporal control.[6-11] Additional upstream sequences often enhance this activity.[2,6,10,11]

TABLE 1
Properties of Seed Storage Proteins from Some Major Crop Species

Species	Proteins	Protein type	Subunit (MW × 10)	Lys (%)	Met (%)	Pro + Glx (%)	Approx proportion of total protein (%)
Wheat (*Triticum aestivum*)	Gliadin	Prolamin	30–70	0.6	1.4	49	40
	Glutenin	Glutelin	up to 133	1.5	1.4	48.8	
Barley (*Hordeum vulgare*)	Hordein	Prolamin	30–55	0.5	0.6	34.2	55
Maize (*Zea mays*)	Zein	Prolamin	15–21	0.3	0.1	63	52
French bean (*Phaseolus vulgaris*)	Phaseolin (G1)	Globulin	45.5–52	7.6	0.5	22.4	50
Soybean (*Glycine max*)	Glycinin (11S)	Globulin	20,40	3.9	1.0	27.9 ⎫	60–80
	β-Conglycinin (7S)		44–90	7.0	0.3	24.8 ⎭	
Pea (*Pisum sativum*)	Legumin (11S)	Globulin	20,40	4.4	0.5	26 ⎫	80
	Vicillin (7S)		30–70	7.9	0.2	22.8 ⎭	
Rice (*Oryza sativa*)	Globulin	Globulin	33–70	3.3	0.9	20.3	10
	Glutenin	Glutelin	16–38	4.8	0.9	24.2	80
Oat (*Avena sativa*)	Endosperm globulin	Globulin	21.7, 31.7	2.6	N	26.7	56

N = not determined.

A. LEGUME STORAGE PROTEINS

In several species of legumes, a number of cultivars have been found to be defective in the expression of specific seed protein genes, and molecular lesions of some of these have been identified.[2] To understand how the expression of seed protein genes is regulated, several authors have characterized the genes for phytohemaglutinin (PHA) conglycinin, glycinin, phaseolin, vicilin, lectin, etc., and studied their expression.[11-13] In the following review a detailed account of these storage proteins and their genes is given.

1. Glycinin and Conglycinin
a. General Characteristics

Among the legumes, soybean seeds contain about 40% protein by weight. In general, four major proteins: glycinin (11S) and α, β, and γ-conglycinin (7S) have been reported from legumes. Glycinin and β-conglycinin are the most abundant and account for about 70% of the total seed protein. The proportions of glycinin and β-conglycinin vary, with ratios of 1:1 to 3:1 in different soybean cultivars. On the average, the ratio of glycinin to conglycinin is 1.6:1.[14]

The isolation of glycinin from soybean seeds is not very difficult. Isolation of these proteins and subsequent analysis revealed that both 11S and 7S proteins are low in sulfur amino acids (1.8% for glycinin and 0.6% for β-conglycinin).[14,15] Glycinin is isolated from seeds as a hexameric complex of M_r 360,000 composed of six nonidentical subunits with M_r averaging 60,000. The subunits, which are not glycosylated, are made up of a polypeptide of approximately M_r 40,000 with an acidic pI and a 20,000 M_r polypeptide with a basic pI, covalently linked by a single disulfide bond. The glycinin hexamer is assembled from five different subunit molecules that fall into two classes: (1) group I contains glycinins, (M_r 58,000) with five to eight methionine residues and consist of two members; and (2) group II glycinins (M_r 62,000 to 69,000) have three methionine residues and consist of two members.[14,15] There is about 80% identity in amino acid sequence between members of the same group, but only 40 to 50% sequence identity between members of different groups.[10,16]

The isolation of glycinin from extracts of total soybean protein is carried out by isoelectric precipitation at pH 6.6 at 4°C.[17] After the protein is denatured in 6 M urea in the presence of 2-mercaptoethanol, the acidic and basic polypeptides are fractionated by anion exchange chromatography. The initial protein fraction contains a mixture of basic polypeptides, while subsequent fractions contain acidic polypeptides. Further separation of the mixture of basic polypeptides by anion exchange chromatography gives five different polypeptides.

The isolation of β-conglycinin is generally done as a trimer of M_r 180,000, which dimerizes to form a hexameric complex at low ionic strength. β-Conglycinin exists in at least seven isomeric forms. In general, it is made up of three major subunits, α', α, and β[18,19] of M_r 76,000, 72,000, and 53,000, respectively,[20] as well as several minor components.[21] In contrast to glycinin, these subunits contain a single polypeptide. The subunit molecules are structurally equivalent, and β-conglycinin complexes arise by random assembly.[15] Unlike glycinin, β-conglycinin is glycosylated with about 5% sugar by weight. The sugar component is covalently linked to one or more asparagine residues and is made up of 2 central N-acetylglucosamine residues to which are attached 7 to 10 mannose residues.[22,23] Expression of genes encoding these subunits is tissue-specific and temporally regulated in soybean plants.[7] Both the α' and β-subunit genes have been isolated and fully or partially characterized by DNA sequence analysis. Each has been transferred to transgenic plants and was found to be expressed in petunia and tobacco plants in a regulated manner as in soybean plants. The results indicate that the 170-bp DNA sequence can function as a strong cis-acting element that controls seed-specific and temporally regulated gene expression.

In order to precipitate β-conglycinin, glycinin is removed first by lowering the pH to 4.8. Lower-molecular-weight proteins are removed by gel filtration.[21] Further purification is obtained on a concanavalin A-Sepharose affinity column to separate β-conglycinin from nonglycosylated protein contaminants. Finally, the three major subunit polypeptides, α', α, and β are separated by a combination of anion- and cation-exchange chromatography.

Genomic clones encoding the α' and β subunit genes of *Glycine max* were isolated and found in divergent orientation.[24] The DNAs were introduced into *Petunia hybrida* by a disarmed Ti plasmid vector.[24] Expression of the introduced genes in transgenic plants was limited to specific stages in developing seeds. The expression was compared with transgenic plants that contained either gene alone. The results suggested that sequences of the α' gene enhanced the level of expression of a nearby seed storage protein gene, and the signal(s) operating in the temporal regulation of seed storage protein genes was (were) common to soybean and petunia and were different from those that govern the final amounts of mRNA and protein.

Another class of globulin, sedimenting at 8.2S, has been characterized in soybean.[23,24] This protein is a relatively minor component and accounts for 5 to 10% of the seed protein, depending on the cultivar. This globulin is a tetramer of M_r 168,000 composed of four identical M_r 42,000 subunits and, in contrast to the 7S and 11S globulins, each subunit is composed of a large polypeptide (M_r 30,000). Unlike the 11S globulin, the large polypeptide is basic and the small polypeptide is acidic. The 8.2S globulin has higher levels of methionine and cysteine than the 11S and 7S proteins of soybean, but is lower in acidic amino acids, accounting for its higher isoelectric point.

b. Organization and Structure

The primary amino acid sequence of the 7S and 11S storage globulins was determined by cloned cDNAs corresponding to mRNAs encoding the proteins. These proteins were found to be encoded by gene families varying from a few to as many as 20 or more members. Group I glycinins of soybean are encoded by 3 or 4 genes. The conglycinin gene family of soybean consists of 15 to 20 members.[25] The other storage proteins are also included in several gene families. In pea, it was estimated that there are 8 genes for legumin, 11 for vicilin, and 1 for convicilin.[26] In French bean, 7 genes appear to encode phaseolin.[27] The oat 12S globulin gene family appears to contain 6 to 8 members.[28]

Soybean globulin genes are scattered throughout the genome. Of the five glycinin genes, two, Gy1 and Gy2, are tightly linked, whereas Gy3 and Gy5 segregate independently. By heteroduplex formation, it was found that the Gy1-Gy2 region differs from Gy3 by a simple deletion or duplication of 4.3 kb of DNA. The group I glycinin genes appear to arise from an ancestral gene that gave rise to the Gy1-Gy2 cluster by duplication. Duplication of the genome (tetraploidization) and deletion of DNA could account for the two glycinin domains, one containing two genes (Gy1-Gy2) and the other one gene (Gy3).

The 15 or more genes encoding β-conglycinin are also in three linkage groups in different regions of the soybean genome. Some genes contain a 556-bp insertion that accounts for the two classes of β-conglycinin mRNA and protein. An interesting feature of the β-conglycinin subunit polypeptides gene in soybean is that it can give rise to multiple charged isomers of the protein.[25] Charge variants of a single-sized polypeptide are often observed after IEF analysis. After introducing a gene encoding a β-conglycinin subunit into petunia, Beachy and co-workers found a number of charged isomers of the soybean protein in seeds of regenerated plants.[4] The formation of these charged variants is not fully known. It is also not known whether the charge variants of these proteins are encoded by different genes or whether they are the result of posttranslational events.

In several species of soybean, the sequence organization of 11S globulins is similar. The 11S globulin genes of pea and soybean have four exons separated by three introns. The introns in these genes vary in length and range up to 600 bp in soybean to less than 100 bp in pea. The three introns are in similar positions and interrupt the coding sequence of the acidic polypeptide twice and the basic polypeptide once.[15,29] A similar arrangement has been found in the 12S globulin genes of oats. Variants of this structure have been found in pea, broad bean, and sunflower.[30] One of the genes in both pea and broad bean lacks the first intron, and the third intron is absent in a gene from sunflower.

Similar structure of the 7S globulin gene has been reported. Genes for an α-subunit, β-conglycinin, and β-subunit of phaseolin consist of six exons separated by five small (100 to 200 bp) introns.[31] Doyle and co-workers also compared the coding sequences of these two genes with cDNA sequences corresponding to a number of other 7S storage proteins and observed that the degree of sequence similarity varies considerably in different regions. Within the exons, regions of high homology are separated by completely diverged regions. On the basis of these patterns, they suggested that evolutionary conservation operates on units smaller than exons. It was thus concluded that legume storage proteins have not arisen by splicing together of exons coding for structural domains.[31,32]

Beachy et al.[4] introduced a soybean gene α'-subunit of β-conglycinin into petunia. Transcripts of the soybean gene were detected in immature embryos but not in leaves of transgenic plants. The soybean α'-subunit accumulated in parallel with the petunia seed storage protein and was estimated to be between 0.1 and 1.0% of the total seed protein. The majority of the α'-subunit polypeptides assembled into multimeric proteins with sedimentation coefficients of 7 to 8S.

2. Legumin

The developing pea seed accumulates large amounts of two major storage proteins (legumin and vicilin) and smaller amounts of four other proteins (pea albumins 1, 2, and 3, and pea seed lectin). These collectively make up about 80% of the storage protein of the seed and approximately 20% of the total weight of the mature seed.[33,34] The genes for storage proteins in pea are expressed during 35 days between pollination and seed maturity and are not expressed at any other stage of plant development. These genes have their own characteristic temporal pattern of expression during seed formation.[35,36] Lycett and co-workers[37,38] analyzed the 5' flanking regions of three pea legumin genes in attempts to delimit nucleotide sequences responsible for regulating the expression of storage globulin genes. An 82-bp sequence was repeated twice in the 5' upstream region, and each repeat contained a pair of 21-bp inverted repeats in one of these genes. This region of the gene was proposed to form either of two mutually exclusive stem-loop structures, one of which having a potentially stable 42-bp stem. Switching between these alternative structures may play a role in the control of the expression of this legumin gene.

In a slightly different approach, Baumlein and colleagues[30] compared the sequences of 11S genes of broadbean, pea, and soybean. They found a region in the 5' flanking DNA of these genes that was more highly conserved than most of the coding region. The region of highest conservation was in a 28-bp sequence about 80 bp upstream of the mRNA cap site in which 25 of the 28 nucleotides were invariant. This consensus sequence, tentatively termed the "legumin box", could not be located in any genes encoding 7S proteins, leading these authors to suggest that it serves a specific regulatory function.

The expression of the legumin genes also appears to be specifically regulated. This is evident from the relative amounts of each storage protein which is characteristic of each pea line under conditions of adequate nutrient supply.[34,39] These amounts change quite dramatically, when peas are subjected

TABLE 2
Legumin Synthesis as a Percentage of Total Protein Synthesis in Cotyledons from Control and Sulfur-Deficient Plants at Two Stages of Seed Development

Development stage	Cotyledon source	Labeling time	[^{14}C] legumin (% of total ^{14}C-protein)
22 DAF	Control	10 min	8.3
		20 h	10.5
	S-deficient	10 min	1.0
		20 h	1.6
25 DAF	Control	10 min	14.8
		20 h	12.8
	S-deficient	10 min	2.9
		20 h	1.9

Note: Cotyledons were pulse labeled for 10 min with ^{14}C amino acids, with or without a subsequent 20-h chase period. DAF = Days after flowering.

From, Spencer, D., Rerie, W. G., Randall, P. J., and Higgins, T. J. V., *Aust. J. Plant Physiol.*, 17, 355, 1990. With permission.

to nutrient stress during seed formation.[35,40,41] Among all the factors which affect the expression of legumin storage proteins, the supply of sulfur (normally as sulfate) has a striking effect (Table 2).[42] Even a mild degree of sulfur deficiency results in a marked reduction in the level of one major and two minor seed storage proteins (legumin and pea albumins 1 and 3). The decrease in these proteins is compensated for by an equivalent increase in the accumulated level of vicilin. In spite of all these major changes, a viable, mature seed is formed.

The deficiency in sulfur supply has also been reported to affect the regulation of legumin genes in lupins,[43] wheat,[44] barley,[45] soybean,[46] cowpea, rape, and sunflower.[42] There was a marked reduction in the constituent polypeptides of napin (M_r 9000 and 4000) with a compensatory increase in a higher molecular weight, 11S cruciferin component under sulfur-deficiency conditions in rape seed. Sulfur deficiency caused a major reduction in the proteins of the 2–4S albumin fraction and this was accompanied by further reduction in some of the proteins of the 11S, helianthinin fraction (Figure 1).[42]

The A subfamily of legumin genes encoded the predominant 11S-class of seed storage proteins in pea (*Pisum sativum* L.). Rerie et al.[47] have determined the complete nucleotide sequence of one such gene (*Leg*A2) located on a 4.2 kb *Eco*RI fragment and encoding a mature subunit with a deduced molecular weight of 56,929 Da. The coding region spans 1825 bp (including three short introns) and exhibits a high degree of identity to *Leg*A1, but

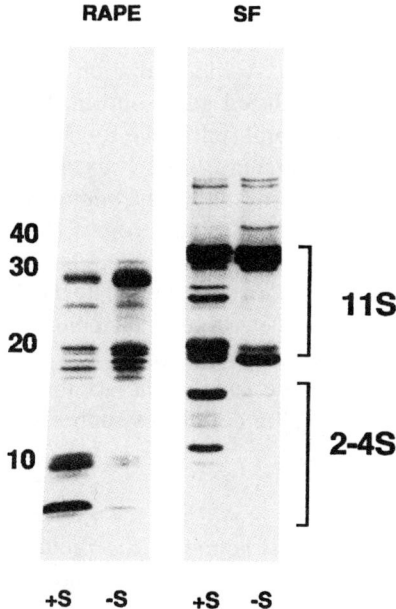

FIGURE 1. Effect of S deficiency on the composition of the seed protein fraction from rapeseed (RAPE) and sunflower (SF). Total soluble protein was extracted from seeds from plants grown under either adequate (+S) or deficient (−S) levels of sulfur. The extracts were fractionated under reducing conditions by electrophoresis on SDS-polyacrylamide gels and detected by staining with Coomassie blue. (From Spencer, D., Rerie, W. G., Randall, P. J., and Higgins, T. J. V., *Aust. J. Physiol.*, 17, 355, 1990. With permission.)

diverges from this gene in both the immediate 5'- and 3'-flanking regions. Nucleotide positions are numbered relative to the proposed cap-site (+1). Potential regulatory sequences are underlined and include the TATA (−26 to −32) and the 'legumin' (−100 to −130) boxes in the 5' region and overlapping polyadenylation signals in the 3' region. The proposed signal peptide is delineated by boldface type preceding the processing site. Regulation at posttranslational, translational, and posttranscriptional level of legumin genes also appears to be affected under reduced sulfur supply, resulting in the reduced level of seed storage proteins. Another legumin gene LeB4 showed seed-specific expression in transgenic tobacco plants.[48] The 2.4-kb upstream sequence alone of this gene, when fused to either the NPTII gene or the GUS gene, led to high enzyme levels in transgenic seeds of both tobacco and *Arabidopsis*. GUS activity was especially intense in the cotyledons, fading out towards the embryonal root tip. An 873 promoter fragment of the *Leg*A gene of pea is known to be fully functional in transgenic plants. This fragment has been enzymatically cleaved and the products examined for the ability to interact specifically with nuclear proteins. The tissue-specific activity of F_1 plants was temporally correlated with legumin gene expression.[49]

Two variants of the seed-specific pea *Leg*A gene encoding legumin were transferred to *Nicotiana plumbaginifolia* plants using the *Agrobacterium tumefaciens* binary vector Bin 19. A wide variation was observed both in the gene copy number of the introduced genes and in the level of pea legumin synthesized in the seeds of several individual transgenic *N. plumbaginifolia* plants.[11] The variation in expression from factors, such as chromosomal position effects of the integrated genes in the *Nicotiana* genome also influenced the *Leg*A genes because the number of gene copies of introduced *Leg*A genes could not be correlated with the expression level in the same plant.

In addition to the cis-acting factors that regulate the developmental expression of storage protein genes, environmental conditions, such as mineral nutrition, have been shown to affect storage protein gene expression. In peas grown with a suboptimal sulfate supply, the accumulation of legumin was greatly reduced.[40] Likewise, sulfur deficiency suppressed the synthesis of 11S globulin of lupine.[50]

3. Vicilin and Convicilin

Vicilin constitutes a major fraction of the protein in pea seed. Native vicilin is a trimeric 7S protein that contains a number of polypeptides, some of which are glycosylated.[33] Vicilin can be divided into two classes of polypeptides on the basis of the size of the primary translation product from which the polypeptides are derived. Sizes of relative molecular mass (M_r) 50,000 or less, result from cleavage, after translation, of a family of M_r-50,000 vicilins at one, both or neither of two internal processing sites.[36,51] Some members of this class are also glycosylated.[52,53] The second class of protein in pea seed, known as convicilin, contains polypeptides of M_r approx. 70,000[54] or 75,000.[55,56] Convicilin is not glycosylated[54] and does not appear to undergo any further processing such as endoproteolytic cleavage.[52]

The two classes of vicilin are distinct gene families that are not genetically linked[57] and are active at different times of seed development.[41] The M_r-50,000 vicilins are encoded by approx. 18 genes[33] that are transcribed early in the deposition of storage reserves. In addition, sulfur deficiency leads to an increase in both the level and the duration of synthesis of M_r 50,000 vicilin polypeptides. The increased accumulation of vicilin correlates with elevated vicilin mRNA in these seeds,[41] leading to the conclusion that control of storage protein synthesis by sulfur availability results from modulation of mRNA levels by transcription, posttranscriptional processing, or turnover. Convicilin, on the other hand, is encoded by a single gene, although subsequent data indicate the presence of at least one other related sequence.[58] Both of these genes are active during the late stages of seed development.[41] The precise relationship between convicilin and vicilin is not known, and there was some uncertainty about the oligomeric nature of convicilin.[59]

Like 7S globulins, convicilin is a trimeric protein. An earlier report indicating that convicilin was a tetramer[54] was based on the relatively low

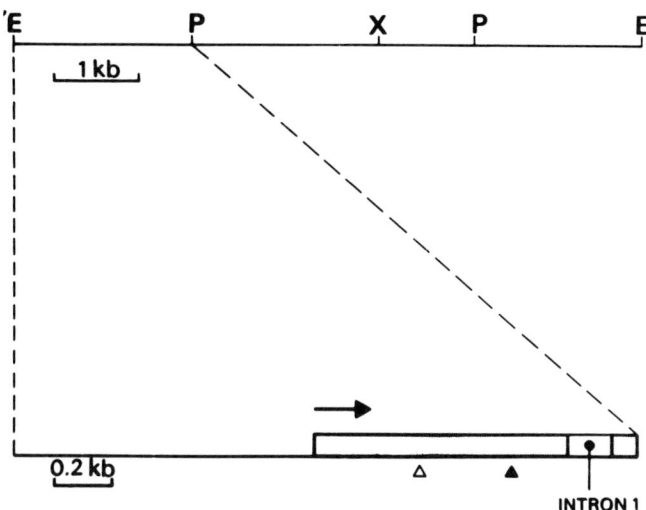

FIGURE 2. The restriction map of a 7.5 kb EcoRI fragment containing the gene for convicilin from pea. The upper line shows the full-102 bp length insert in p5.1. The lower line shows the 2.1-kb EcoRI to PstI fragment (pEN6), which was sequenced with the coding region shown by the heavy line. The open region shows the position of the first intron. An arrow indicates the direction of transcription. The open arrowhead shows the beginning of the convicilin cDNA clone, pPS15-28. The solid arrowhead marks the position at which convicilin and vicilin begin to show sequence similarity. The letters E, P and X refer to the restriction enzymes, EcoRI, PstI, and XbaI, respectively. (From Newbegin, E.J., de Lumen, B. O., Chandler, P. M. et al., *Planta*, 180, 461, 1990. With permission.)

resolving power of gel filtration chromatography. Subsequently the utilization of both sedimentation equilibrium centrifugation and chemical cross-linking indicated that convicilin has a trimeric structure.[56] The close serological relationship between vicilin and convicilin[51,54] was explained by the fact that convicilin represents vicilin with an N-terminal extension.[33,58] Such extended proteins are also found in other seeds, such as the α' of β-conglycinin of soybean[31] and α-globulins of cotton.[60] The entire nucleotide sequence of the convicilin gene was deduced by Newbegin et al.[56] The flanking regions of the convicilin gene showed the features common to RNA polymerase transcription units from plants and animals in the form of conserved "CAAT" and "TATA" boxes upstream of the coding region and repeated polyadenylation signals 3' to the end of translation (Figure 2). A single intron of 151 bp was found in the partially sequenced genomic clone at the corresponding position relative to the deduced amino acid sequence.[58,61-64]

The complete gene encoding convicilin (Figures 3 and 4) was also transferred to tobacco (*Nicotiana tabacum* L.), and the characteristics of its expression in the seeds of transgenic plants were studied.[56] An unprocessed polypeptide, which was found only in the seeds of the transgenic plants, was identical in size to pea convicilin (which does not undergo post-translational

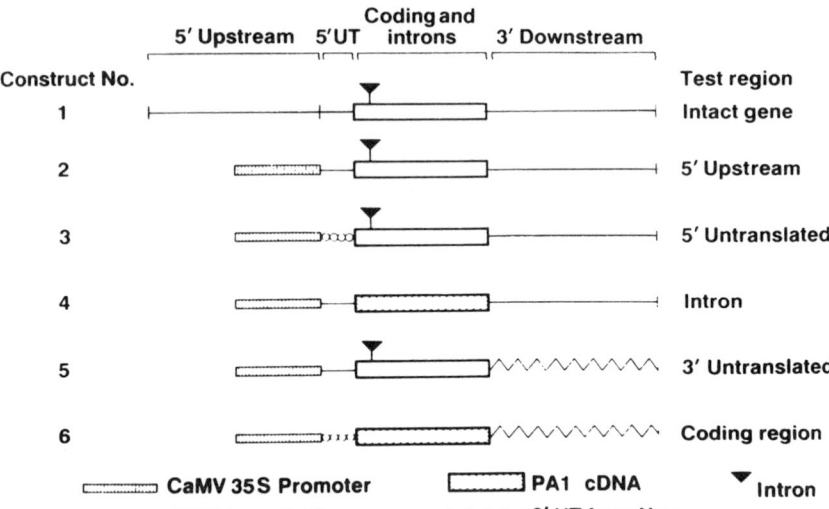

FIGURE 3. Diagrammatic representation of six genes designed to locate the sulfur-responsive element(s) in the PA1 gene. 5' UT from Rubisco refers to the sequence between the cap site and the first ATG of the gene for the small subunit of ribulose biphosphate carboxylase; 3' UT refers to the 3' flanking sequence from the nopaline synthetase gene from *Agrobacterium tumifaciens*. (From Spencer, D., Rerie, W. G., Randall, P. J., and Higgins, T. J. V., *Aust. J. Plant Physiol.*, 17, 355, 1990. With permssion.)

cleavage in peas), was partially processed to polypeptides of a relative molecular mass (M_r) of approx. 50,000 in transgenic tobacco seeds. There was a twofold variation in the level of convicilin accumulated by the mature seeds of a number of transgenic plants and this was well correlated with the number of gene copies incorporated into the different transformants. In the seeds of tobacco plants that contained a single copy of the transferred gene, it was estimated that convicilin comprised up to 2% of the seed protein.

The contents of sulfur-containing amino acids vary considerably in convicilin, vicilin, and lectin,[51] legumin,[29,37] and PA1 (pea albumin 1)[62] and PA2.[51] To explain this phenomenon Spencer et al.[42] proposed that all storage proteins would continue to be synthesized at a normal rate, even during sulfur-deficient conditions, but there are posttranslational changes such as that more sulfur-rich proteins are degraded faster to become available again for the synthesis of other essential proteins. However, there are exceptions to this recycling model and it was not supported by studies carried out in legumin synthesis.[35] In fact, reduced levels of legumin were found to be due to a reduced rate of synthesis and not due to an increased rate of degradation after synthesis. Subsequent studies by Spencer et al.[42] revealed that the effect is at the transcriptional level, because only trace amounts of mRNAs corre-

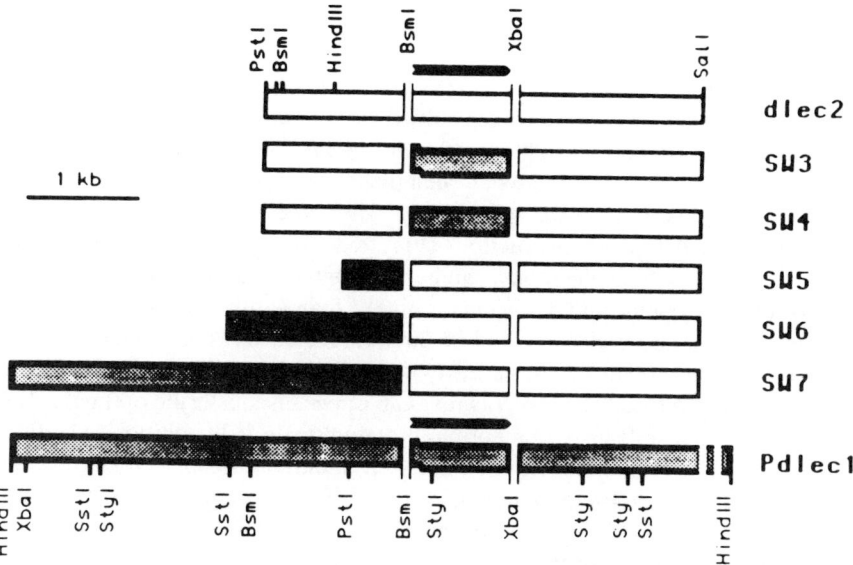

FIGURE 4. Scheme of the swap constructs. The top bar (open) represents the restriction enzyme map of a 3.4-kb *PstI/SalI* fragment carrying the PHA-L gene *dlec2* of *Phaseolus vulgaris* cv. Greensleeves. The extent of the coding sequence (816 bp, no introns) is shown by the rightward-pointing solid arrow-bar. The bottom bar (shaded) represents the major portion of the 7.5-kb *Hind*III fragment that carries the pseudogene *Pdlec*1 from *P. vulgaris* cv. Pinto. The gene carries a coding region (right-pointing arrow-bar) of 824 bp that has no introns and would be a PHA coding sequence if it were not interrupted by a single-base pair deletion in the 11th codon (shown by the break in the bar). The coding regions of *dlec2* and *Pdlec*1 are delimited by *Bsm*I and *Xba*I sites, which are shown as interruptions. *Bsm*I is located at −12 relative to the translation stop at the end of the coding region. The structure of all chimeric genes, SW3 to SW7, is depicted as assemblies of portions of the two parent genes. Repair of the coding sequence of *Pdlec*1, as shown in SW4, was achieved by swapping the otherwise identical regions between the *Bsm*I and *Sty*I restriction sites of the two Pinto PHA genes. (From Voelker, T.A., Moreno, J., and Chrispeels, M. J., *Plant Cell,* 2, 255, 1990. With permission of the American Society of Plant Physiologists.)

sponding to legumin and PA1 are synthesized under sulfur deficiency conditions. However, the possibility that the regulation is also affected at the translational and posttranslational levels cannot be ruled out.

According to the model proposed by Spencer et al.,[42] under sulfur deficient conditions there is a reduction in the pool of amino acids such as cysteine and methionine and the mRNA available for protein synthesis is not completely utilized because of this deficiency and degraded faster. There are several evidences against this model which have been discussed in detail by Spencer et al.[42] Further, this model is also not applicable to vicilin as the analysis of amino acid sequence and translational events revealed that this storage protein is synthesized from a precursor protein containing a signal peptide which is processed in the endoplasmic reticulum to yield a functional storage protein. The mechanisms for regulating the level of expression of such a sensitive gene are common in general and do not appear to be dependent on plant and legumin genes which are known to be expressed with great degree of fidelity into tobacco.[56,62,65] These genes were regulated in a similar way in tobacco as in pea. Under sulfur stress, very little legumin was accumulated in tobacco seeds.[65] In addition, the PA1 storage protein was found to accumulate in leaves, and when such plants were grown under sulfur stress the level of this protein was reduced by 90% in the leaves.[42] Spencer et al.[42] further modified gene constructions (Figure 3) to test these possibilities.

Legumin polypeptides have higher amounts of sulfur-containing amino acids (1.0% cysteine and 0.7% methionine)[66] than vicilin polypeptides, which have very few or no cysteine and methionine residues. Thus, a lack of sulfur shifts storage protein synthesis away from relatively sulfur-rich proteins to sulfur-poor proteins. The effect is reversible, because restoring optimal sulfur levels causes a shift back to a normal protein profile.[41] In addition to sulfate, compounds such as methionine, cysteine, glutathione, and mercaptoethanol also cause increased legumin synthesis in sulfur-deficient seeds. The mechanism by which sulfur availability modulates mRNA levels is unknown, but apparently the effect is not regulated by cysteinyl- and methionyl-tRNA levels, which are unchanged under sulfur-deficient conditions.

4. Phytohemagglutinin and Phaseolin

The seeds of common bean *Phaseolus vulgaris* L. contain two such proteins, phaseolin and phytohemagglutinin (PHA), which represent about 50% and 10% of the total seed proteins, respectively. While no catalytic activity has been so far detected for phaseolin, the ability of PHA to bind to carbohydrates and its strong erythro-agglutinating and lymphocyte-mitogenic activities has led to its classification as a lectin. The low content of methionine in phaseolin, the absence of sulfur amino acids in PHA and its antimetabolic properties greatly contribute to the lowering of the nutritional value of the bean seeds. Recent advances in genetic engineering techniques give rise to the possibility of the production of nutritionally improved bean varieties, by *in vitro* modification of the genes coding for both proteins. To this end, the organization and expression of these genes have been investigated in detail. The mechanisms of synthesis and processing of these proteins and, most important, processing events during intracellular transport, accumulation, and biological properties are also well known.

a. Synthesis and Glycosylation

Recent studies have elucidated the steps in the synthesis of phaseolin and PHA. Phaseolin is a glycoprotein with M_r of about 150,000 and sedimentation coefficient of 7.1S and belongs to the vicilin class of storage proteins. Early work has shown that it is synthesized by polysomes bound to the endoplasmic reticulum (ER) of the cotyledonary cells of developing bean seeds, and then accumulates in the protein bodies (PB).[67] During synthesis, a signal peptide is cleaved from the nascent polypeptide, then high-mannose oligosaccharide side chains are added co-translationally.[68] Analysis by 2D polyacrylamide gel electrophoresis (PAGE) of a mixture of unglycosylated and glycosylated phaseolin polypeptides of each of the electrophoretic variants studied showed that each unglycosylated polypeptide precursor corresponds to two glycosylated ones. Peptide mapping analysis of these glycosylated polypeptides demonstrated that the same polypeptide can acquire either one or two of such oligosaccharide side chains.[69,70] Since the phaseolin genes contain two putative glycosylation sites,[71] the different extents of glycosylation might be due to different accessibility for the processing enzymes at the site.[71,72] The finding that expression of a phaseolin cDNA gene in yeast yields two glycosylated polypeptides[73] favors this hypothesis.

As mentioned earlier, PHA is a glycoprotein with a M_r of about 115,000 and a sedimentation coefficient of 6.2S. In its native form it is a tetramer which contains, in all possible combinations, two different polypeptides, termed E (erythro-agglutinating) and L (lymphocyte-mitogen). Synthesis and accumulation of PHA in the developing seeds is synchronous with that of phaseolin. After cleavage of the signal peptide from the nascent chain, two high-mannose oligosaccharide side chains are co-translationally added to each subunit.[74,75] A small amount of subunit E can also acquire a third side chain. This is in agreement with the nucleotide sequence of a PHA-E subunit gene which indicates the existence of three putative glycosylation sites.[76-78] This site seems more accessible when PHA is synthesized and glycosylated *in vitro*.

b. Intracellular Transport and Processing

Newly synthesized and glycosylated phaseolin and PHA are transiently accumulated in the ER, as precursor forms, before their transport and accumulation in the PB,[69] and both radioactive proteins disappear from the ER with a half-life of 1.2 h.

Early work on cereal storage proteins indicated that PHA proteins are synthesized directly at their site of accumulation. In corn, polysomes actively engaged in zein synthesis are bound both to the membranes of the ER and to the membranes surrounding the PB.[79] Zein is also synthesized on the ER and then is probably transported directly to the PB, which are connected to the ER.[80] In legumes, however, no polysomes are bound to the PB membrane and, furthermore, no connection between the site of synthesis, the ER, and the PB has been found.

The finding that mature PHA contains oligosaccharide side chains of the modified type, containing fucose and xylose, helped to shed light on the process.[81-83] Fucose and xylose incorporation have been shown to be part of an extensive processing of one of the two carbohydrate side chains of PHA. This process, which occurs during passage of PHA through the Golgi apparatus and in the intracellular transport of seed storage proteins, has also been demonstrated for phaseolin and for the castor bean lectin.[84,85] The oligosaccharide side chains of phaseolin, however, do not incorporate fucose and xylose at appreciable levels during transport of the protein through the Golgi apparatus and maintain a high-mannose structure.

Further modification to mature forms of legume storage proteins and lectins occurs in the PB.[32] In bean PB this process takes about 24 hours for completion. Processing of phaseolin is confined to the protein moiety of the molecule. In fact, when glycosylation is inhibited *in vivo*, the two unglycosylated phaseolin polypeptides that accumulate in the PB have higher electrophoretic mobilities than their counterparts associated with the ER. On the contrary, in the case of PHA, the protein moiety is not involved in the final processing step.

The two oligosaccharide side chains of phaseolin and one of the side chains of PHA are not modified in the Golgi apparatus and remain of the high-mannose type, while the other side chain of PHA is extensively modified. Endoglycosidase H (endoH) is an enzyme which *in vitro* removes the high mannose chains from the polypeptide backbone of glycoproteins, leaving only the inner GlcNAc residue linked to asparagine; the enzyme is, however, not effective on modified chains. Phaseoline, when in its native form, is unaffected by endo H, while the side chains are readily hydrolyzed when the protein is denatured. However, one of the two high-mannose chains of the ER-associated PHA can be hydrolyzed by the enzyme, even when the protein is in the native form, whereas in the case of phaseolin both chains are hydrolyzed from the denatured PHA. The side chain is susceptible to endo H digestion, and the ER-associated precursor of PHA (which contains one modified and one high-mannose chain) is digested by endo H only when the protein is denatured. Thus the side chains which are not modified during intracellular transport of bean storage proteins cannot be digested with endo H *in vitro*, unless the quaternary structure of these glycoproteins is altered by denaturation.

c. Transport and Biological Activities

The function of processing events which storage proteins undergo during intracellular transport and accumulation is not clearly understood. There is direct evidence that unglycosylated PHA synthesized *in vivo* by developing cotyledons, treated with tunicamycin, does accumulate in the PB and that the rate of transport from the ER to the PB is not negatively affected by the lack of oligosaccharide side chains. In addition, unglycosylated PHA isolated from such cotyledons maintains erythroagglutinating and mitogenic activities at levels comparable to those of normal PHA.[68] Glycosylation of PHA therefore is not a prerequisite for correct intracellular transport and maintenance of biological activities. Unglycosylated pea vicilin, synthesized *in vivo* in the presence of tunicamycin, was found to be chased out of the ER with a half life comparable to that of the normally glycosylated form.[53] When the castor bean lectin reaches the PB, its protein moiety is processed to yield smaller subunits. This cleavage takes place also when the precursor is in its unglycoslylated form, indicating that it has been transported inside the PB.[54] Thus it appears that glycosylation is not a prerequisite for correct intracellular transport of legume storage proteins. It might be, however, possible, that glycosylation is important to preserve a particular conformation which makes the protein less susceptible to the hydrolases known to be present in the PB.[86] In some cases, proteins have been shown to increase susceptibility to proteases *in vivo*, when synthesized in an unglycosylated form.[87,88]

d. Phytohemagglutinin and Phaseolin Genes

There exist sequences homologous to PHA genes in cultivars which lack PHA. The level of hybridization suggests that the absence of PHA is not due to large deletions of PHA genes. The lack of mRNA homologous to PHA genes in developing cotyledons of this cultivar indicates that the mutation is at the transcriptional level. The restriction patterns of the DNA of PHA-deficient cultivars were identical, suggesting that they might have originated from a common ancestor. These cultivars also contain small amounts of a protein (which can be recognized by PHA antibodies) which is synthesized and processed like PHA, has an M_r similar to PHA-L and has mitogenic activity comparable to normal PHA of the TG2 type.[74] Whether this protein represents low levels of all PHA-L polypeptides, or normal levels of one of the PHA-L polypetides, is not known. Alternatively, this protein could be a PHA-related protein present in all bean cultivars, the presence of which is masked by the more abundant PHA. The complete absence of PHA-E polypeptides and the presence of these PHA-L, or PHA-like proteins suggest that lectin genes might be subjected to different regulatory mechanisms.

Several bean cultivars have been identified that are deficient in PHA accumulation.[89] For example, the Pinto cultivar contains very low levels of PHA-L (about 5% of the normal level) and no PHA-E.[90] In the *Pdlec2* gene, which encodes a PHA-1 polypeptide,[91,92] a quantitative element has been deleted from the promoter, resulting in a down-regulation of the mRNA and protein levels, compared with a normal cultivar such as Greensleeves. The *Pdlec1* gene that encodes a PHA-E polypeptide is a pseudogene that has a single-base pair deletion in codon 11. This deletion results in a translational shift into another frame leading into a stop at codon 53. Only a truncated, aberrant, 52-amino acid polypeptide could be sythesized instead of the 275-amino acid full-length PHA molecule. Besides this frameshift, the nucleotide sequence of *Pdlec1* is more than 90% identical to a normal PHA-E allele including the 5' promoter region and the poly(A) signal 3' of the coding frame. No other obvious lesions could be found in the pseudogene. With gene-specific probes, in Pinto cotyledons a 600-fold reduction in cytoplasmic poly(A) RNA for PHA-E, compared with the normal cultivar Greensleeves, was noticed.[13] According to this description, *Pdlec1* is comparable with the

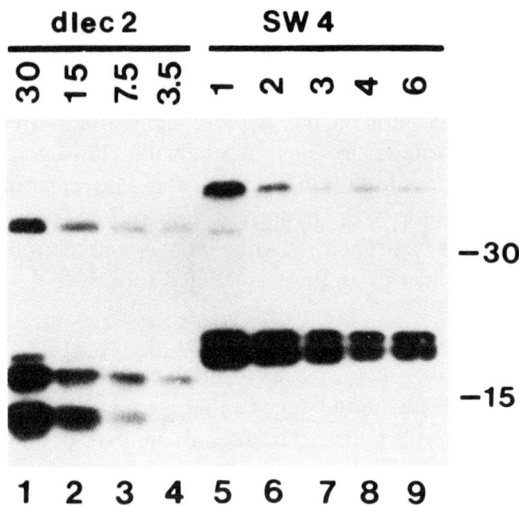

FIGURE 5. Immunoblot analysis of the SW4 transformants. Mature tobacco seeds were extracted with a low-salt, nondenaturing buffer and the proteins were separated by SDS-PAGE, electroblotted to nitrocellulose, and developed with rabbit PHA antiserum. Lanes 1 to 4: dilution series of seed extracts of *dlec*2-transformed tobacco plants; the numbers at the top indicate the micrograms of protein loaded per lane. Lanes 5 to 9: seed extracts of five independent plants transformed with the SW4 constructs (30 μg of protein per lane). Molecular size markers on the right are in kDa. (From Voelker, T.A., Moreno, J., and Chrispeels, M. J., *Plant Cell,* 2, 255, 1990. With permission of the American Society of Plant Physiologists.)

human γ-globulin pseudogene, which is nearly identical to its functional neighbor, but has its sixth codon mutated to a chain termination signal. In this case, after conversion to a functional frame, no transcripts could be detected in vivo, i.e., this pseudogene must carry additional regulatory mutations besides the nonsense mutation.[93] The frameshift mutation results in mRNA destabilization and leads to reduced accumulation levels, because the soybean trypsin inhibitor psuedogene greatly reduced levels of poly(A) mRNA *in vivo* accompanied by near-normal rates of transcription in isolated nuclei.

Voelker et al.[91] have shown that a PHA gene from the normal cultivar Greensleeves of *Phaseolis vulgaris* is expressed in a conserved manner in the seeds of transgenic tobacco (Figures 4 and 5). This made it possible to introduce the *Pdlec*1 pseudogene into this heterologous system. Voelker et al.[94] further found that its promoter functioned normally. The pair of PHA-E coding frames of the *Pdlec*1 pseudogene resulted in the accumulation of much higher levels of mRNA in the seeds of transgenic tobacco.

Further analysis of the transgenic seeds showed that the *Pdlec*1 promoter is fully functional (Figure 6). Voelker et al.[94] also repaired the *Pdlec*1 coding frame *in vitro* and inserted the repaired and unrepaired versions into a PHA gene expression cassette. In transgenic tobacco, both constructs showed *Pdlec*1

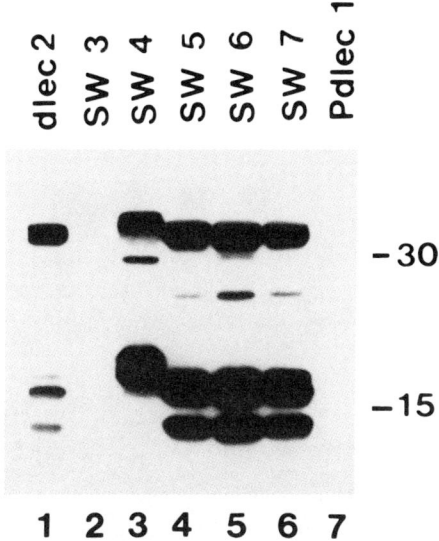

FIGURE 6. Immunoblot analysis of PHA accumulation in the seeds of transgenic tobacco. Proteins of fully mature seeds were extracted with a low-salt/Triton X-100-containing buffer and equal proportions of protein from four to five independently transformed plants mixed before SDS/PAGE (30 μg of total protein per sample). Molecular size markers on the right are in kDa. (From Voelker, T.A., Moreno, J., and Chrispeels, M. J., *Plant Cell*, 2, 255, 1990. With permission of the American Society of Plant Physiologists.)

transcript accumulation in the second half of seed maturation (Figure 7). The single-base frame repair boosted the peak transcript levels by a factor of 40 and resulted in the synthesis of PHA-E at normal levels. Voelker et al.[94] proposed that the premature translation stop caused by the frameshift leads to faster breakdown of the *Pdlec*1 mRNA, thereby preventing this transcript from accumulating to high levels.

Phaseolins display a type of regulation described for legumin of pea. Phaseolin genes exhibit peak expression when the cotyledons are 17 mm in length. Very little or no phaseolin protein or mRNA occur in vegetative organs.[5] Accumulation of storage proteins and their mRNAS is confined to the cotyledons and the embryonic axis (excluding the root meristem). Developmentally regulated expression of bean phaseolin in embryos of transgenic tobacco seed revealed the evolutionary conservation of regulatory and processing mechanisms for mRNA (transcription and intron splicing) and seed storage protein (transit signal cleavage and correct glycosylation). These findings have subsequently been confirmed for bean lectin and many other seed protein genes from legumes, *Brassica, Arabidopsis,* and sunflower in a variety of transgenic plants.[2]

Phaseolin 5'-flanking sequences (-795 to $+20$) are sufficient to confer correct spatial and temporal regulation upon a bacterial gene coding for GUS

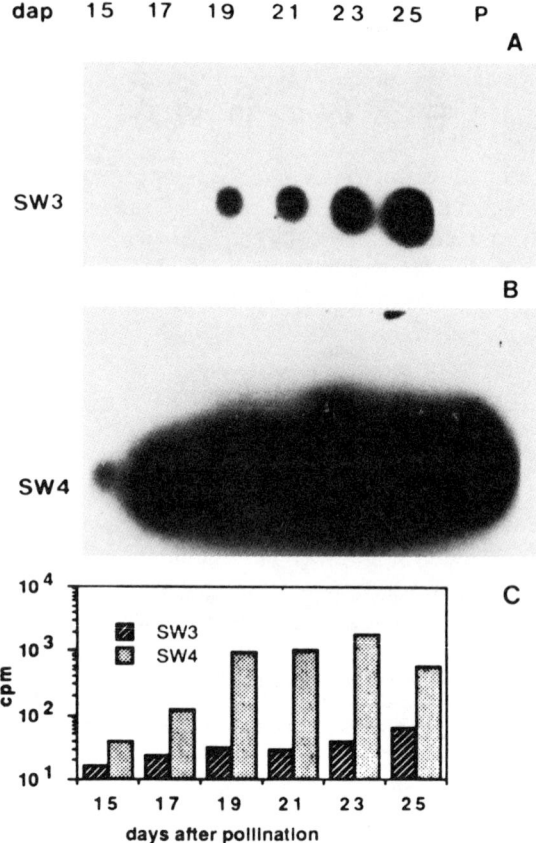

FIGURE 7. Quantitation of *Pdlec*1 mRNA levels in the developing seeds of transgenic tobacco. Total RNA was isolated from developing seeds at 2-day intervals after pollinating (DAP). Equals numbers of capsules from independently transformed plants from the same stage were pooled before extraction to integrate for plant-to-plant variation. For RNase protection, 10 µg of total RNA pools was used. The *Pdlec*1 mRNA was detected with a 700-base antisense RNA probe (1 µCi) derived from an *in vitro* runoff transcription with bacteriophage T7 RNA polymerase using ^{32}P-UTP at 200 Ci/mM. This probe is protected by the *Sty*I/*Xba*I 3' reading frame sequence of *Pdlec*1. After RNase digestion, the samples were electrophoresed in a 1% agarose-formaldehyde gel. All protected probes migrate as a 700-base piece in the same position as original probe (lane P). (A) SW3 pool RNA. (B) SW4 pool RNA. (C) Quantitation of the signals in (A) and (B). Appropriate pieces of the original gel were cut out, and the ^{32}P activity was measured. (Baseline, 20 cpm, obtained with RNA from seeds of untransformed from all values in the plot). (From Voelker, T.A., Moreno, J., and Chrispeels, M. J., *Plant Cell*, 2, 255, 1990. With permission of the American Society of Plant Physiologists.)

in transgenic tobacco plants. Similarly regulated expression of reporter genes has been demonstrated for upstream sequences from soybean β-conglycinin.[7] Therefore, although posttranscriptional regulation of phaseolin mRNA stability can affect the level of phaseolin accumulation,[12] primary control of the time and location of phaseolin expression is at the level of gene transcription.

The complexity of the phaseolin upstream region was indicated by the fact that an upstream A/T-rich sequence (-682 to 628) enhances GUS expression from a minimal CaMV 35S promoter in many vegetative tissues of tobacco, but when present in the context of the full-length phaseolin promoter, is silent in all tissues other than the developing embryo.[95] By means of gene transfer experiments in tobacco and transient gene expression assays in bean cotyledon cells, Bustos et al.[95] showed that the β-phaseolin upstream region (-795 to $+20$) has a modular organization of at least five regulatory domains that activate or decrease GUS expression *in vivo*.

In vitro mutagenesis was used to supplement the sulfur amino acid codon content of a gene encoding β-phaseolin, storage protein.[78] The number of methionine codons in the phaseolin gene was increased from three to nine by insertion of a 45-bp synthetic duplex. Although similar levels of phaseolin RNA were detected in seeds of plants transformed with either the normal or modified (hilmet) gene, the quantity of hilmet protein was consistently much lower than normal β-phaseolin. Hilmet phaseolin is expressed in a temporal- and organ-specific fashion, and is N-glycosylated and assembled into trimers in the manner of normal phaseolin.

Mutations affecting the spatial and temporal regulation of the β-phaseolin gene encoding the major storage protein of bean (*Phaseolus vulgaris*) were analyzed by stable and transient transformation approaches.[96-98] Spatial information is specified primarily by two upstream activating sequences (UAS). UAS1 (-295 to 109) was sufficient for seed-specific expression from both homologous and heterologous (CaMV 35S) promoters. *In situ* localization of GUS expression in tobacco embryos demonstrated that UAS1 activity was restricted to the cotyledons and shoot meristem. A second positive domain, UAS2 (-468 to -391), extended gene activity to the hypocotyl. Temporal control of GUS expression was found to involve two negative regulatory sequences, NRS1 (-391 to -295) and NRS (-518 to -418), as well as the positive domain UAS1. The deletion of either of the negative elements caused premature onset of GUS expression. These findings also provided the first example of the involvement of negative elements in the temporal control of gene expression in higher plants.

Sengupta-Gopalan et al.[5] introduced a gene encoding the β-subunit of the 7S protein of French bean into tobacco and immunologically detected phaseolin in seeds of regenerated plants, where it accounted for about 1% of the total seed protein. By immunocytochemical techniques, phaseolin was found in protein bodies in six of seven tobacco embryos but in only one of seven endosperms.[96] The phaseolin was glycosylated, but it was proteolytically cleaved differently than in bean seeds. Very little of the protein was found in transformed calli or seedlings, indicating that expression of the gene was both tissue specific and developmentally regulated. A single copy of the phaseolin gene was present in the transformed plants, and it was inherited as a simple Mendelian dominant trait.

5. Arcelin

Arcelin is a novel seed storage protein first found in several accessions of the wild Mexican bean *Phaseolus vulgaris* L. var. *aborigenecus*.[97,98] It is closely related to bean lectin, both at the amino acid and nucleotide level.[99] There are five arcelic variants of arcelin.[100] The Arc 1 protein migrates has two bands on polyacrylamide gels with M_r of 35,000 and 38,000. These two bands are glycosylation variants because a single band is seen following chemical deglycosylation.[101] The presence of arcelin correlates with reduced expression of phaseolin.[98] In these seeds, arcelin accounts for about 10% of the total seed protein.[93] It is interesting to note that arcelin has insecticidal properties for the bruchid beetles, a prominent bean pest.[102] The arcelin cDNA sequences have been worked out recently.[101,103,104]

B. CEREAL STORAGE PROTEINS

The major storage proteins in most cereals are alcohol-soluble prolamines. In general, as in *Triticum* (wheat), *Hordeum* (barley), *Secale* (rye), *Zea* (maize), *Sorghum* (sorghum or milo), and *Pennisetum* (millet), the prolamines account for 50 to 60% of the total endosperm protein. Rice and oats are unusual because their prolamine fractions account for only 5 to 10% of the endosperm protein, and most of the storage protein is similar to the 11S globulins.

Cereal prolamines are characterized by a high content of proline and glutamine (hence the name prolamine) and a low level of charged amino acids, especially lysine. The low lysine content is primarily responsible for the poor nutritional quality of high prolamine-containing cereals. The prolamines of most cereals are complex mixtures of polypeptides that occur in protein bodies. Some of these proteins can be cross-linked by disulfide bonds, thus forming high-molecular-weight complexes. Differences in the solubility and the potential of some to cross-link and associate with other proteins have made it difficult to isolate and characterize prolamine proteins by traditional protein chemistry. The analysis of recombinant DNA clones corresponding to prolamine mRNAs and genes, however, has allowed the deduction of the primary amino acid sequences of many prolamines.

NH_2-terminal sequence analysis of mixtures of prolamines, shows that while prolamines from a given species are complex, there is homology among them. Furthermore, there is conservation of prolamine NH_2-terminal protein sequences among members of the various cereal tribes. For example, prolamine sequences of wheat, rye, and barley are more closely related to one another than they are to the Oryzae (rice) and the Avenae (oats). The prolamines of the Andropogoneae (maize, *Tripsacum,* and *Sorghum*) are more closely related to one another than to those of the Paniceae (millet). The sequence relationships among this heterogeneous group of proteins have prompted several reviewers to suggest that prolamines originated through duplication and subsequent mutation of an ancestral prolamine gene, and this hypothesis has been largely supported by molecular analyses of storage protein genes.

1. Storage Proteins of Wheat and Barley

Wheat prolamine storage proteins are traditionally divided into gliadins and glutenins, which together form a gluten complex responsible for the breadmaking properties of wheat. Glutenin is believed to impart elasticity to dough, while gliadin makes it viscous and provides extensibility. Typically, wheat storage proteins consist of about 50% gliadin, 10% high-molecular-weight (HMW) glutenin, and 40% low-molecular-weight (LMW) glutenin subunits.

Traditionally, gliadins are distinguished as being soluble in 70% ethanol at room temperature, while glutenins are insoluble. Glutenins dissolve in dissociating media such as dilute acid, urea, or ionic detergents. The two prolamine groups are also classified according to their state of aggregation in dissociating media rather than solely on their solubility: the gliadins are a mixture of single polypeptides, while the glutenins are aggregated, primarily through disulfide bonds. Treatment of glutenin with a reducing agent, such as 2-mercaptoethanol, causes dissociation into two groups of polypeptides classified as low-molecular-mass (LMM) and high-molecular-mass (HMM) subunits. With this system of separation, about 10% of the gliadin (variously referred to as HMM gliadin, aggregated gliadin, LMM glutenin, and glutenin III) is transferred to the glutenin fraction.

Two-dimensional PAGE techniques separate gliadins and glutenins into subunits.[44] Most of the gliadin fraction consists of polypeptides of M_r 32,000 to 44,000 that are distinguished as α, β, ω, and γ-gliadins, based on decreasing mobility at acidic pI. The ω-gliadins have the lowest mobility at acidic pI and exhibit M_r of 50,000 to 70,000. The size and charge heterogeneity of gliadin proteins is variable among different wheat varieties, and this provides one method to differentiate between them.

Following reduction of disulfide bonds, the HMM glutenins can be resolved into polypeptides with M_r ranging from 80,000 to 150,000. Under similar conditions, the LMM glutenins migrate with M_r similar to the γ-gliadins. These proteins are optimally separated from the γ-gliadins by a combination of equilibrium and nonequilibrium IEF. The LMM glutenins can then be resolved into two sets of polypeptides with either acidic or basic pIs.

Miflin and co-workers[45] proposed an alternative method for the classification of wheat prolamines based not on aggregation, but rather on M_r and sulfur amino acid content. These investigators distinguished three groups of proteins called sulfur-rich (S-rich), sulfur-poor (S-poor), and high molecular-weight (HMW). The S-rich wheat prolamines, which contain 2.5 to 3.5% cysteine plus methionine, consist of the α-, β-, and γ-gliadins as well as the "aggregated gliadins". The S-poor wheat prolamines, containing 0 to 0.2% cysteine plus methionine, are composed of the ω-gliadins. In addition to having higher M_r than the S-rich prolamines, the S-poor proteins contain higher percentages of glutamine, proline, and phenylalanine. These three amino acids are present in a ratio of 4:3:1 and account for about 80% of the total. The HMW prolamines have a sulfur amino acid content that is intermediate between the S-rich and S-poor fractions (0.5 to 1.9%) and are cross-linked by disulfide bonds.

At least partial amino acid sequence information is available for many of the gliadin and glutenin proteins, and the complete amino acid sequence is known for examples of the α-, β-, and γ-gliadins[105] and several HMW glutenins.[106] The α-, β-, and γ-gliadins are structurally related and can be divided into several distinct domains or modules.[107] The gliadins that have been characterized to date are around 300 amino acids long. The mature protein is preceded by a signal peptide of approximately 20 amino acids that directs the synthesis of the protein into the lumen of the RER. Immediately following the signal sequence is a region of 90 to 100 amino acids that contains tandemly repeated peptides composed mainly of proline and glutamine. Because these repeats are tandem and slightly degenerate, different investigators have deduced somewhat different consensus repeats. Sumner-Smith et al.[108] recognized nine copies of a 12-amino acid repeat (Pro-Gln-Pro-Gln-Pro-Phe-Pro-Pro-Gln-Gln-Pro-Tyr) in an α, and β-type gliadin, although these authors differed in the assignment of the first amino acid in the repeat. A shorter 7- or 8-amino acid repeat (Pro-Gln-Gln-Pro-Phe-Pro-Leu/Gln-Gln) was determined in this region of the γ-gliadin.[109] The remainder of the protein is divided into four distinct regions. The first and third are segments of 25 to 30 residues that are mostly polyglutamine. The second and fourth are regions of 60 to 70 amino acids that are designated "unique". These regions are deficient in three of the four amino acids that constitute the repetitive peptide region (Phe, Gln, and Pro) and contain the majority of the Thr, Ala, Ser, Cys, Val, Arg, and Ile.

There also appears a somewhat different organization of the primary amino acid sequence of the S-rich wheat prolamines. Following the signal peptide, the protein is divided into a proline-rich and a proline-poor domain. The proline-rich region, which constitutes the first one-third of the molecule, corresponds to the region of tandemly repeated peptides. The proline-poor domain consists of seven regions that are designated I_1, A, I_2, B, I_3, C, and I_4. Regions A, B, and C are strongly conserved among all the S-rich wheat prolamines, while regions I_1 to I_4 are more variable. The I_1 region corresponds to the first polyglutamine-rich tract, and I_3 corresponds to most of the second polyglutamine tract.

The complete primary amino acid sequences of several HMW glutenin subunits have also been determined.[106] The length of the mature proteins vary from 504 to 817 amino acids, but each appears to be organized into three domains: an NH_2-terminal region of around 100 residues, a middle repetitive region of 400 to 670 residues, and a COOH-terminal region of around 40 residues. Two consensus peptides appear in the repetitive domain of these molecules. One is a hexapeptide (Pro-Gly-Gln-Gly-Gln-Gln) and one is a nonapeptide (Gly-Tyr-Tyr-Pro-Tyr-Ser-Leu-Gln-Gln). The repetitive peptides are interspersed such that copies of the nonapeptide are located between regions containing several tandemly arranged copies of the hexapeptide repeat. The nonrepetitive regions have some homology to the nonrepetitive regions of the S-rich gliadin genes. These proteins contain relatively low amounts of cysteine, four to six residues, which are clustered in the NH_2-terminal and COOH-terminal regions of the protein. Interestingly, several lysine residues are present in the NH_2-terminal and COOH-terminal regions.

Most of the information pertaining to the secondary and tertiary structures of wheat prolamines comes from circular dichroism (CD) measurements or from computer-generated models based on the conformation of model proteins whose structures are known from crystallographic analyses. However, there are limitations to interpretations of structural predictions with either approach. To analyze the CD spectrum, the proteins must be dissolved in alcohol solutions or dissociating media, and whether or not the polypeptides assume the same conformation in these solutions as they do in the cell is unknown. Computer-generated models are incorrect because most of the model proteins upon which the structural predictions are based are hydrophilic rather than hydrophobic. Presumably, predictions based on a combination of these procedures have a measure of validity.

Based on CD analysis predictions, the α-gliadins contain predominantly α-helix (35%) and β-turn (35%) with the remainder being β-sheet and random coil.[109] Most of the α-helical regions are presumed to be associated with polyglutamine stretches, while the turns are in the proline-rich regions: the repetitive domain and the COOH-terminal domain. The CD spectra of the β- and γ-gliadins are similar to the α-gliadins, suggesting that these proteins have similar conformations.

The CD analysis of ω-gliadins reveals little α-helix or β-sheet structure, with most of the molecule composed of β-turn structure.[110] A preponderance of β-turn structure is also deduced from CD analysis of the HMW subunits of glutenin, and it is thought to result from the many repeated peptides in the central region of the protein.[110] The NH_2-terminal and COOH-terminal regions, which contain most of the cysteine residues, are thought to be α-helical. It is proposed that cross-linking of cysteine residues in these regions causes the HMW glutenins to assemble into linear polymers,[110] which are responsible for the elastic nature of bread dough. The presence of multiple cysteines in the NH_2-terminal region may allow branching and cross-linking with the LMW subunits of glutenin.

Although complete amino acid sequences of the ω-gliadins are unknown, partial amino acid sequencing and comparison with similar proteins in barley suggest that these proteins contain a significant proportion of repeated peptides similar to the HMW glutenins.

The prolamines of barley, hordeins, are structurally related to the wheat prolamines. Hordeins are classified into four main groups of polypeptides designated A, B, C, and D, based on separation by SDS-PAGE. The A group is heterogeneous and does not show characteristics of prolamine proteins. The group B hordeins, which have M_r of 30,000 to 50,000 and account for 80 to 90% of the prolamine fraction, are structurally similar to the S-rich wheat prolamines. Like the gliadins, the region of the protein following the signal peptide is proline rich and contains a series of tandemly repeated peptides with the sequence Pro-Gln-Gln-Pro-Pro making up the core of the repeat.[106] Several variations of this repeat can be recognized, all consisting of the tetrapeptide Pro-Gln-Gln-Pro with the addition of Phe-Ile, Val, Pro, Phe-Pro, or Gln-Pro-Tyr. This region encompasses approximately the first one-fourth of the polypeptide. The remainder of the protein is relatively deficient in proline, has no repeated peptides, and contains most of the cysteine residues. Within the COOH-terminal portion of the protein are three regions, designated A, B, and C, that are highly conserved between the B hordeins and the α-, β-, and γ-gliadins. The extent of similarity between these regions clearly indicates a common origin for the proteins.

Only partial amino acid sequences are available for the C hordeins,[106] but these proteins appear to be structurally related to the S-poor wheat prolamines, the ω-gliadins. The C hordeins, which have M_r of 60,000 to 80,000 and comprise 10 to 20% of the prolamine fraction, contain a repeated peptide (Pro-Gln-Gln-Pro-Phe-Pro-Gln-Gln) that is thought to account for most of the protein sequence. The first four amino acids of this repeat are identical with the repeated peptides in the NH_2-terminal portion of the B hordein, and it may be that they have a common origin.

The D hordein accounts for about 5% of the barley prolamine fraction and consists of a few polypeptides with apparent M_r of 105,000. The amino acid sequences of the D hordeins have not been determined, but they are thought to be related to the HMW subunits of wheat glutenin. Barley flour does not have the same rheological qualities for bread-making as wheat flour, so there may be some structural differences between these proteins.

2. Rice Storage Proteins

Glutelin is the major storage protein of rice endosperm, comprising up to 80% of the rice seed protein.[111,112] Despite the relative insolubility characteristics of these proteins, elucidation of their primary sequences has revealed that they are homologous to the 11S globulin proteins, typically accumulated in embryonic seed tissues of dicotyledonous plants.[109,113,114] Glutelins are encoded by a multigene family that contains at least three distinct subfamilies, each comprised of approximately five to eight gene copies.[109] Genomic clones representing each of these three subfamilies (designated Gt1, Gt2, and Gt3) have been isolated.[109] Gt1 and Gt2 share 95% nucleotide sequence homology in their coding sequences, whereas the coding regions of Gt3 share only about 83% identity with those of Gt1 and Gt2. The close relationship between Gt1 and Gt2 is also evident in the 5' and 3' flanking regions and in the introns.

Inspection of the 5' flanking sequences reveals that all three genes are relatively conserved from -125 to -1 bp from the translational start.[109] This region contains the transcriptional initiation site as well as the ubiquitous TATA and CCAAT boxes. Sequences upstream to -900 bp are highly conserved in Gt1 and Gt2 except for the presence of several insertions and deletions of DNA segments. The promoter sequence of Gt3, however, diverges significantly from Gt1 and Gt2 upstream of position -186.[109] The significant divergence between the upstream promoter regions of Gt3 compared to Gt1 and Gt2 suggests that the Gt3 subfamily may be regulated differently from the other glutelin genes.

In order to gain an understanding of the mechanism by which the genes are expressed, Leisy et al.[115] have transformed tobacco with Gt3 promoter-CAT chimeric constructs (Figure 8) and have monitored the expression of CAT activity directed by the Gt3 promoter in these plants. Their results indicate that these constructs direct the highest levels of CAT enzyme activity in seeds (Figure 9), with maximal levels occurring about 16 days after flowering (DAF). In some, but not all, plants, lower but significant levels of CAT enzyme activity were also detected in certain nonseed tissues as well, indicating that the expression of this monocot-derived gene is not always strictly controlled in a heterologous tobacco host.

3. Maize Storage Proteins
a. Maize Prolamines

The endosperm of maize (*Zea mays*), in common with that of other cereals, is a tissue primarily devoted to the accumulation of starch and proteins which later provide the nitrogen and energy sources for the germinating seedling. In this tissue, between days 15 and 40 after fertilization, the major storage proteins, the zeins, are synthesized by membrane-bound polyribosomes and transported into the lumen of the endoplasmic reticulum where they are assembled into protein bodies. At maturity about 55% of the total protein content of the tissue consists of zeins.

b. Molecular Organization and Structure

The zeins (also called maize prolamines) are composed of several proteins that differ in alcohol solubility, depending on the presence of reducing agents. Various investigators have used slightly different procedures for isolating zeins, leading to a somewhat complicated nomenclature.[116] The total zein

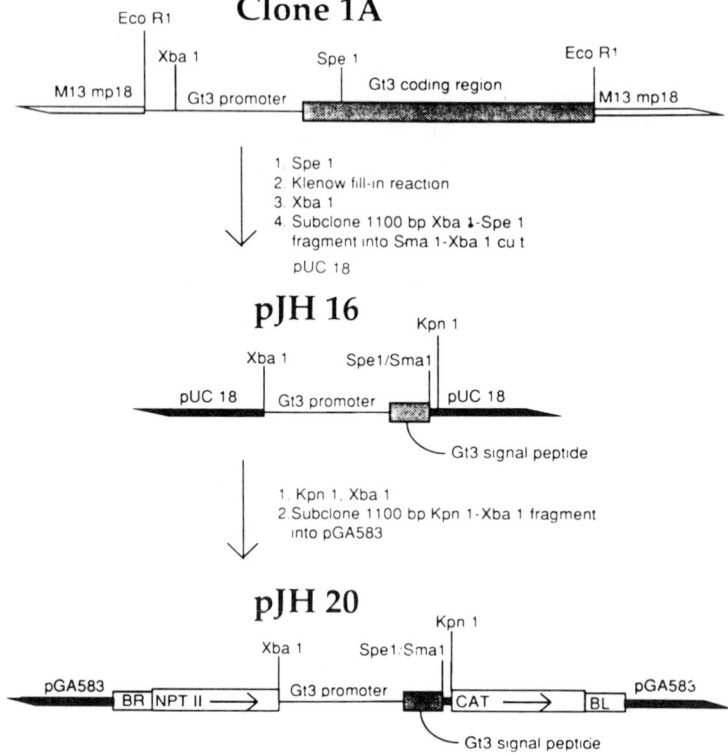

FIGURE 8. Cloning of the 5' flanking sequence of the Gt3 gene from the M13mp18 clone 1A into pGA583 to form pJH20. (From Leisy, D.J., Hnilo, J., Zhiao, Y., and Okita, W., *Plant Mol. Biol.*, 14, 41, 1989. With permission.)

fraction can be extracted from maize endosperm in 70% ethanol or 55% 2-propanol containing 1% 2-mercaptoethanol. Separation by SDS-PAGE resolves components with apparent M_r 27,000, 22,000, 19,000, 15,000, 14,000, and 10,000. Estimates of molecular weight for these groups of polypeptides vary, but the patterns of protein separation are very similar. Esen[116] fractionated zeins into groups designated as α, β, and γ on the basis of their solubility in 2-propanol in the presence or absence of reducing agent. The proteins within these groups are structurally distinct, so this appears to be a valid basis on which to distinguish them.

The α-zeins, which typically account for about 70% of the total zein fraction, are composed of four or five polypeptides with apparent M_r ranging from 19,000 to 24,000. Each of these can be further separated into differently charged species by IEF. The IEF heterogeneity is genotype specific and is inherited codominantly in a simple Mendelian fashion.[117] The heterogeneity has been useful in distinguishing among various maize genotypes.

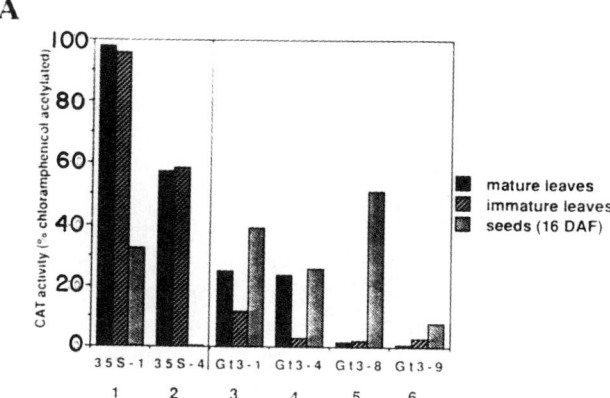
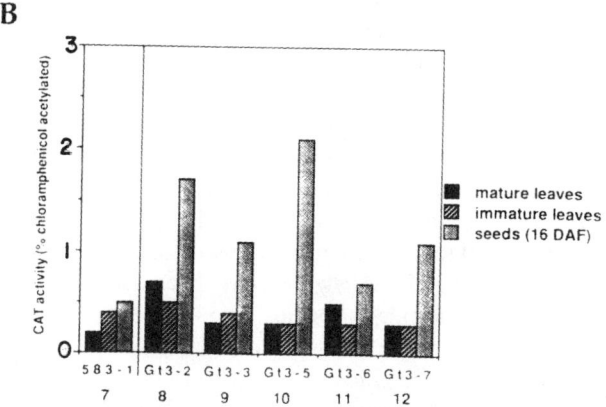

FIGURE 9. Expression of chimeric Gt3-promotor-CAT constructs in leaves and seeds of transgenic tobacco. Each CAT enzyme assay contained extract equivalent of 30 μg of total protein. Mature leaves were taken from the central part of the plant and were approximately 20 to 25 cm in length. Immature leaves were taken from near the top of the plant and were approximately 5 to 7 cm in length. Only the most apical portions of the leaves were used for making extracts. Seeds were harvested at 16 days after flowering. Panel (A): data from plants exhibiting high levels of CAT enzyme activity in seed extracts. Panel (B): data from plants exhibiting low levels of CAT enzyme activity in seed extracts. Plants were transformed with CaMV 35S promoter-CAT constructs (lanes 1 and 2), a promoterless CAT construct (pGA583, lane 7), and a Gt3 promoter-CAT construct (lanes 3 to 6 and 8 to 12). (From Leisy, D.J., Hnilo, J., Zhiao, Y., and Okita, W., *Plant Mol. Biol.*, 14, 41, 1989. With permission.)

The α-zeins are primarily preproteins with an NH_2-terminal signal peptide of 20 or 21 amino acids.[118] The length of the mature protein is variable and ranges from 210 to 245 amino acids. All of the zeins have high contents of glutamine (25%), leucine (20%), alanine (15%), and proline (11%), and none has been identified that contains lysine. A distinguishing feature of the α-zeins is the presence of tandemly repeated peptides of approximately 20 amino acids in the central region of the protein.[118] These repeats vary slightly in length but can be averaged to obtain a consensus-repeated peptide sequence. On the basis of CD analysis and computer-generated structural models,[119] it was proposed that the repeated peptides are α-helices that interact through hydrogen bonding of polar amino acids on the surface of the repeats, as well as between glutamines at the ends of the repeats. This may contribute to aggregation of α-zeins within protein bodies.[119]

The β-zeins are proteins of M_r 14,000 to 16,000 that account for around 15% of the zein fraction. These polypeptides show less charge heterogeneity than the α-zeins on IEF analysis,[120] with each component composed of only one or two polypeptides. Amino acid sequence analysis of an M_r 15,000 zein showed that the mature protein following the signal peptide contains 160 amino acids. This protein has less glutamine (16%), leucine (10%), and proline (9%) than the α-zeins but contains significantly more of the sulfur amino acids, methionine (11%) and cysteine (4%).[116] Unlike the α-zeins, the β-zeins contain no repetitive peptides. Based on CD and computer-generated structural analysis, the proteins have very little α-helical structure and are mostly composed of strand and turn structure.

The γ-zein also referred to as the "reduced-soluble protein", alcohol-soluble reduced glutelin or glutelin-2,[121] generally accounts for 20% of the total zein,[116] although it can make up as much as 50% of this fraction. In addition to being soluble in alcohol solutions, the γ-zein is soluble in saline solution, and this has led to its extraction in several different protein fractions. Although charge variants of the γ-zein have been isolated,[116] the protein appears to be primarily a single species.[121] Prat et al.[121] recognized five distinct, consecutive regions in the 180-amino acid protein sequence: an NH$_2$-terminal segment of 11 amino acids, a repetitive peptide region composed of 8 tandem copies of the hexapeptide, an alternating Pro-X sequence between residues 70 and 91, and a COOH-terminal part rich in Gln that can be divided into two parts. The fourth region, comprising residues 92–148, is cysteine-rich and has some internal homology: the heptapeptide is repeated twice and the related sequence Gln-Cys-Gln-Ser-Leu-Arg also appears once in this region. The fifth region, beginning with residue 149, has no apparent internal homology. While the γ-zein is structurally distinct from both the α- and β-zeins, it does have two regions of internal homology to the M_r 15,000 β-zeins: there are matches of 10 residues each between amino acids 34 to 49 of the β-zein and 112 to 125 of the γ-zein, and between amino acids 61 to 73 of the β-zein and 130 to 141 of the γ-zein. These regions of homology probably account for the small degree of cross-reactivity between the M_r 15,000 β-zein and antisera directed against the γ-zein. Little is known about the structure of the γ-zein. The fact that it migrates more slowly on SDS-PAGE than the larger α-zeins may be a reflection of an unusual secondary structure associated with its high proline content (25%). Another fraction of M_r 10,000 zein protein contains a high proportion of sulfur amino acids and has an amino acid composition resembling that of the M_r 15,000 β-zein. However, it differs somewhat in alcohol solubility from the β-zeins[116] and may represent a structurally distinct fourth type of maize storage protein.

The study of the regulation of the gene expression in the developing maize endosperm is facilitated by the existence of viable mutants, many of which affect storage protein synthesis. Genetic analysis suggests that some of these mutants may correspond to regulatory genes, which control the timing and level of zein expression. Salamani and his colleagues[122] mapped loci controlling zein synthesis to several maize chromosomes. The genes controlling a number of the α-zeins are on chromosomes 4, 7, and 10. Seven genes corresponding to zeins of M_r 19,000 and 20,000 have been mapped to the short arm of chromosome 7 and linked with two regulatory loci, opaque-2 and DeB-30, that exert qualitative and quantitative effects on zein synthesis. A set of nine genes corresponding to M_r 20,000 and 22,000 zeins was mapped to the short and long arms of chromosome 4, and associated with these is the regulatory locus *floury*-2. A third locus corresponding to an M_r 22,000 zein is on the long arm of chromosome 10, and associated with it is the regulatory locus *opaque*-7. Although the gene or genes encoding the γ-zein have not been mapped, a gene encoding an M_r 15,000 β-zein has recently been mapped to the long arm of chromosome 6.[28] The gene corresponding to the M_r 10,000 zein protein has been mapped to the short arm of chromosome 7.

The complexity among zein proteins detected by IEF analysis has also been demonstrated among the genes that encode these proteins. There are multiple families of cDNA clones corresponding to zein mRNAs that can be distinguished from one another by different hybridization criteria.[120] The heterogeneity among these sequences results from variation in the number of elements encoding repetitive peptide and from high frequencies of base substitutions, insertions, and deletions. The cDNA clones hybridize to multiple coding sequences in the genome, leading to estimates of between 75 and 150 genes;[123] however, there appear to be only one or two copies of genes encoding the α-, β-, and γ-zeins.

Although little is known about the molecular organization of the zein genes, there is evidence of close linkage among some of the zein genes. Several genomic clones have been isolated that contain multiple genes spaced within a few kb.[120] The genes are tandemly arranged 5' to 3', and the DNA sequences flanking the coding regions are conserved. Analysis of one gene family of zeins suggested that the region of homology extends from about 1 kb on the 5' end to 0.3 kb on the 3' end.[124] Outside these conserved regions the DNA sequences are moderately to highly repetitive. The conserved regions flanking the genes contain eukaryotic promoters and polyadenylation sequences. The coding sequences of zein genes contain no introns. Several genomic clones have been isolated in which the coding sequences have altered initiation codons or premature stop codons.[120] These are thought to be pseudogenes, but it has been suggested that transcripts from genes with premature stop codons may be translated.

The zeins appear to be encoded by at least four major gene families. Two families are reponsible for the synthesis of the 19–20-kDa and 22-kDa zeins to which 15 to 22 polypeptides belong.[123] In wild-type endosperm, expression of zein genes is under the control of several regulatory genes.[122] One of these, as mentioned earlier, is the Opaque-2 (O2) locus, located on chromosome 7, which exerts its major effect on the level of the 22-kDa zein polypeptide for which there are several structral genes located on chromosome 4.[122] In homozygous O2 mutants this class of zeins is greatly reduced,[122] whereas the levels of other zein classes are only weakly affected. The effect of the mutation is reduction of the overall zein content to a level of 50 to 70% of wild type, depending on the genetic background in which the mutation is present.[122] The O2 mutation results in a reduction in the level of 22-kDa zein transcripts as shown by *in vitro* nuclear run-on studies and is therefore likely to be involved in transcriptional activation.

The O2 mutant also lacks a number of non-zein polypeptides present in the wild-type endosperm, the most abundant of which is a 32-kDa albumin termed b-32. The b-32 protein is linked to and may be regulated or encoded by the Opaque-6 (O6) locus.[122] It is also known that the expression of b-32 during seed development is temporally and quantitatively coordinated with the deposition of storage proteins, and that in the O2 mutant, the level of b-32 mRNA is reduced to <5% of wild-type. The gene encoding the b-32 protein has been cloned and the complete amino acid squence of this protein has been derived.[125] The function of b-32 is currently undefined. However, a recent search for homology to available protein sequences revealed a region of strong homology (33.5%) in a 173-amino acid long region. While it remains to be unequivocally demonstrated that the b-32 protein is the product of the O6 gene, it is nevertheless clear that the activation of the b-32 gene requires an O2-encoded function.

The O2 gene has been cloned by using a transposon tagging strategy with the help of the mobile Spm[126] and Ac.[127] Translation of the corresponding cDNA results in the production of a 48-kDa protein, which possesses characteristic features in common with the "leucine zipper" class of trans-acting factors. Members of this class include the proteins c-jun, c-fos, C/EBP, GCN-4, and CREB.[128] O2 shares with these trancription factors the presence of a region rich in basic amino acids immediately preceding the leucine repeat.

c. Transgenic Plants

Reports concerning the activity in transgenic plants of genes encoding the maize endosperm protein zein have started to appear. Matzke et al.[129] reported the first evidence for transcription of a monocot gene in a dicot host plant. A genomic clone for the maize storage protein zein, introduced along with T-DNA into sunflower cells, is transcribed from its own promoter. The zein mRNA made in sunflower can be translated *in vitro* to yield an immunoprecipitable protein. However, they have been unable to detect the presence of the zein protein in the engineered sunflower tissue. Similarly, several other workers, have transferred genes encoding 15-, 19-, and 23-kDa zein genes into sunflower callus tissue and, although they were transcribed, zein proteins could not be detected.[129] Zein proteins have also been synthesized in the giant green alga *Acetabularia mediterranea* following microinjection of the corresponding genes into the nuclei of *Acetabularia* cells.[130] A 15-kDa zein protein has been shown to be synthesized and accumulated in transgenic tobacco seeds. However, this gene, under the control of the β-phaseolin promoter, was also shown to be active in tobacco seeds.[5]

In petunia plants transformed with a zein gene (Z19ab1) encoding a 19-kDa protein, expression was not tissue specific and low levels of transcripts were observed in both leaves and seeds.[130] With respect to the activity of zein gene promoters in transgenic plants, a recent report indicated that the expression of a zein gene encoding a 19-kDa zein protein was not developmentally regulated in transgenic petunia: zein RNA was found not only in seeds but also in other organs; zein protein could not be detected in any organ. In contrast, Jefferson et al.[131] have obtained results using a sensitive reporter gene, GUS, which demonstrates the tissue (endosperm)-specific activity of a zein gene (Z4) promoter in transgenic tobacco plants.

Transgenic tobacco plants containing a maize gene (Z4) were obtained, using a modified Ti plasmid vector.[132] Although a polyadenylated transcript homologous to the Z4 gene was present in the seeds of some of these transgenic plants, zein protein could not be detected in any of the plants tested (35 total). To simplify the analysis of the tissue specificity of the Z4 promoter ($Z4_{pro}$) in different organs of transformed tobacco plants, additional transgenic plants containing the chimeric genes $Z4_{pro}$-CAT and $Z4_{pro}$-GUS were produced. Very weak seed-specific CAT activity was observed in one out of ten $Z4_{pro}$-CAT-transformed plants. When the more sensitive GUS assay system was used to evaluate $Z4_{pro}$ activity in tobacco, it could be shown in all 11 transgenic tobacco seeds. In addition, when the protein-coding region of the Z4 gene and a related zein cDNA clone were placed under the control of the 35S promoter from cauliflower mosaic virus (CaMV), the corresponding zein proteins were synthesized and detected in leaves, roots and endosperm tissue, but not embryos, of seeds from transgenic tobacco plants.[132]

Although the Z4 gene encodes a 23 kDa protein, the promoter region is almost identical to that of the Z19ab1.[133,134] Promoters of other monocot genes which are normally expressed in endosperm have also maintained the same tissue specificity in transgenic tobacco plants.[8,9] In a number of cases, therefore, tobacco has proven to be a suitable heterogeneous system for investigating the upstream sequences required for tissue-specific activity of monocot seed protein genes.

A zein gene (Z4) promoter containing 886 bp upstream from the transcription start site has been shown previously to be active specifically in the endosperm of transgenic tobacco seeds.[135] To investigate the region required

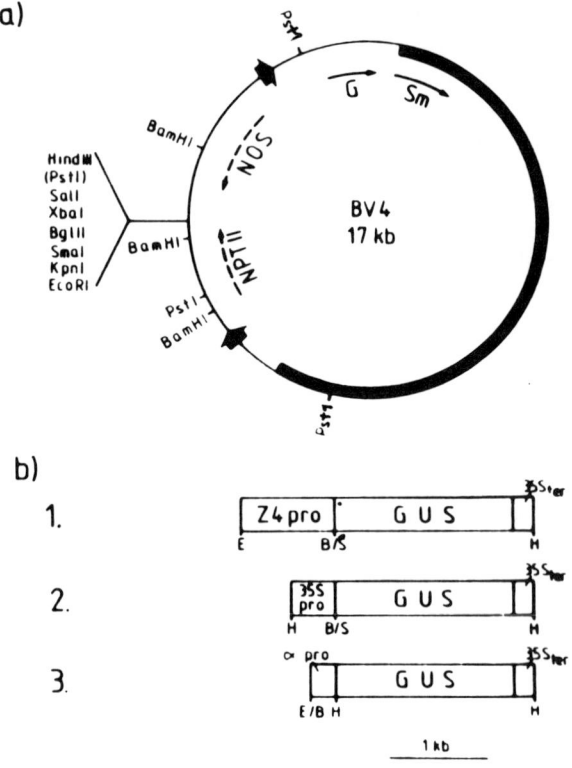

FIGURE 10. Various promoters used to drive GUS expression in transgenic tobacco plants. Chimeric GUS genes under the control of the 866 bp $Z4_{pro}$, as well as the 420-bp promoter of the 35S transcript of cauliflower mosaic virus ($35S_{pro}$) and the 257-bp α'_{pro} (b) were inserted into the polylinker of the binary vector, BV4 (a). Only the restriction enzyme sites necessary for understanding the constructions are shown. H, *Hind*II; S, *Sma*I. (From Matzke, A. J. M., Stoger, E. M., Schernthaner, J. P., and Matzke, M. A., *Plant Mol. Biol.*, 14, 323, 1990. With permission.)

for this tissue-specific activity, deletions of the Z4 promoter were constructed and placed upstream of the GUS reporter gene (Figure 10). When these deletions were tested in transgenic tobacco plants, seed-specific GUS activity, which reached a peak between 15 and 19 DAP, was observed for promoters extending from -886 to -174 (Table 3). Interestingly, the 174-bp promoter lacked the complete 15-bp consensus sequence found in the same position in all zein genes so far sequenced. With the next shorter promoter in the deletion series (79 bp), which just included the CAAT and TATA elements, negligible GUS activity was observed in seeds. The results demonstrated that 174 bp upstream of the transcription start site are sufficient for tissue-specific and temporally regulated activity of the Z4 promoter in tobacco.

TABLE 3
Time Course of GUS Activity in Transgenic Tobacco Seeds

Promotor	\multicolumn{9}{c}{Days after pollination (DAP)}
	8–9	10–11	12–13	14–15	16–17	18–19	20–21	22–23	24–25	Ripe
ZA_{pro}										
886 bp [23]		420	730	2,500	2,300	610	180	22	7	4
591 bp	48	125		118	185	240	182	50	31	5
491 bp	15		29	57	124	205	159	108		10
246 bp	5		82	120	140	220		40	20	6
174 bp		50	80	600	1,400	1,250	820	78	50	5
79 bp	69		57		22	41	20		2	
Normal tobacco	12		15	12	9	9	6	3	2	2
α'_{pro}		700	1,500	3,900	16,100	37,000	40,000	43,000		112,000
$35S_{pro}$	8,000		47,000	107,000	111,000	86,000	68,000	41,000	21,000	9,000

Note: Activities are given in pmol 4-methyl umbelliferone (MCS) per mg protein per min. The plants shown are those with the highest activity obtained with each construction. The peak of activity is underlined.

From, Matzke, A. J. M., Stoger, E. M., Schernthaner, J. P., and Matzke, M. A., *Plant Mol. Biol.*, 14, 323, 1990. With permission.

A maize genomic clone containing a Z4 gene was inserted into the T-DNA of the Ti plasmid pTiT37. An *Agrobacterium tumefaciens* strain harboring this modified Ti plasmid was used to infect stem sections of young plants or explants of dicotyledonous *Solanum nigrum*.[136] Axenic transformed calli active in nopaline synthesis were obtained and transgenic plants were differentiated from them. DNA Southern hybridization and RNA dot-hybridization analyses showed that the zein gene is really transferred integrated and transcribed into mRNA in the transformed calli and shoots. But the presence of the zein protein cannot be detected in either the transformed calli or the transgenic shoots.

III. PRODUCTION OF SMALL PEPTIDES OF 2S ALBUMIN

With a few exceptions, most seed storage proteins seem to have no enzymatic function, making them promising targets for modification, either for composition or for inserting sequences enclosing bioactive peptides. The production of small peptides in 2S albumins is one such example.

The 2S albumins are among the smallest seed storage proteins, consisting of two subunits of approximately 9 and 3 kDa linked by disulfide bridges.[137] As with other classes of seed proteins, comparisons of the peptide sequences of 2S albumins revealed both highly conserved and variable regions. It is interesting to note that the number and position of eight cysteine residues is highly conserved.[138] One intercysteine region showed variation in both length and sequence between species. 2S albumin seed storage proteins undergo a complex series of posttranslational proteolytic cleavages. In order to determine if this process is correctly carried out in transgenic plants, the gene AT2S1 encoding an *Arabidopsis thaliana* 2S albumin has been expressed in transgenic tobacco.[139] Initial experiments using a reporter gene demonstrated that the AT2S1 promoter directs seed specific expression in both transgenic tobacco and *Brassica napus* plants. The entire AT2S1 gene was then transferred into tobacco plants, where it showed a tissue-specific and developmentally regulated expression. *Arabidopsis* 2S albumin accumulates up to 0.1% of the total high-salt extractable seed protein. Vondekerckhove et al.[140] modified the portion of an *Arabidopsis thaliana* 2S albumin gene encoding this variable region by subsituting part of it for sequences encoding the five amino acid neuropeptide leu-enkephalin and flanking it with codons containing tryptic cleavage sites. The modified gene was transformed into both *Arabidopsis* and *Brassica napus* (oil seed rape), and 2S albumins from the seeds of transformed plants were isolated; due to their smaller size and solubility in low-salt solutions, purification of this class of storage proteins is particularly straightforward. After tryptic cleavage, the peptide, still with an extra lysine residue on the carboxyl terminus, was isolated using reversed-phase HPLC. The extra lysine was subsequently removed using carboxypeptidase B.

The yields from this first example were extremely promising. Even in the diminutive *Arabidopsis,* 206 nmol peptide (113 μg peptide per gram of seed) were obtained, suggesting that a single greenhouse filled with such plants could, for some peptides, produce enough peptide to meet demand. Whether this would be more economical than fermentation techniques is not yet clear. In *Brassica,* a widely grown crop plant producing 3000 kg of seed per hectare, 50 nmol peptide per gram of seed were reported, and plants with significantly higher yields have since been generated. This would make the isolation of tens of hundreds of grams per hectare possible, depending on the size of the peptide. A variety of questions must be answered first, however, before such production is possible for commercially interesting peptides.

IV. CONCLUSIONS AND FUTURE PROSPECTS

The seeds of higher plants contain large quantities of storage proteins which serve as a source of nutrition for germinating seedlings. The proteins found in the seeds of different species have been classified on the basis of size and their solubility in different solvents. Proteins of the same class are highly conserved among species. Some, but not all, are glycosylated *in vivo.* Some proteins may represent 60% or more of the total seed protein. Seeds are natural storage organs and can be stored for a long periods of time. The storage proteins are aggregated in specialized vacuoles called protein bodies. With few exceptions, most seed storage proteins seem to have no enzymatic function, making them important targets for modification, either for altering the overall amino acid composition or for inserting sequences encoding bioactive peptides.

Two basic types of storage proteins can be distinguished in seeds: the globulins, which are found in the embryos of most seeds, and the prolamines, which occur in the endosperms of many cereals. It would not be surprising if they evolved from common ancestral genes, but as they now exist there are significant structural differences between these proteins and between the genes that encode them. Although both globulins and prolamines have a high content of amide amino acids, their primary and secondary structures are quite different. Subunits of storage globulins are spherical and composed of three domains. Insertions and deletions of short peptides account for most of the structural variation among globulin storage proteins of different species. Although less is known about the tertiary and quaternary structure of prolamine proteins, the best-studied examples have more extended and rod-like conformations. Duplication of short peptides appears to have occurred in the evolution of prolamine proteins. The segmented nature of some prolamines, e.g., those of wheat and barley, suggests they may have evolved from parts of other proteins. There is insufficient evidence to evaluate the role that may have been played in the evolution of these proteins by intron/exon structures. All of the globulin genes examined thus far contain introns, while the prolamine genes completely lack introns. The size of introns in the 7S and 11S globulin genes in legumes is variable, but their number and organization are generally conserved.

The mechanisms by which storage proteins are synthesized and deposited in cells of embryo or endosperm tissue are conserved among seed plants. In all cases they are made as preproteins that are transported through the endomembrane system; they become concentrated and associate into aggregates called protein bodies. In maize, sorghum, and rice, aggregation occurs directly within the lumen of the RER, but for the prolamines of other cereals and the globulins of most dicots, transport continues through the ER and Golgi apparatus to the vacuole. During their transport, globulin storage proteins are subject to proteolytic processing. Cleavage of 11S globulins occurs between the acidic and basic parts of subunit polypeptides and may play a role in the association of 7–8S complexes into 11S complexes. It is less clear what function proteolytic digestion plays in the processing and assembly of 7S globulins. Cleavage of these proteins is more variable among species. The fact that the same protein can undergo digestion at different sites in different plants suggests that the cleavage may not be important in processing and assembly. Since the 7S complexes form within the RER immediately following synthesis, the peptide cleavages may not affect the structure of the molecule.

The role of glycosylation in the transport and assembly of storage proteins is also unclear. Glycosylation is apparently not involved in transport or assembly of 11S globulins or cereal prolamines. However, many, though not all, 7S storage globulins undergo a series of sugar addition and subtraction reactions. Present evidence suggests that these carbohydrate moieties are not necessary for protein transport. It is possible that they are involved in assembly within the protein body or in proteolytic digestion during seed germination. The storage proteins are strictly regulated, expression usually occurring only in developing seeds and following a distinct temporal pattern.

The ability to transform plants with storage protein genes provides a tool to address important questions pertaining to gene regulation as well as the synthesis and processing of these proteins. The DNA sequences responsible for developmental regulation of storage globulins appear to be relatively short cis-acting elements that precede the genes by several hundred base pairs. Although equivalent sequences remain to be identified in cereals, the conservation of DNA sequences flanking prolamine genes suggests they are similarly situated. There must nevertheless be some difference in the regulation of prolamine and globulin genes, since one group is expressed in the embryo and the other in the endosperm.

The genes coding for a number of these storage proteins have now been cloned and extensively characterized. A considerable amount is known about the structure of the genes, but, as yet, it is not understood how their expression is regulated in the seed. It has been suggested that nutritional improvement could be achieved by using site-directed mutagenesis with the objective of introducing more lysine or methionine codons into the gene sequences. The engineered gene would then be replaced in the plant and, when expressed in the seed, should considerably increase in the levels of the deficient amino acids. However, such an approach would still entail an extensive program of breeding and testing, since this would be essential in order to ensure the production of an acceptable cultivar.

These aims are currently far from being fulfilled although some steps have been achieved. Various seed storage protein genes have been cloned and expressed in transgenic plants. The genes have been cloned and methods are available for engineering the appropriate changes in them. Several reports of successful replacement of a storage protein gene into a crop plant, have begun to appear. Rapid progress in the area of plant gene vectors will probably make this type of manipulation possible in a wider range of dicotyledonous plants in the near future. There are still problems, however, for cereal crops as, except for controlling elements, there are no natural vector systems available. Liposome-mediated DNA transfer may be possible. To produce a plant in which all cells carry the engineering genes, the vector must be used to place the gene into protoplasts which can be regenerated somatically to produce whole plants. Unfortunately, many crop plants, particularly cereals, have proved recalcitrant as regards this type of manipulation, although some progress has now been made in regenerating cereals from embryogenic callus cultures, and in a few cases from protoplasts.

The idea of using recombinant DNA technology to improve seed crops is attractive in that it may allow very specific and controlled changes to be carried out. It can be argued that this approach is more likely to achieve improved nutritional quality along with the production of well filled grains, because the aim is to subtly alter the nature of the proteins while maintaining the balance between the different types of protein in the seed. Most mutant lines selected so far have an altered amino acid composition by virtue of an overall reduction in storage protein synthesis, more often than not resulting in small grains.

The problems outlined so far are purely physical ones of achieving the placement of an engineered gene into a plant, but there are many more factors to be considered. First, the regulation of these genes is maintained over the inserted genes. Production of these genes is tightly controlled in the plant and it is essential that this control be maintained over the inserted genes. Production of these storage proteins is usually encoded in multigene families, so the introduction of one engineered gene may have little effect on the overall amino acid composition. Finally, assuming it is possible to introduce large amounts of an engineered protein into the seed, there may be associated detrimental effects. For instance, if a legume seed requires a larger supply of sulfur-containing amino acids, from which source are these going to be supplied? In the case of cereals, the storage proteins are packaged into endoplasmic reticulum-associated protein bodies by virtue of the hydrophobicity of the proteins themselves. The presence of extra lysine residues in the storage proteins may reduce their hydrophobicity and interfere with their packaging. Such interference may alter subsequent processes involved in seed maturation. Inefficient packaging is likely to be conducive to the production of a well-filled grain. Until these types of specific gene transfers have been achieved, one can only guess at the biological consequences for any given crop species.

The time-scale for the use of recombinant DNA to improve seed proteins is impossible to predict, although some workers suggested that it could be achieved over a period as short as 3 to 5 years. It is likely that such approaches will take longer than that, since breeding and testing the new cultivars at the field level will take the usual passage of several seasons. The grower, as well as the molecular biologist, must be convinced that a new cultivar produced is of a superior and reliable quality.

Despite several difficulties there are significant grounds for optimism. The experiments done so far reveal that several seed storage proteins can be stably accumulated and experiments are underway to used modifed genes in transformation experiments. Further increasing knowledge of the structure and assembly of seed storage proteins and the intracellular targeting mechanisms involved should make design of modified proteins less of a "hit and miss" operation in the future. In view of the difficulties encountered, it can be suggested that new methods with increasing efficiency of extraction and purification should be developed, and expression levels should be enhanced by using suitable promoters.

While seed storage proteins seem to have no enzymatic functions, many undergo differing degrees of posttranslational processing and all (with the exception of the maize zeins), must pass through the intracellular transport systems via the endoplasmic reticulum and the Golgi complex to the proteins bodies. Any modification which interferes with these processes, or simply produces an unstable protein, is likely to result in failure of development of protein bodies. The feasibility and economics of producing a small peptide through the 2S albumin gene have not been worked out. As the peptide increases in length, the chance increases that unwanted cleavage sites for the more inexpensive proteases (or chemical cleavage sites) will be present. This would necessitate the use of more expensive reagents for the cleavage step. Peptides requiring special modification, such as amidation of the termini, will neccessitate further processing. In such cases, synthesis may still be the most viable option.

REFERENCES

1. **Hermann, R. G. and Larkins, B.**, *Plant Molecular Biology,* Plenum Press, New York, 1991.
2. **Goldberg, R. B., Barker, S. J., and Perez-Grau, L.**, Regulation of gene expression during plant embryogenesis, *Cell,* 56, 149, 1989.
3. **Higgins, T. J. V., Newbigin, E. J., Spencer, D., Llewellyn, D. J., and Craig, S.**, The sequence of a pea vicilin gene and its expression in transgenic tobacco plants, *Plant Mol. Biol.,* 11, 683, 1988.
4. **Beachy, R. N., Chen, Z. L., Horsch, R. B., Rogers, S. B., Hoffmann, N. J., and Fraley, R. T.**, Accumulation and assembly of soybean β-conglycinin in seeds of transformed petunia plants, *EMBO J.,* 4, 3047, 1985.
5. **Sengupta-Gopalan, C., Reichert, N. A., Barker, R. F., Hall, T. C., and Kemp, J. D.**, Developmentally regulated expression of the bean β-phaseolin gene in tobacco seed, *Proc. Natl. Acad. Sci. U.S.A.,* 82, 3320, 1985.
6. **Chen, Z. L., Schuler, M. A., and Beachy, R. N.**, Functional analysis of regulatory elements in a plant embryo-specific gene, *Proc. Natl. Acad. Sci. U.S.A.,* 83, 8560, 1986.
7. **Chen, Z. L., Naito, S., Nakamura, I., and Beachy, R. N.**, Regulated expression of genes encoding soybean β-conglycinins in transgenic plants, *Dev. Genet.,* 10, 112, 1989.
8. **Colot, V., Robert, L. S., Kavanagh, T. A., Bevan, M. W., and Thompson, R. D.**, Localization of sequences in wheat endosperm protein genes which confer tissue-specific expression in tobacco, *EMBO J.,* 6, 3559, 1987.
9. **Robert, L. S., Thompson, R. D., and Flavell, R. B.**, Tissue-specific expression of a wheat high molecular weight glutenin gene in transgenic tobacco, *Plant Cell,* 1, 569, 1989.
10. **Shirsat, A., Wilford, N., Croy, R., and Boutler, D.**, Sequences responsible for the tissue specific promoter activity of a pea legumin gene in tobacco, *Mol. Gen. Genet.,* 215, 326, 1989.
11. **Shirsat, A., Wilford, N., and Croy, R. R. D.**, Gene copy number and levels of expression in transgenic plants of a seed specific gene, *Plant Sci.,* 61, 75, 1989.
12. **Chappell, J. and Chrispeels, M. Z. J.**, Transcriptional and posttranscriptional control of phaseolin and phytohemagglutinin gene expression in developing cotyledons of *Phaseolus vulgaris, Plant Physiol.,* 81, 50, 1986.
13. **Voelker, T., Staswick, P., and Chrispeels, M. J.**, Molecular analysis of two phytohemagglutinin genes and their expression in *Phaseolus vulgaris* cv. Pinto, a lectin-deficient cultivar of the bean, *EMBO J.,* 5, 3075, 1986.
14. **Nelson, N. C., Dickson, C. D., Cho, T. J., Thanh, V. H., Scollon, B. J., Fischer, R. L., Drews, G. N., and Golberg, R. B.**, Characterization of the glycinin gene family in soybean, *Plant Cell,* 1, 313, 1989.
15. **William, J. P. L., Scott, M. P., and Nielson, N. C.**, Assembly properties of modified subunit in the glycinin subunit family, *Plant Mol. Biol.,* 2, Herrman R.G. and Larkin B., Eds., Plenum. Press, New York.
16. **Moreira, M. A., Hermodson, M. A., Larkins, B. A., and Nielsen, N. C.**, Comparison of the primary structure of the acidic polypeptides, *Arch. Biochem. Biophys.,* 210, 633, 1981.
17. **Moreira, M. A., Hermodson, M. A., Larkins, B. A., and Nielsen, N. C.**, Partial characterization of the acidic and basic polypeptides of glycinin, *J. Biol. Chem.,* 254, 9921, 1979.
18. **Thanh, V. H. and Shibasaki, K.**, Heterogeneity of β-conglycinin, *Biochem. Biophys. Acta,* 439, 326, 1976.
19. **Thanh, V. H. and Shibasaki, K.**, β-Conglycinin from soybean proteins. Isolation and immunological and physiochemical properties of the monomeric forms, *Biochem. Biophys. Acta,* 490, 370, 1977.

20. **Meinke, D. W., Chen, J., and Beachy, R. N.**, Expression of storage-protein genes during soybean seed development, *Planta*, 153, 130, 1981.
21. **Coates, J. B., Mederos, J. S., Thanh, V. H., and Nielsen, N. C.**, Characterization of the subunits of β-conglycinin, *Arch. Biochem. Biophys.*, 243, 184, 1985.
22. **Yamauchi, F. and Yamagishi, T.**, Carbohydrate sequence of soybean 7S protein, *Agric. Biol. Chem.*, 43, 505, 1979.
23. **Yamauchi, F., Saio, K., and Yamagishi, T.**, Isolation and partial characterization of salt extractable glubulin from soybean seeds, *Agric. Biol. Chem.*, 48, 645, 1984.
24. **Naito, S., Dube, P. H., and Beachy, R. N.**, Differentiation expression of conglycinin α' and β subunit genes in transgenic plants, *Plant Mol. Biol.*, 11, 109, 1989.
25. **Ladin, B. F., Doyel, J. J., and Beachy, R. N.**, Molecular characterzation of a deletion mutant affecting the α'-subunit and β-conglycinin of soybean, *J. Mol. Appl. Genet.*, 2, 372, 1984.
26. **Domoney, C. and Casey, R.**, Measurement of gene number for storage proteins in *Pisum*, *Nucleic Acids Res.*, 13, 687, 1985.
27. **Talbot, D. R., Adang, M. J., Slightom, J. L., and Hall, T. C.**, Size and organization of the multigene family encoding phaseolin, the major seed storage protein of *Phaseolus vulgaris* L., *Mol. Gen. Genet.*, 198, 42, 1984.
28. **Shotwell, M. A., Afonso, C., Davies, E., Chesnut, R. L. S., and Larkins, B. A.**, Molecular characterization of oat seed globulins, *Plant Physiol.*, 87, 698, 1988.
29. **Lycett, G. W., Delauney, A. J., Zhao, A. J., Gatehouse, J. A., Croy, R. R. D., and Boutler, D.**, Two genes coding for the legumin protein of *Pisum sativum* L. contain sequence repeats, *Plant Mol. Biol.*, 3, 91, 1984.
30. **Baumalein, H., Wobus, U., Pustell, J., and Kafatos, F. C.**, The legumin gene family: structure of a B type gene of *Vicia faba* and a possible legumin gene specific regulatory element, *Nucleic Acids Res.*, 14, 2707, 1986.
31. **Doyle, J. J., Schuler, M. A., Godette, W. D., Zenger, V., Beachy, R. N., and Slightom, J. L.**, The glycosylated seed storage proteins of *Glycine max* and *Phaseolus vulgaris*, *J. Biol. Chem.*, 261, 9228, 1986.
32. **Higgins, T. J. V.**, Synthesis and regulation of major proteins in seeds, *Annu. Rev. Plant Physiol.*, 75, 191, 1984.
33. **Casey, R., Domoney, C., and Ellis, N.**, Legume storage proteins and their genes, *Oxford Surv. Plant Mol. Cell Biol.*, 3, 1, 1986.
34. **Schroeder, H. E.**, Major albumins of *Pisum* cotyledons, *J. Sci. Food Agric.*, 35, 191, 1984.
35. **Chandler, P. M., Higgins, T. J. V., Randall, P. J., and Spencer, D.**, Regulation of legumin levels in developing pea seeds under conditions of sulfur deficiency. Rates of legumin synthesis and levels of legumin mRNA, *Plant Physiol.*, 71, 47, 1983.
36. **Gatehouse, J. A., Lycett, G. W., Delauney, A. J., Croy, R. R. D., and Boutler, D.**, Sequence specificity of the post-translational proteolytic cleavage of vicilin, a seed storage protein of pea (*Pisum sativum* L.), *Biochem. J.*, 212, 427, 1983.
37. **Lycett, G. W., Croy, R. R. D., Shirsat, A. H., and Boutler, D.**, The complete nucleotide sequence of the legumin gene from pea (*Pisum sativum* L.), *Nucleic Acids Res.*, 12, 4493, 1984.
38. **Lycett, G. W., Croy, R. R. D., Shirsat, A. H., Richards, D. M., and Boutler, D.**, The 5'-flanking regions of three legumin genes, comparison of the DNA sequence, *Nucleic Acids Res.*, 13, 6733, 1985.
39. **Davies, D. R.**, The r_a locus and legumin synthesis in *Pisum sativum*, *Biochem. Gen.*, 18, 1207, 1980.
40. **Randall, P. J., Thompson, J. A., and Schroeder, H. E.**, Cotyledonary storage proteins in *Pisum sativum*. IV. Effects of sulfur, phosphorus, potassium and magnesium deficiency, *Aust. J. Plant Physiol.*, 6, 11, 1979.

41. **Chandler, P. M., Spencer, D., Randall, P. J., and Higgins, T. J. V.**, Influence of sulfur nutrition on developmental patterns of some major pea seed proteins and their mRNAs, *Plant Physiol.*, 75, 651, 1984.
42. **Spencer, D., Rerie, W. G., Randall, P. J., and Higgins, T. J. V.**, The regulation of pea seed storage protein genes by sulfur stress, *Aust. J. Plant Physiol.*, 17, 355, 1990.
43. **Blagrove, R. J., Gillespie, R. M., and Randall, P. J.**, Effect of sulfur supply on the seed globulin composition of *Lupinus angustifolius*, *Aust. J. Plant Physiol.*, 3, 173, 1976.
44. **Wrigley, C. W., duCross, D. L., Archer, M. J., Downie, P. G., and Roxburgh, C. M.**, The sulfur content of wheat endosperm proteins and its relevance to grain quality, *Aust. J. Plant Physiol.*, 7, 755, 1980.
45. **Shewry, P. R., Franklin, J., Parmar, S., Smith, S. J., and Miflin, B. J.**, The effects of sulfur starvation on the amino acid and protein compositions of barley grain, *J. Cereal Sci.*, 21, 1983.
46. **Gayler, K. R. and Skyes, G. F.**, Effects of nutritional stress on the storage proteins of soybeans, *Plant Physiol.*, 78, 582, 1985.
47. **Rerie, W., Whilecross, M. I., and Higgins, T. J. V.**, Nucleotide sequence of an A-type legumin gene from pea, *Nucleic Acids Res.*, 18, 655, 1990.
48. **Baeumlein, H., Boerjan, W., Nagy, I., Panitz, R., Inze, D., and Wobus, U.**, Upstream sequences regulating legumin gene expression in heterologous transgenic plants, *Mol. Gen. Genet.*, 225, 121, 1991.
49. **Meakin, P. J. and Gatehouse, J. A.**, Interaction of seed nuclear proteins with transcriptionally-enhancing regions of the pea (*Pisum sativum* L.) *leg*A gene promoter, *Planta*, 183, 471, 1991.
50. **Shotwell, M. A. and Larkins, B. A.**, The biochemistry and molecular biology of plants, *Biochem. Plants*, 15, 297, 1989.
51. **Spencer, D., Chandler, P. M., Higgins, T. J. V., Inglis, A. S., and Rubira, M.**, Sequence interrelationships of the subunits of vicilin from pea seeds, *Plant Mol. Biol.*, 2, 259, 1983.
52. **Chrispeels, M. J., Higgins, T. J. V., and Spencer, D.**, Assembly of storage protein oligomers in the endospermic reticulum and processing of the polypeptides in the protein bodies of developing cotyledons, *J. Cell Biol.*, 93, 306, 1982.
53. **Chrispeels, M. J., Higgins, T. J. V., Craig, S., and Spencer, D.**, Role of the endoplasmic reticulum in the synthesis of reserve proteins and the kinetics of their transport to protein bodies in developing pea cotyledons, *J. Cell Biol.*, 93, 5, 1982.
54. **Croy, R. D., Gatehouse, J. A., Tyler, M., and Boutler, D.**, The purification and characterization of a third storage protein (convicilin) from the seeds of pea (*Pisum sativum* L.), *Biochem. J.*, 191, 509, 1980.
55. **Thomson, J. A. and Schroeder, H. E.**, Cotyledonary storage proteins in *Pisum sativum*. II. Hereditary variation in components of the legumin and vicilin fractions, *Aust. J. Plant Physiol.*, 5, 281, 1978.
56. **Newbigin, E. J., deLumen, B. O., Chandler, P. M., Gould, A., Blagrove, R. J., March, J. F., Kortt, A. A., and Higgins, T. J. V.**, Pea convicilin: structure and primary sequence of the protein and expression of a gene in the seeds of transgenic tobacco, *Planta*, 180, 461, 1990.
57. **Ellis, T. H. N., Domoney, C., Castleton, W. C., and Davies, D. R.**, Vicilin genes of *Pisum*, *Mol. Gen. Genet.*, 205, 164, 1986.
58. **Bown, D., Ellis, T. H. N., and Gatehouse, J. A.**, The sequence of a gene encoding convicilin from pea (*Pisum sativum* L.) shows that convicilin differs from vicilin by an insertion near the N-terminus, *Biochem. J.*, 2511, 7171, 1988.
59. **Casey, R. and Sanger, E.**, Purification and some properties of a 7S seed storage protein from *Pisum* (pea), *Biochem. Soc. Trans.*, 8, 658, 1980.
60. **Chlan, C. A., Pyle, J. B., Legocki, A. A., and Dure, L., III**, Development biochemistry of cotton seed embryogenesis and germination XVIII. cDNA and amino acid sequence of members of the storage protein family, *Plant Mol. Biol.*, 7, 475, 1986.

61. **Higgins, T. J. V.**, Synthesis and regulation of major proteins in seeds, *Annu. Rev. Plant Physiol.*, 35, 191, 1984.
62. **Higgins, T. J. V., Chandler, P. M., Zurawski, G., Button, S. C., and Spencer, D.**, The biosynthesis and primary structure of pea seed lectin, *J. Biol. Chem.*, 258, 9544., 1986.
63. **Higgins, T. J. V., Chandler, P. M., Randall, P. J., Spencer, D., Beach, L. R., Blagrove, R. J., Kortt, A. A., and Inglis, A. S.**, Gene structure, protein structure and regulation of the synthesis of a sulfur-rich protein in pea seeds, *J. Biol. Chem.*, 261, 11124, 1989.
64. **Higgins, T. J. V., Beach, L. R., Spencer, D., Chandler, P. M., Randall, P. J., Blagrove, R. J., Kortt, A. A., and Gurthire, R. E.**, cDNA and protein sequence of a major pea seed albumin (PA2: M_r = 26,000), *Plant Mol. Biol.*, 8, 37, 1987.
65. **Rerie, W. G.**, The Structure of Pea Legumin Genes and Their Expression in Tobacco, Ph.D. thesis, Australlina National University, Australlina, 1989.
66. **Casey, R. and Short, M. N.**, N-Terminal amino acid sequence of the β-subunit of legumin from *Pisum sativum, Phytochemistry*, 20, 21, 1981.
67. **Bollini, R. and Chrispeels, M. J.**, The rough endoplasmic reticulum is the site of reserve-protein synthesis in developing *Phaseolus vulgaris* cotyledons, *Planta*, 146, 487, 1979.
68. **Bollini, R., Ceriotti, A., Daminati, M. G., and Vitale, A.**, Glycosylation is not needed for intracellular transport of phytohemagglutinin developing *Phaseolus vulgaris* cotyledons and for the maintenance of its biological activities, *Physiol. Plant.*, 65, 15, 1985.
69. **Bollini, R., Van der Wilden, W., and Chrispeels, M. J.**, A precursor of the reserve protein phaseolin is transiently associated with the endoplasmic reticulum of developing *Phaseolus vulgaris* cotyledons, *Physiol. Plant.*, 55, 82, 1982.
70. **Bollini, R., Vitale A., and Chrispeels, M. J.**, *In vivo* and *in vitro* processing of seed reserve protein in the endoplasmic reticulum: evidence for two glycosylation steps, *J. Cell Biol.*, 96, 999, 1983.
71. **Silightom, J. L., Sun, S. M., and Hall, T. K. C.**, Complete nucleotide sequence of a French bean storage protein gene: phaseolin, *Proc. Natl. Acad. Sci. U.S.A.*, 80, 1897, 1983.
72. **Lioi, L. and Bollini, R.**, Contribution of processing events to the molecular heterogeneity of four banding types of phaseolin, the major storage protein of *Phaseolus vulgaris, Plant Mol. Biol.*, 3, 345, 1984.
73. **Harris, C. J., Lea, K., and Slightom, J. L.**, Expression of phaseolin cDNA genes in yeast under control of natural plant DNA sequences, *Proc. Natl. Acad. Sci. U.S.A.*, 82, 334, 1985.
74. **Vitale, A., Ceriotti, A., Bollini, R., and Chrispeels, M. J.**, Biosynthesis and processing of phytohemagglutinin in developing bean cotyledons, *Eur. J. Biochem.*, 141, 97, 1984.
75. **Ueng, P., Galili, G., Sapanara, V., Goldsbrough, P. B., Dube, P., Beachy R. N., and Larkins, B. A.**, Expression of maize storage protein gene in *Petunia* plants is not restricted to seeds, *Plant Physiol.*, 86, 1281, 1988.
76. **Hoffman, L. M.**, Structure of a chromosomal *Phaseolus vulgaris* lectin gene and its transcript, *J. Mol. Appl. Genet.*, 2, 447, 1984.
77. **Hoffman, L. M. and Donaldson, D. D.**, Characterization of two *Phaseolus vulgaris* phytohemaglutinin genes closely linked on the chromosome, *EMBO J.*, 4, 883, 1985.
78. **Hoffman, L. M., Donaldson, D. D., and Herman, E. M.**, A modified storage protein is synthesized, processed, and degraded in the transgenic plants, *Plant Mol. Biol.*, 11, 711, 1988.
79. **Viotti, A., Sala, E., Alberi, P., and Soave, C.**, Heterogeneity of zein synthesized *in vitro, Plant Sci. Lett.*, 13, 365, 1978.
80. **Larkins, B. A. and Kurkman, W. J.**, Synthesis and deposition of zein in protein bodies of maize endosperm, *Plant Physiol.*, 62, 256, 1978.

81. **Chrispeels, M. J.**, Incorporation of fucose into the carbohydrate moiety of phytohemagglutinin in developing *Phaseolus vulgaris* cotyledons, *Planta,* 157, 454, 1983.
82. **Chrispeels, M. J.**, The Golgi apparatus mediates the transport of phytohemagglutinin to the protein bodies in bean cotyledons, *Planta,* 158, 140, 1983.
83. **Chrispeels, M. J., Higgins, T. J. V., Craig, S., and Spencer, D.**, The role of the endoplasmic reticulum in the synthesis of reserve proteins and the kinetics of their transport to protein bodies in developing pea cotyledons, *J. Cell Biol.,* 93, 5, 1980.
84. **Lord, M. J.**, Precursor of *Ricin* and *Ricinus communis* agglutinin. Glycosylation and processing during synthesis and intracellular transport, *Eur. J. Biochem.,* 146, 411, 1985.
85. **Vitale, A. and Chrispeels, M. J.**, Transient N-acetylglucosamine in the biosynthesis of phytohemagglutinin: attachment in the Golgi apparatus and removal in protein bodies, *J. Cell Biol.,* 99, 133. 1984.
86. **Van der Wilden, W., Herman, E. H., and Chrispeels, M. J.**, Protein bodies of mung bean cotyledons as autophagic organelles, *Proc. Natl. Acad. Sci. U.S.A.,* 77, 428, 1980.
87. **Olden, K., Pratt, R. M., and Yamada, K. M.**, Role of carbohydrate in biological function of the adhesive glycoprotein fibronectin, *Proc. Natl. Acad. Sci. U.S.A.,* 76, 3343, 1979.
88. **Polonof, E., Machida, C. A., and Kabat, D.**, Glycosylation and intracellular transport of membrane glycoproteins encoded by murine leukemia viruses, *J. Biol. Chem.,* 257, 14023, 1982.
89. **Brucher, O.**, Absence of phytohemaglutinin in wild and cultivated beans from South America. *Proc. Trop. Region Am. Soc. Hortic. Sci.,* 12, 68, 1968.
90. **Pusztai, A., Grant, G., and Stewart, J. C.**, A New type of *Phaseolus vulgaris* (cv. pinto III) seed lectin: isolation and characterization, *Biochim. Biophys. Acta,* 671, 146, 1981.
91. **Voelker, T., Sturm, A., and Chrispeels, M. J.**, Differences in expression between two seed lectin alleles obtained from normal and lectin-deficient beans are maintained in transgenic tobacco, *EMBO J.,* 6, 3571, 1987.
92. **Riggs, C. D., Voelker, T. A., and Chrispeels, M. J.**, Cotyledon nuclear proteins bind to DNA fragments harboring regulatory elements of phytohemagglutinin genes, *Plant Cell,* 1, 609, 1989.
93. **Hill, A. V. S., Nicholis, R. D., Thein, S. L., and Higgs, D. R.**, Recombination within the human embryonic zeta-globin locus: A common zeta-zeta chromosome produced by gene conversion of the psi-zeta gene, *Cell,* 42, 809, 1985.
94. **Voelker, T. A., Moreno, J., and Chrispeels, M. J.**, Expression analysis of a pseudogene in transgenic tobacco: A frameshift mutation prevents mRNA accumulation, *Plant Cell,* 2, 255, 1990.
95. **Bustos, M. M., Begum, D., Kalkan, F. A., Battraw, M. J., and Hall, C. T.**, Positive and negative cis-acting DNA domains are required for spatial and temporal regulation of gene expression by a seed storage protein promoter, *EMBO J.,* 10, 1469, 1991.
96. **Greenwood, J. S. and Chrispeels, M. J.**, Correct targeting of the bean storage protein phaseolin in the seeds of transformed tobacco, *Plant Physiol.,* 79, 65, 1985.
97. **Romero, J.**, Genetic Variability in the Seed Protein of Nondomesticated Bean (*Phaseolus vulgaris* L. var. *aborigenecus*) and the Inheritance and Physiological Effects of Arcelin, a Novel Seed Protein, Ph.D. thesis, University of Wisconsin, Madison, 1984.
98. **Romero, J., Yandell, B. S., and Bliss, F. A.**, Bean arcelin. I. Inheritance of a novel seed protein of *Phaseolus vulgaris* L. and its effect on seed composition, *Theor. Appl. Genet.,* 72, 123, 1986.
99. **Osborn, T. C., Alexander, D. C., Sun, S. S. M., Cardona, C., and Bliss, F. A.**, Insecticidal activity and lectin homology of arcelin seed protein, *Science,* 240, 207, 1988.
100. **Osborn, T. C., Blake, T., Gepts, P., and Bliss, F. A.**, Bean arcelin. II. Genetic variation, inheritance and linkage relationships of a novel protein of *Phaseolus vulgaris* L., *Theor. Appl. Genet.,* 71, 847, 1986.

101. **Osborn, T. C., Burrow, M., and Bliss, F. A.,** Purification and characterization of arcelin seed protein from common bean, *Plant Physiol.*, 86, 399, 1988.
102. **Lioi, L. and Bollini, R.,** Identification of a new arcelin variant in wild bean seeds, *Bean Improv. Crop.*, 32, 28, 1989.
103. **Anthony, J. L., Haar, R. A. V., and Hall, T. C.,** Nucleotide sequence of a genomic clone encoding arcelin, a lectin-like seed protein from *Phaseolus vulgaris*, *Plant Physiol.*, 97, 839, 1991.
104. **John, M. E. and Long, C. M.,** Sequence analysis of arcelin. 2. A lectin-like plant protein, *Gene*, 86, 171, 1990.
105. **Kasarda, D. D., Onkita, T. W., Bernardin, J. E., Baecker, P. A., Nimmo, C. C., Lew, E. J.-L., Dieler, M. D., and Green, F. C.,** Nucleic acid (cDNA) and amino acid sequence of α-type gliadins from wheat (*Triticum aestivum*), *Proc. Natl. Acad. Sci. U.S.A.*, 81, 4712, 1984.
106. **Forde, J., Malpica, J. M., Halford, N. G., Shewry, P. R., Anderson, O. D., Greene, F. C., and Miffin, B. J.,** The nucleotide sequence of HMW glutenin subunit gene located on chromosome IA of wheat (*Triticum aestivum* L.), *Nucleic Acids Res.*, 13, 68171, 1985.
107. **Reeck, G. R. and Hedgcoth, C.,** Amino acid sequence alignment of cereal storage proteins, *FEBS Lett.*, 180, 291, 1985.
108. **Sumner-Smith, M., Rafalski, A., Sugiyama, T., Stoll, M., and Soll, D.,** Conservation and variability of wheat α/β gliadin genes, *Nucleic Acids Res.*, 13, 3905, 1985.
109. **Okita, T. W., Hwang, Y. S., Hnilo, J., Kim, W. T., Ayan, A. P., Larsen, R., and Krishan, H. B.,** Structure and expression of the rice glutein multigene family, *J. Biol. Chem.*, 264, 12573, 1989.
110. **Tatham, A. S., Shewry, P. R., and Miffin, B. J.,** Wheat gluten elasticity: a similar molecular basis to elastin?, *FEBS Lett.*, 177, 205, 1984.
111. **Hoekema, A., Hirsch, P. R., Hooykaas, P. J. J., and Schilperoort, R. A.,** A binary vector strategy based on separation of *vir*-and T-region of the *Agrobacterium tumefaciens* Ti-plasmid, *Nature*, 303, 179, 1983.
112. **Takaiwa, F., Ebinumaa, H., Kikuchi, S., and Oono, K.,** Nucleotide sequence of a rice glutelin gene, *FEBS Lett.*, 221, 43, 1987.
113. **Higuchi, W. and Furukawa, C.,** A rice glutelin gene and soybean glycin have evolved from a common ancestral gene, *Gene*, 55, 245, 1987.
114. **Zhao, W. M., Gatehouse, J. A., and Boutler, D.,** The purification and partial amino acid sequence of a polypeptide from the rice glutelin fraction of rice grains: homology to the pea legumin, *FEBS Lett.*, 162, 96, 1983.
115. **Leisy, D. J., Hnilo, J., Zhao, Y., and Okita, W.,** Expression of a rice glutelin in transgenic tobacco, *Plant Mol. Biol.*, 14, 41, 1989.
116. **Esen, A.,** Separation of alcohol-soluble protein (zeins) from maize into three fractions by different solubility, *Plant Physiol.*, 80, 623, 1986.
117. **Righetti, P. G., Gianazza, E., Viotti, A., and Soave, C.,** Heterogeneity of storage proteins in maize, *Planta*, 136, 115, 1977.
118. **Geraghty, D., Peifer, M. A., Rubenstein, I., and Messing, J.,** Primary structure of plant storage proteins: zein, *Nucleic Acids Res.*, 9, 5163, 1981.
119. **Argos, P., Pedersen, K., Marks, M. D., and Larkins, B. A.,** A structural model for maize zein proteins, *J. Biol. Chem.*, 257, 9984, 1982.
120. **Marks, M. D., Lindell, J. S., and Larkins, B. A.,** Quantitative analysis of the accumulation of zein mRNA during maize endosperm development, *J. Biol. Chem.*, 260, 16445, 1985.
121. **Prat, S., Cortadas, J., Pugdomenech, P., and Palau, J.,** Nucleic acid (cDNA) and amino acid sequences of the maize endosperm protein glutelin-2, *Nucleic Acids Res.*, 13, 1493, 1985.

122. **Soave, C. and Salamani, F.**, Organization regulation of zein genes in maize endosperm, *Phil. Trans. R. Soc. London Ser. B*, 304, 341, 1984.
123. **Hagen, G. and Rubenstein, H.**, Complex organization of zein genes in maize, *Gene*, 13, 239, 1981.
124. **Kriz, A., Boston, R. S., Slightom, J. L., and Larkins, B. A.**, Structural and transcriptional analysis of DNA sequences flanking genes that encode 19 kilodalton zeins, *Mol. Gen. Genet.*, 207, 90, 1987.
125. **Di Fonzo, N., Hartings, H., Brembilla, M., Motto, M., Soave, C., Navarro, E., Palau, J., Rohde, W., and Salamani, F.**, The b-32 protein from maize endosperm and albumin sequences regulated by the O2 locus: nucleic acid (cDNA) amino acid sequences, *Mol. Gen. Genet.*, 212, 481, 1988.
126. **Schmidt, R. J., Burr, F. A., and Burr, B.**, Transposon tagging and molecular analysis of the maize regulatory locus opaque-2, *Science*, 238, 960, 1987.
127. **Motto, M., Maddalonim, M., Ponzani, G., Brembilla, M., Matotta, R., Di Fonzo, N., Soave, C., Thompson, R., and Salamani, F.**, Molecular cloning of O2-m5 allele of 8 *Zea mays* using transposon marking, *Mol. Gen. Genet.*, 212, 488, 1988.
128. **Bohmann, D., Bos, T. J., Admon, A., Nishimura, T., Vogt., P. K., and Tjian, R.**, Human proto oncogene c-jum encodes a DNA binding protein with structural and functional properties of transcription factor AP-1, *Science*, 238, 1386, 1987.
129. **Matzke, M. A., Susani, M., Binns, A., Lewis, E. D., Rubenstein, I. and Matzke, A. J. M.**, Transcription of a zein gene introduced into sunflower using a Ti plasmid vector, *EMBO J.*, 3, 1525, 1984.
130. **Landridge, W. H. R., Czernilofsky, A. P., and Szalay, A. A.**, EMBO Course Lab Manual, *Transfer and Expression of Genes in Higher Plants*, Max-Planck-Institut für Züchtungsforschung, Cologne, 1985, 106.
131. **Jefferson, R. A., Kavanagh, T. A., and Bevan M. W.**, GUS fusions: β-glucuronidase as a sensitive and versatile gene fusion marker in higher plants, *EMBO J.*, 6, 3901, 1987.
132. **Schernthaner, J. P., Matzke, M. A., and Matzke, A. J.**, Endosperm-specific activity of a zein gene promoter in transgenic plants, *EMBO J.*, 7, 1249, 1988.
133. **Hu, N.-T., Peifier, M. A., Heidecker, G., Messing, J., and Rubenstein, I.**, Primary structure of a genomic zein sequence of maize, *EMBO J.*, 1, 1337, 1982.
134. **Pedersen, K., Devereux, J., Wilson, D. R., Sheldon, E., and Larkins, B. A.**, Cloning and sequence analysis reveal structural variation among related zein genes in maize, *Cell*, 29, 1015, 1982.
135. **Matzke, A. J. M., Stoger, E. M., Schernthaner, J. P., and Matzke, M. A.**, Deletion analysis of zein gene promoter in transgenic tobacco plants, *Plant Mol. Biol.*, 14, 323, 1990.
136. **Deng, W., Shao, Q., and Jiang, X.**, Expression of Zein gene (Z4) introduced into dicotyledonous *Solanum nigrum*, *Sci. Sin. (B) Chem. Life Sci. Earth Sci.*, 31, 1085, 1988.
137. **Youle, R. and Huang, A. H. C.**, Occurrence of low molecular weight and high cysteine containing albumin storage protein in oil seeds of diverse species, *Am. J. Bot.*, 68, 44, 1981.
138. **Krebbers, E., Herdies, L., De Clercq, A. et al.**, Determination of the processing sites of an *Arabidopsis* 2S albumin acid characterization of the complete gene family, *Plant Physiol.*, 87, 859, 1988.
139. **De Clercq, A., Vandewiele, M., De Rycke, R., Van Damme, J., Van Montagu, M., Krebbers, E., and Vandekerckhove, J.**, Expression and processing of an *Arabidopsis* 2S albumin in transgenic tobacco, *Plant Physiol.*, 92, 899, 1990.
140. **Vandekerckhove, J., Van Damme, J., Van Lijsebettens, M. et al.**, Enkephalins produced in transgenic plants using modified 2S seed storage proteins, *Bio/Technology*, 7, 929, 1989.

Chapter 6

IMPACT OF GENETICALLY MODIFIED CROPS IN AGRICULTURE AND ECOLOGICAL RISK

I. INTRODUCTION

During the last 5 years, availability of gene transfer systems has catalyzed a major refocusing on plants as a biological system. The use of genetically engineered plants as an analytical tool to explore unique aspects of gene regulation and development and the potential to produce novel commercial crop varieties has created a high level of scientific excitement and has driven research into many new areas. The breadth of information to be gained from the study of transgenic plants is serving as an important focus for unifying basic plant scientific research in plant breeding, pathology, biochemistry, and physiology with molecular biology. In addition, the elucidation of regulation of gene expression, which is the fundamental basis for manipulating cellular metabolism, has offered the possibility of extending physiological and genetic observations to the mechanism level. Through genetic engineering, genes can now be removed from one organism and placed into the genome of another, where they are expressed along with the recipient's own genes. The relatedness of donor and recipient is no longer a limiting factor for genetic modification. The promise of genetic engineering for crop improvement arises from this vast new reservoir of characteristics.

The development of genetically engineered plants (also called transgenic plants) is progressing rapidly. Research has moved from the laboratory and greenhouse to the field, where increasing numbers of small-scale experiments on transgenic plants are now being conducted throughout the U.S. and Europe.[1] The major crops that can currently be improved with genetic techniques are soybean, cotton, rice, and alfalfa, and commercial introductions of genetically engineered varieties are likely in the mid 1990s. Rapid progress is being made in the genetic engineering of corn, and it is likely that genetically engineered corn hybrids carrying traits for resistance to herbicides, insects, and viral diseases will reach the marketplace by the year 2000.

Genetic engineering of plants also offers exciting opportunities for the food-processing industry to develop new products and more cost-effective processes. While many of the early successful examples of genetically engineered plants have focused on agronomic genes, it is possible that the food processing and specialty chemical industries may represent the greatest commercial opportunity for biotechnology. Examples of such applications include production of (1) larger quantities of starch or specialized starches with various degrees of branching and chain length to improve texture and storage properties; (2) higher quantities of specific oils or the elimination of particular fatty acids in seed crops; and (3) proteins with nutritionally balanced amino acid composition. The ability to reduce processing costs by the elimination of antinutritive or off-flavor components in foods is quite feasible with antisense nucleic acid technology. The enzymes and genes involved in the biosynthesis of some of these compounds have been hampered by the low quantities of enzymes present in the producing cells, but new techniques based on gene tagging may overcome these difficulties.

Enormous opportunities lie in the succesful use of crops for both commodity and specialty chemical products. Plants have traditionally been a source of a wide range of polymeric materials. These range from starch and celluloses, which are carbohydrate based, to polyhydrocarbons such as rubber and waxes. Many of these polymers have been replaced in the last two to three decades by synthetic materials derived from petroleum-based products. However, the cost, supply, and waste-stream problems often associated with petroleum-based products are issues that are focusing renewed attention on the use of biological polymers. Genetic engineering will significantly enlarge the spectrum and composition of available plant polymers.

Plants also offer the potential for production of foreign proteins with various applications to health care. Proteins such as neuropeptides, growth hormones, and blood factors could be produced in plant seeds, and this may ultimately prove to be an economical means of production. Several mammalian proteins have been produced in genetically engineered plants,[2] and expression of pharmaceutical peptides in oilseed rape plants has been reported.[3]

II. TRANSGENIC PLANTS AND THEIR COMMERCIAL VALUE

Of all the biotechnologies, plant agricultural R and D may have the most realistic timetable for commercial pay-offs. The technology for inserting good expressing genes into plant cells is in hand, and scientists are identifying more and more genes and the traits they control. Successful field trial results are enabling companies to address questions of technological feasibility, costs, and the timing of commercial product introduction with more precision. And the established nature of the agro-marketplace allows for reasonable calculations of an investment payback period based on the amount of value added to crop yields and the potential gain in market share.

Another feature distinguishing plant agricultural commercialization in molecular biology and genetic engineering from other biotech endeavors is the relative straightforwardness of obtaining patent protection on unique new plant varieties. No sociological roadblocks exist equivalent to patenting transgenic animals and microbes. Incorporating genetic modifications in plants helps assure that royalties will be true to the patent holder or licensee.[3] Added to this is the recent, positive attitude shown by regulators — and even environmental groups — towards careful field testing, and the sum total shows that plant agriculture (via molecular biology) may prove one of the most predictable ways of corporations to profit from biotechnology in the 1990s.

Apart from the availability of *Agrobacterium tumefaciens* transformation system, another reason for the rapidity of the race toward agrobiotechnology commercialization is the availability of a wide assortment of selectable and scoreable markers that can be inserted easily into genetic constructs. These marker genes allow for multiple transformations of selected lines and optimization of the selection process for different species. In addition, scoreable markers are valuable aids for demonstrating the targeting of transformation to particular cell types in tissue explants, as well as for analyzing expression and heritability of foreign DNA inserts. The ease of using tissue explants — including leaves, stems, and cotyledons — instead of depending on older, protoplast-based methods, enables researchers to best utilize *Agrobacterium*'s ability to transform intact plant tissues for scaling up to conduct field tests and, eventually, for commercial production.

As mentioned earlier, the genetic modification of crops using new techniques of manipulating DNA and introducing it into plant cells opens up many possibilities for the production of improved varieties of crop plants. Before the potential of this technological advance can be realized, however, it will be necessary to carry out extensive field testing of these modified crops in the same manner as any new breeding material. To date, genetically modified plants have been produced in well over 50 species,[3] which include those showing evidence that a particular character is stably transformed and inherited. It is unfortunate that several of the reported instances of "transformation", even with annual species, do not fulfill this criterion. Indeed, strict criteria which demand a combination of genetic, phenotypic and physical data[4] must be met before any putative genetically modified plant could be included in a breeding program. There is sufficient precedent from the animal field to support such a caution. For example, there are several reports of intramolecular, nonintegrated DNA being present several months after the transformation experiments and even of passing through meiosis into progeny as extrachromosomal sequences. It is not known whether such a phenomenon occurs in plants.[3]

It must be stressed that most of the reports of genetic modification have been on a single (or a few) varieties which are unlikely to be typical of the species. Herein lies a question of strategy, namely, whether in order to be commercially useful, a technique for DNA introduction must be capable of being applied to any genotype of a species, or whether it would be practicable to transfer and introduce a gene from a "model" (i.e., easy to transform) variety to varieties of greater commercial value by crossing. At present there are very few, if any, important crop species which can be transformed at will. This means that reports of new successes, although to be applauded, may not lead directly to commercial success.

The initial successes in genetic modification have been in the introduction of herbicide, insect, and disease resistance into plants. These examples are simple traits that can be easily studied, and from them much basic information on genetically modified plants is being obtained.

A. HERBICIDE RESISTANCE

The development of herbicide-resistant varieties has generated not only interest but also considerable controversy over recent years. However, much of this controversy is based upon misconceptions of the reasons for this work. Herbicide resistance, particularly where mode of action studies had identified a sensitive enzyme, was an obvious candidate for investigation. Such interest has led to the isolation of genes providing resistance/tolerance to several herbicides. However, the development of such resistant/tolerant plants, useful as they might be for scientific study, will not lead to their automatic commercialization and is not an evidence of any determined policy on behalf of chemical companies to increase herbicide usage. It is illogical to suppose that the availability of a crop variety resistant to a particular herbicide will force the farmer to purchase supplies of the compound. He or she will make a decision on economic grounds and the herbicide regime chosen will be that which can provide effective control at the lowest possible cost, as has been argued by Eckes et al.:[5] "Because of the possibility of using herbicides only when required, i.e., not until weed infestation has gone beyond an acceptable level, and because of the replacement of several herbicides by a single nonselective one, the use of herbicide-resistant plants should result in less use of herbicides." It will also be possible to replace older products with products ecologically more favorable with fewer carry-over problems. The potential problem of herbicide-resistant weeds will be made neither better nor worse by the availability of herbicide-resistant crops. Such contingencies will be prevented or managed, as now, by sensible rotation of crops and herbicides.

B. INSECT RESISTANCE

The major emphasis in this area has been on the development of plants containing a protein isolated from the soil-living bacterium *Bacillus thuringiensis* (Bt). Protein crystal/spore preparations of Bt have been used as commercial preparations for several years and it is estimated that over 2000 tonnes of this preparation have been applied world-wide with no undesirable effect. Several genes encoding for insecticidal Bt proteins have now been inserted into plants and proved to give some level of protection against lepidopteran insects.[6] More recently, additional genes have been isolated and these are claimed to be specifically active against Coleoptera, Diptera, or even nematodes. It is conceivable, therefore, that precise targeting to an individual pest could be successful. Also, induction of an introduced gene in response to insect attack could be possible.[7] Additional routes to insect control (such as the use of cowpea trypsin inhibitor) are also being actively pursued, and in the long term it is possible that combinations of genes might be preferable (see Chapter 2).

C. DISEASE RESISTANCE

As with the examples of herbicides and insect resistance cited above, the first strategies to obtain disease-resistant plants have involved the use of relatively simple traits coded by single genes. A number of molecular approaches to obtaining virus resistance are being examined (see Chapter 4). For example, it has been found that by inserting coat protein genes from a virus into a plant, infection by the virus is inhibited. This had given promising results in several crops, and field trials have been carried out.[3,8] It should be stressed that there is no effective chemical method to control virus disease in plants and so virus-resistant cultivars are needed in many crops. Progress is also being made in modifying plants for resistance to fungal diseases, for example by introducing a gene coding for chitinase, and field trials have already been carried out with such plants.

D. FIELD PERFORMANCE OF TRANSGENIC PLANTS

To date, little quantitative data on the performance of the genetically modified plants described above are publicly available. It is, however, possible to list certain requirements which must be achieved if such material is to have any value in breeding programs, rather than remain an interesting curiosity. First, the material must show a stable and predictable expression of the inroduced gene(s), both over generations and over time during development. In other words, the gene(s) must not be subject to rapid (or gradual) inactivation or deletion. It must be stable in response to enviromental stress, and for a seed crop or hybrid variety it must be stable in different genetic backgrounds.

Among the most extensive of field trials examining the levels of expression of an introduced gene is that conducted recently with tobacco in which a DNA sequence encoding the bacterial enzyme β-glucuronidase (GUS) had been inserted.[8] The level of activity of this enzyme can be easily measured by a fluorometric assay. Individual plants showing low and high levels of expression were selected. Samples were taken from plants at various times throughout the growing season and assayed in the laboratory. Although variation in GUS activity between plants was high, there was no change in expression levels from one generation to another, and the relatively high or low levels of expression were maintained during the growth season in both locations.

The second requirement is that the gene should show an adequate level of expression. This is most easily considered with reference to herbicide resistance, where some of the plants produced to date do not show high levels of "resistance" and would not survive under conventional field applications. Third, the overall agronomic performance of the modified plant should be comparable with the nonmodified cultivars with which it is likely to compete; only in a few extreme cases, would a farmer be willing to sacrifice yield for the potential benefit of an extra characteristic. From the evidence available to date, it is not clear whether existing examples of modified crops meet these criteria.

In addition to the need to confirm the agronomic performance of genetically modified crops, it is of course also necessary to demonstrate that these novel types of plant have no unacceptable effect on the environment. For this reason, all trials of such plants have to be strictly controlled and permission has to be sought from the relevant government authorities. In the U.S., this currently consists of a joint system comprising the Intentional Introduction Subcommittee of the Advisory Committee on Genetic Manipulation — a Health and Safety Executive responsibility — and the Interim Advisory Committee on Introductions, within the Department of Environment.[3] These committees are composed of independent experts to advise on the environmental, medical, and scientific aspects of each proposal. In addition, permission may be required from other government departments (such as the Ministry of Agriculture, Fisheries and Food), according to the nature of the experiment. Particular environmental issues which are addressed during the course of each assessment include:

1. Could the plant become invasive of natural habitats to the detriment of native species or the local environment in total?
2. Could it become a noxious or pernicious weed?
3. Could the introduced gene, by natural outcrossing in the field, be transferred to other species and produce new hybrids which could then present ecological or environmental problems?

Only a few small-scale trials have been carried out in the U.K. to date, though world-wide the number has already crossed 100 and many more trials are planned for the future.[3] No trial yet conducted has given any cause for concern in any of these areas.

III. ECOLOGICAL AND ENVIRONMENTAL IMPACTS AND LEGISLATION

Concerns have been raised regarding the safety of environmental release of transgenic organisms. The most visible of these early controversies was over the field test of a deletion mutant of *Pseudomonas syringae* (called ice minus), designed to reduce frost damage to certain crops. Partly in response to public concerns, in 1986 the Executive Office of the President in the U.S. established the Coordinated Frame Work for Regulation of Biotechnology to ensure the safety of biotechnology research and products. Various executive agencies were given regulatory authority to oversee particular categories of organisms created through genetic engineering.

A novel British research program called PROSAMO (Planned Release of Selected and Modified Organisms) has produced its batch of results on the ecological behavior of a genetically manipulated variety of oilseed rape (known to Americans as canola). Their preliminary data indicate that these plants do not outgrow their competitors in the wild, nor is there any evidence that they pass on their foreign genes to other species. PROSAMO is designed to provide scientific data on the fate of both engineered plants and microbes released into the environment. Ecological data are currently being collected on a variety of oilseed rape.

Until recently, the predominant ecological concern was that plant taxa with novel engineered traits would become pests in the environment. Studies of crop-weed comparisons have shown that plants can evolve invasive genotypes, based on a few major gene polymorphisms, which are usually followed by physiological or life-history changes. The probability that this change to weediness will occur with transgenic crops is unknown. The direct evolution of transgenic weeds, however, is not the only or even a primary concern with the release of plants. Crop plants are capable of transferring genes, by hybridization, over relatively long distances to related plants that differ markedly in their life-history characteristics. Many of these relatives have the weedy characteristics of high reproductive output and seed dispersal. Thus the genes carefully inserted by the molecular biologists may be further transferred into new organisms by the plants themselves.

How, then, can the benefits of the approaches described above be best confirmed and integrated into agriculture? First, the processes involved must be systematized. Accurate, unbiased information must be freely available.

Within agriculture, plant breeders and agronomists should be brought into the process of developing genetically modified crops. There is a need for exchange of information on the goals to be set and the most beneficial way to use genetically modified crops in plant breeding. This applies especially to the development of plants containing characteristics designed to improve plant breeding itself. For example, plants are being produced in which it is possible to control male fertility in such a way that hybrid seed can be obtained on crops where this was not practicable previously.[3]

In addition, education must be accompanied by a careful adherence to the regulatory demands of each country. This is an area of some concern at present in that independent and often inconsistent regulations governing the trials of genetically modified organsisms are being, or have been, drafted in different countries. Within Europe, some countries (notably Denmark) have already legislated on this subject, while others such as the U.K. have drafted legislation which is not yet law (although guidelines are in force). Yet others have no published regulations at this time. Above the national jurisdictions, the European Community has drafted a separate set of directives and the Organization and Development has issued specimen guidelines.

Outside Europe, detailed regulations exist in some countries (U.S., Australia), while in the majority there is no regulatory framework. Since the agricultural sector is the core of international businesses, and commodity trading is indispensable to providing reliable food supplies, the determination of common standards is of vital importance. These standards must be based on coherent, scientifically rigorous rules, if they are to achieve the respect of those affected.

Another area that is undergoing review at present is patent legislation in relation to genetically modified plants.[4] Appropriate property protection for the results is needed if the work is to be carried out in the private sector. The U.S. now allows patents on plants, and the European Community is moving towards this position with the draft directive on the protection of biological inventions.[3] Amendment of The International Committee on Plant Variety Protection to take into account developments in biotechnology represents well-based hope for a resolution of this matter that will take into account the interests of innovators, breeders, and the public.

Molecular biologists working to improve crops often state that the ecological risks from releasing genetically engineered (transgenic) organisms are small and point out that breeders have introduced new genes into organsims by crossing with wild relatives for decades with no apparent harm. There is also a feeling that biotechnology in fact reduces risk because traits are more tightly controlled in these crosses, that is, single genes instead of blocks of genes are transferred into the crop.

Traditional breeding techniques usually rely on crosses with closely related plants, either subspecies or wild ancestral species of the crop, because incompatibility barriers often prevent crossing between more distantly related gene pools. Incompatibility presents no barrier with molecular engineering techniques and, in fact, genes can be introduced from a closely related species, distantly related species, or even an organism from outside the plant kingdom.

The set of human health and environmental issues related to the effects of release of genetically modified organisms on the environment has been placed under the aegis of the Environmental Protection Agencies (EPA) and the U.S. Department of Agriculture (USDA).[3] Biotechnology application in agriculture and forestry is now regulated by the USDA. However, the field testing of transgenic plants is undertaken by the Animal and Plant Health Inspection Service (APHIS).[3] APHIS, as required under the National Environmental Policy Act, conducts the environmental assessment of each plant considered for release. After studying all the parameters, APHIS certifies that no risk is involved in the release of the plants and finally clears the release of the transgenic plants to the environment.[9-11]

Now the USDA completely assumes the responsibility of releasing genetically engineered plants.[1,3] Most of the transgenic plants are produced through *Agrobacterium tumefaciens* as the vector, and have been declared safe since a nonpathogenic vector is used in the system. In addition, promoters derived from CaMV 35S especially have also been considered seriously while releasing the transgenic plants.

What might be the environmental consequences of widespread acceptance of new varieties containing traits that confer environmental tolerance or pest resistance? Plants with these traits, acquired either by the escape of the engineered crop or from wild plants that have crossed with the crop, could have serious impacts on man-made plant communities.

The most obvious threat comes from plants that might become more serious weeds in agricultural systems. For example, herbicides considered environmentally safe would no longer be effective against weeds that had captured a gene for herbicide resistance, forcing the use of more dangerous chemicals.[12] Similarly, widespread use of insect resistance traits may lead to the rapid evolution of pest species for which alternate resistance factors or chemical control would be neccessary. Impact on natural communities would be less predictable, but might have important consequences. For example, weeds with a novel trait that makes them less susceptible to their usual herbivores (such as weedy tomatoes that express Bt toxin) might exhibit greater reproductive success. Rare plant species might be eliminated through competitive displacement, and the genetic diversity of nonspecific plants might be reduced. Insect communities might also be affected. If lepidopteran herbivores were removed from plant species, insects eating these plant species may face greater competition.

Plant communities distant from the site of the novel weed might also be affected. Janzen[13] notes that some lepidoptera from Guanacaste National Park in northwestern Costa Rica cross the central mountain range to spend the dry season in the San Carlos Valley, an area that is becoming increasingly converted to agriculture. The extent to which the migrating species use areas that contain the engineered crop plant or weed could affect insect population dynamics in the distant natural community. In addition, although lepidoptera are herbivores as larvae, many are important pollinators of wild species as adults. Therefore, if larval mortality is high, plant communities might also be disrupted by the decreased availability of pollinators. Further disruption of the pollinator and plant communities could occur if the toxin is expressed in plant nectar.

In fact the environmental assessments address four issues involving dissemination of genetic material: the extent of movement of genetic material of plants of the same species, the determination of adequate isolation distances to prevent outcrossing, the extent of dissemination of genetic material to closely related plant species, and methods to prevent the dispersal of genetic material from seeds and vegetative plant parts surviving in the field tests.

The problem of intra- and interspecific cross-pollination among transgenic plants was analyzed critically with some specific examples such as potatoes by Wrubel et al.[3] They suggested that plants like Russet Burbank potatoes do not produce viable pollen and hence their risk of outcrossing is essentially zero. The situation is, however, different for other potato varieties such as Lemhi Russet, which do produce viable pollens, and therefore enough data during field trials should be collected for such plants. Wrubel et al.[3] also described the determination of adequate distances which should be maintained in field tests so that pollens from transgenic plants do not fertilize nontransformed species. However, APHIS does not require applicants to determine the extent and frequency of pollen movement nor the effectiveness of border rows in linking the transmission of pollen during field tests.

The possibility of interspecific cross-pollination is greatest when two sexually compatible species exist in close proximity.[12] Thus, genes might be exchanged from crop and plant and weedy relative. Most plants do not readily cross with related species without human intervention; if they can cross, the plants with which they are compatible do not exist in the region of the test site. Wrubel et al.[3] cited two examples where transgenic cotton plants would be allowed to be released and flower in Hawaii. In one assessment, wild-type plants in the area were absent from the test site thus avoiding the hybridization risk. But they suggested that immediately after the field trial the transgenic plants must be disposed of and the seed bank should be destroyed. There are several methods of disposal of plants and there is no explanation in the environmental assessments why a particular method is preferable or used in one instance and not in another. However, APHIS should be convinced that disposal methods proposed in the application are adequate to prevent the escape of genetic material. In addition, there are no recommendations to generate data on the cold-hardness of seeds and potential movement of seeds out of test sites by birds, rodents, ants, and beetles.

Although there have been discussions and debate on genetically engineered plants becoming weeds when released in the environment, so far there is no direct evidence to prove this.[9-11,14,15] However, APHIS takes the position that if an untransformed crop plant is not considered a weed, then changing a single gene cannot transform the plant into a weed.[3] It is also thought that genetic modifications of plants might increase the susceptibility of the plants to disease or increase the palatability of the plant to herbivores. However, this concept, although included in the environmental risk assessment, has not been considered very seriously. In addition, the transmission of genetic material from the transgenic plants to any plant in the local community was also thought to be highly unlikely.

Wrubel et al.[3] suggested that although USDA/APHIS have considered several aspects for which data should be generated during field tests, the APHIS personnel must base their conclusions on empirical data. Experiments on gene flow between conspecifics and between transformed plants and close relatives should be included. The potential for weediness and enhanced competitiveness of transgenic plants should be assessed carefully.

Over the last decades, researchers discovered how to engineer the DNA of many plants so that they can express Bt toxins themselves (see Chapter 2). The simple bacterium has triggered a multimillion dollar research effort in agricultural industry to reproduce self-protecting crops that do not require chemicals to fight off pests. Now at the verge of a revolution in agriculture, researchers are afraid that Bt's miracles may be coming to an end. In nearly 12 generations Bt-resistant pests can be generated in the laboratory, and if we are not careful, the insects which remain susceptible to Bt will be lost, especially when Bt-engineered plants are used. However, officials at the Monsanto Chemical Corporation, which is running one of the largest programs to insert Bt resistance into crop plant DNA, say that its engineered plants are less likely to produce resistant pests. They explain that the toxins will come from inside the plants rather than from outside, where they can wash away. They argue that the major advantage of such plants will be in making the insects vulnerable to Bt as soon as they begin to eat the plants. However, researchers want the regulatory agencies to issue guidelines which can help companies and farmers to use engineered plants with care.

Brassica napus ssp. *olifera* is being engineered for resistance to Basta at four experimental stations in Britain. Assuming that the plant itself is safe, nothing can be said about the safety of the plant genes that it carries. If several weeds which are close relatives of rape pick up the herbicide-resistant genes from their engineered relative, it will pose even a greater problem. Several other crops where transgenic plants have been created are bound to differ more from oilseed rape. For example, sugar beet (*Beta vulgaris*) is wind pollinated, so its genes will travel much further, and beet has several wild relatives with which it is known to hybridize.

REFERENCES

1. **McCammon, S. L. and Medley, T. L.**, Certification for the planned introduction of transgenic plants into the environment, in *Molecular and Cellular Biology of Potato*, Vadya, M. E. and Park, W. D., Eds., C.A.B. International, Wallingford, U.K., 1990, 233.
2. **Sigmons, P. C., Dekker, B. M. M., Schrammeijer, B., Verwoered, T. C., van den Elzen, P. J. M., and Hoekema, A.,** Production of correctly processed human serum albumin in transgenic plants, *Bio/Technology*, 8, 217, 1990.
3. **Wrubel, R. P., Krimsky, S., and Wetzer, R. E.**, Field testing transgenic plants, *BioScience*, 42, 280, 1992.
4. **Crawley, M. J.**, The ecology of genetically engineered organisms: assessing the environmental risks, in *Introduction of Genetically Modified Organisms into the Environment*, Mooney, H. A. and Bernardi, G., Eds., John Wiley & Sons, New York, 1990, 133.
5. **Eckes, P., Donn, G., and Wengenmayer, F.**, Genetic engineering with plants, *Angew. Chem. Intl. Eng. Ed.*, 26, 382, 1987.
6. **Delannay, X., LaVallee, B. J., Proksch, R. K., Fuchs, R. L., Sims, S. R., Greenplate, J. T., Mattone, P. G., Dodson, R. B., Augustinem, J. J., Layton, J. G., and Fischoff, D. A.**, Field performance of transgenic tomato plants expressing the *Bacillus thuringiensis* var. *kurstaki* insect control protein, *Biotechnology*, 7, 1265, 1989.

7. **Thornberg, R. W., Kerman, A., and Molin, L.**, Chloramphenicol acetyl transferase (CAT) protein is expressed in transgenic tobacco field tests following attack by insects, *Plant Physiol.*, 92, 500, 1990.
8. **Dunwell, J. M. and Paul, E. M.**, Impact of genetically modified crops in agriculture, *Outlook Agric.*, 19, 103, 1990.
9. **Colwell, R. K.**, Ecology and biotechnology: expectations and outliers, in *Safety Assurance for Environmental Introductions of Genetically-Engineered Organisms into the Environment,* Springer-Verlag, New York, 163, 1988.
10. National Academy of Sciences (NAS), *Introduction of Recombinant DNA-Engineered Organisms into the Enviroment: Key Issues,* National Academy Press, Washington, D.C., 1987.
11. National Research Council (NRC), *Field Testing Genetically Modified Organisms: Framework for Decisions,* National Academy Press, Washington, D.C., 1989.
12. **Ellstrand, N. C. and Hoffman, C. A.**, Hybridization as an avenue of escape for engineered genes, *BioScience,* 40, 438, 1990.
13. **Janzen, D. J.**, Insect diversity of a Costa Rican dryforest: why keep it and how?, *Biol. J. Linn. Soc.,* 30, 343, 1987.
14. **Uchimiya, H., Handa, T., and Brar, D. S.**, Transgenic plants, *J. Biotechnol.,* 12, 1, 1989.
15. **Tidje, J. M., Colwell, R. K., Grossman, Y. L., Hodson, R. E., Lenski, R. E., Mack, R. N., and Regal, P. J.**, The planned introduction of genetically engineered organisms: ecological considerations and recommendations, *Ecology,* 70, 298, 1989.

INDEX

A

Acetabularia mediterranea, zein protein synthesis in, 203
Acetohydroxy acid synthase, see Acetolactate synthase
Acetolactate synthase
  genes for, see ALS genes
  isozymes of, 93
  as target for herbicide-resistant plant production, 89
Aedes aegypti, *cryB1* gene product effect on, 58
Agrobacterium tumefaciens
  binary T-DNA plasmid replication in, 3, 11, 126
  coat protein gene transfer using, 118, 128, 133, 135, 136, 139
  cocultivation techniques for, 9–12, 35
  disarmed strains of, 5
  DNA transfer using, 2–12, 13, 19, 51, 54, 65, 73, 91, 103, 221, 226
  pRK290 stability in, 5
  satellite RNA transfer using, 122
  seed protein gene transfer using, 180, 206
  spheroplasts, use in transformation studies, 13
  wild-type, plant tissue transformed by, 3
Alachlor
  detoxification by glutathione-S-transferase, 101
  resistance to, 91
Albumins, as seed proteins, 171
2S Albumins, production of, 206–207
Alfalfa
  engineered phosphinothricin resistance in, 97
  engineered sulfonylurea resistance in, 97
Alfalfa mosaic virus
  CP gene of
    introduction into tobacco, 121, 125, 136
    resistance mediated by, 131, 133, 136–139, 149, 150, 155, 157
  in CP(+) plants
    field trials on, 159, 161

  resistance in, 127, 128, 129–130, 139, 140, 146, 147
  use in Bt endotoxin gene introduction to tobacco, 64
Alnus incana, GUS gene expression in, 8
AL1 protein, transcription of, 125
ALS gene(s), 91–97
  DNA hybridization studies on, 95–96
  expression after biolistics bombardment, 33
  mutants
    in microbes, 93–95
    in plants, 95–97, 99
ALS isozymes
  amino acid residues in, 94–95
  isolation and properties of, 94
Amaranthus hybrid, *psbA* mutant gene of, use in herbicide-resistance engineering, 90–91
Amino acid biosynthetic pathways, as herbicide targets, 92
Aminoglycoside phosphotransferase type II gene, plasmid carrying, 13
Amitrole
  cross-resistance to, 86
  plant resistance to, 86
Anabaena 7120, ALS gene in, 96
Anagasta kuehniella, P1 protoxin effects on, 58
Anthocyanin pigment, molecular studies of gene for, 29
Antisense nucleic acid technology
  food off-flavor removal by, 220
  use in engineered plant virus protection, 118, 123–124
APHIS, role in biotechnology regulation, 226, 228
Arabidopsis
  EPSPS protein of, 99–100
  pOCA18 delivery of DNA of, 20
  seed protein genes of, 189
  yeast artificial chromosome studies on, 21
Arabidopsis thaliana
  ALS gene in, 95, 96–97
  2S albumin gene of, transfer to tobacco, 206
  sulfonylurea-resistant mutants of, 93, 96–97

Arcelin, properties and cDNA of, 192
AroA gene mutant, in herbicide-resistance engineering, 89, 97–98
Arthrobacter, glyphosate degradation by, 102
Aspartyl-proteinase inhibitors, 49
Atrazine
  detoxification by glutathione-S-transferase, 101
  engineered plant resistance to, 89, 91
  resistant Q_B protein of, 90
Avena sativa, see Oat
Azidesatrazine, use in herbicide-binding studies, 89

B

Bacillus cereus, B. thuringiensis protoxin gene transfer to, 61, 62, 63
Bacillus cereus subsp. fowler and lewin, P1 crystalline inclusion of, 56
Bacillus medusa, inclusion formation in, 61–62
Bacillus popilliae, P1 crystalline inclusion of, 56
Bacillus sphaericus subspecies, toxin genes of, 60
Bacillus subtilis
  sporulation genes of, 63
  use in BT endotoxin gene studies, 60, 62
Bacillus thuringiensis
  chromosomal coded genes of, 59–60
  endotoxins of, see Bt endotoxins
  genes of
    Agrobacterium-mediated transfer of, 65, 73
    expression in transgenic plants, 65, 66, 74
    selection and cloning of, 75
Bacillus thuringiensis aizawai, endotoxin of
  sequences, 58
  toxicity to insects, 75
Bacillus thuringiensis flagella, P1 crystalline inclusion of, 56
Bacillus thuringiensis subsp. dendrolimus, protoxin genes of, 60
Bacillus thuringiensis subsp. finitimus
  P1 crystalline inclusion of, 56
  plasmids of, 59, 62, 63
  protoxin genes of, 59
Bacillus thuringiensis subsp. sotto, endotoxin genes of, 61
Bacillus thuringiensis subsp. subtoxicus, plasmids of, 59
Bacillus thuringiensis subsp. wuhanensis, protoxin genes of, 60
Bacillus thuringiensis var. berliner
  endotoxin of
    genes for, 61, 62, 74
    sequences, 58
    use in pest control, 55, 58
  plasmids of, 59
  protoxin genes of, 58, 60, 62, 63
Bacillus thuringiensis var. dendrolimus, P1 protoxin of, 57
Bacillus thuringiensis var. galleriae, protoxin genes of, 60
Bacillus thuringiensis var. israelensis
  endotoxin of
    genes for, 60, 61, 74
    use in pest control, 55
  inclusions of, 56
  plasmids of, 58
Bacillus thuringiensis var. kurstaki
  endotoxin of, 74
  genes for, 58, 60, 61, 62, 63, 66
  use in pest control, 55, 57
  plasmids of, 59
  protoxin genes of, 58, 59–60
Bacillus thuringiensis var. tenebrionis, insecticidal crystal protein of, 58, 74
Bacillus thuringiensis var. thuringiensis, plasmids of, 58
Bacteria
  herbicide-detoxifying genes from, 101–104
  plant cell cocultivation with, 9
  sulfometuron methyl toxicity to, 93–95
bar gene
  in herbicide-resistance engineering, 89
  pat gene comparison to, 103, 107
  as plant selection marker, 5, 32, 33
  PPT and bialophos resistance conferred by, 102, 103
Barley
  chymotrypsin inhibitors in, 50
  dicamba tolerance in, 101
  legumin gene regulation in, 178
  mutant with high lysine content, 172
  protoplast electroporation of, 17
  seed-storage protein properties of, 173, 192, 193–196

Index

Bean
  seed storage protein properties of, 173, 184–192
  TMV strain infection of, 130–131
Beet army worm, see *Spodoptera exigna*
Beta vulgaris, see Sugar beet
Bialophos, engineered plant resistance to, 102, 103
Bin19, as *Agrobacterium* binary vector, 10–11
Binary method, for separationof T-DNA and *vir* genes, 3
Biolistics technique, application of, 29, 33, 36
Black nightshade, see *Solanum nigrum*
Black spruce, electroporation studies on, 24
Blood factors, production by genetic engineering, 220
Brassica spp., seed protein genes of, 189
Brassica campestris, triazine resistance in, 91
Brassica campestris var. *rapa*, genetic transformation of, 2
Brassica juncea, CAT as unsuitable reporter gene for, 7
Brassica napus
  CAT as unsuitable reporter gene for, 7
  2S albumin expression in, 206, 207
  transgenic plants from, by electroporation, 25
  triazine resistance in, 91
Brassica napus ssp. *olifera*, genetically engineered, 229
Breadmaking, wheat prolamine role in, 193
Breadwheat
  electroporation studies on, 24
  high-velocity microprojective bombardment delivery of DNA to, 34
Broad bean, globulin genes in, 176, 177
Bromoxynil
  engineered plant resistance to, 89, 102
  tolerance to, 88
Bruchid beetles, arcelin toxicity to, 192
Bt endotoxins, 49, 55–71
  characterization of, 55–58, 74
  engineered development of, 49, 228–229
  genes for
    characteriziation and cloning, 60–61
    regulation, 61–63
  insect resistance to, 49, 222–223
  transgenic plants expressing, 64–70, 222, 228–229
bt2 gene, Ti plasmid containing, 65–66
bt884 gene, Ti plasmid containing, 65
bt:neo23 gene, Ti plasmid containing, 65
bt:neo860 gene, Ti plasmid containing, 65
bxn gene
  in engineered herbicide resistance, 89
  role in bacterial herbicide metabolism, 102

C

Cabbage looper, see *Trichoplasia ni*
Callosobruchus maculatius, cowpea resistance to, 55
Callus transformation, using microprojectile bombardment, 33, 34
CaMV, see Cauliflower mosaic virus
Canola, see Rapeseed
Carbonilides, thylakoid membrane binding of, 88
CAT enzyme, use as monitor for RNA or DNA transfer, 5–6, 28–29
CAT gene
  disadvantages of, 8
  electroporation use in transformation studies of, 17, 19, 21, 24
  expression after microprojectile bombardment, 29
  expression in woody species, 8, 52
  as plant selection marker and reporter gene, 5, 6, 7, 25, 54
CAT protein, accumulation in wounded tobacco tissue, 55
Cauliflower mosaic virus
  cross-protection mediated by, 148
  replication of, 153
  35S promoter from, use in gene transfer, 203, 204, 226
  use as biological vector, 1–2, 13, 25, 50, 64
Caulimoviruses, genomes of, 162
C58C1 Rif^R pGV2260, as Bt endotoxin vector, 65
Cell cycle, effects on electroporation, 23
Celluloses, engineered production of, 220
Cereals, transgenic, 2, 7, 9
  production by electroporation, 25
Cereal storage proteins, 192–206
Chenopodium amaranticolor, use in satellite RNA infectivity test, 122

Chitinase, use in fungal disease resistance, 223
Chlamydomonas
  herbicide resistance genes in, 88
  sensitivity to sulfometuron methyl in, 93
Chlamydomonas reinhardtii, transformation by microprojectile methods, 30
Chloramphenicol acetyl transferase, see CAT enzyme
  gene encoding, see CAT gene
Chlorate, plant resistance to, 86
Chlorotoluron, metabolism of, 87
Chlorsulfuron
  mode of action of, 93
  plant resistance to, 86
Chromosomes, artificial, from yeast, 20
Chymotrypsin inhibitors, plant storage sites of, 50
Citrus spp., cross-protection role in diseases of, 162
Clover mosaic virus, RNA encoding coat protein of, 135
Coat protein
  genes encoding, see CP genes
  resistance mediated by, 131–147
    mechanisms of, 147–156
  role in virus replication, 153–154
Cocultivation techniques, for *Agrobacterium tumefaciens*, 9–12, 35
ColE1 origin of replication, in plant virus vectors, 4, 5
Colorado potato beetle, serine proteinase inhibitors and, 49
Commercial value, of transgenic plants, 220–224
Conglycinin
  genes for, 174, 176
  properties of, 174
α-Conglycinin, properties of, 173, 174
β-Conglycinin
  genes for, 176, 191
    transfer, 175, 176, 177
  isolation of, 174–175
  properties of, 173, 174, 175, 181
γ-Conglycinin, properties of, 173, 174
Convicilin
  genes for, 176, 181
    transfer, 181–182
  properties of, 180–181
  sulfur-containing amino acids in, 182
Corn, genetically engineered hybrids of, 219

Cotton
  Bt endotoxin gene introduction into, 64–65, 66, 67, 74
  α-globulins of, 181
  transgenic plants from, 36, 228
    using microprojectile bombardment, 33
Cowpea
  legumin gene regulation in, 178
  trypsin inhibitor gene of, transfer to tobacco, 54–55
Cowpea chlorotic mottle virus, symptoms of, 154
Cowpea mosaic virus, infection mechanism of, 117
Cowpea trypsin inhibitor, use in insect control, 222
CP-antisense, expression in transgenic plants, 124
CP genes
  chimeric genes encoding, 124–126
  comparison of amino acid sequences of, 131
  3' end sequence of, 125–126
  isolation of, 124–125
  nontranslated regions of, 124–125
  plants expressing, see CP(+) plants
  plants lacking expression of, see CP(−) plants
  promoters for, 125
  protection conferred by, detection, 126–131
  role in cross-protection, 147–150
  use in engineered plant virus resistance, 118–121
CP(+) plants
  development of disease symptoms in, 127–128
  field trials on, 159–161
  specificity of protection in, 129–131
  systemic infection delay in, 154–155
CP(−) plants, development of disease symptoms in, 127
Crop plants, genetic engineering of, 1–47, 221, 228–229
Cross-pollination, among transgenic plants, 228
Cross-protection
  nonstructural protein role in, 151–153
  role of CP genes in, 147–150
  use against plant virus infection, 118, 162–163

Index

Cruciferin, sulfur deficiency effects on level of, 178
cryB1 and *cryB2* genes, encoding of *Bacillus thuringiensis* var. *kurstaki* HD-1 endotoxin by, 58
cryIA gene(s)
　DNA seqeunce of, 68
　for insect control proteins in transgenic plants, 67–68, 70
Cucumber mosaic virus
　resistance to, in CP(+) plants, 127, 130, 131, 139, 146, 147, 161
　use in tobacco plant transformation, 122, 123

D

2,4-D
　engineered plant resistance to, 89
　mixed-function oxidases in metabolism of, 87, 101
Dactylis glomerata, transgenic plants from, by electroporation, 25
DCM1, resistant Q_B protein of, 90
Delta endotoxin, of *Bacillus thuringiensis*, 56
Diabrotica sp., trypsin inhibitor toxicity to, 55
3,5-Dibromo-4-hydroxy benzonitrile, bacterial degradation of, 102
Dicamba, tolerance to, 101
Dichlofop-methyl, glucose conjugate of, 87
2,4-Dichlorophenoxyacetic acid, see 2,4-D
Dicots, transgenic, production by electroporation, 25
Dihydrofolate reductase, gene for, replacement in cauliflower mosaic virus, 1–2
Dipteran vectors of disease, Bt endotoxin control of, 55, 56, 74, 222
Disease-resistant plants, genetically engineered, 117–169, 223
DNA
　chemically mediated uptake of, 12–14
　delivery by electroporation, 14–23
　delivery by microinjection, 27–28
　delivery by microprojectile bombardment method, 28–34
DNA viruses, use as biological vectors, 1
D_1 protein, see Herbicide-binding protein
Drug resistance marker, in plant virus vectors, 3

E

Ecology, engineered plant effects on, 71–73, 219–230
Electric discharge particle acceleration method, of gene transfer, 34, 36
Electropermeabilization, of bilayer membranes, 16
Electroporation, 8, 12, 14–23, 35
　definition of, 14, 35
　disadvantages of, 35–36
　electrical parameters affecting, 16–19
　factors affecting, 15–23, 35
　mechanism of, 15, 35
　media and temperature effects on, 22–23
　plasmid DNA concentration and carrier DNA effects on, 21–22
　plasmid-size effects on, 19–21
　protoplast physiology effects on, 23
　transient and stable transformation by, 24–26
ELISA
　of chloramphenicol acetyl transferase, 6
　use in detection of coat protein antibodies, 126, 127
Endopolygalacturonase, role in proteinase inhibitor induction, 50
γ-Endotoxins, of *Bacillus thuringiensis*, 55, 74
　monoclonal antibodies against, 75
　promoter regions of genes for, 61
　properties of, 56
5-Enol pyruvylshikimate phosphate, microprojectile bombardment delivery of gene for, 34
5-Enol pyruvylshikimate-3-phosphate synthase, see EPSPS
Environmental impacts, of genetic engineered organisms and plants, 71–73, 224
Environmental Protection Agency, role in biotechnology regulation, 226
EPSPS
　gene amplification for, 99, 100
　mutant *AroA* genes and, 97–98
　proteins coded by, 34, 99
　as target for herbicide-resistant plant production, 89, 97–98
Escherichia coli
　ALS isozymes in, 93, 94–95, 96
　AroA gene of, 98
　binary T-DNA plasmid replication in, 3, 4, 11

pRK290 stability in, 5
Ptt^R expression in, 105
spheroplasts, use in transformation studies, 13
use in Bt endotoxin gene studies, 60, 61, 62, 65, 74–75
yeast artificial chromosome maintenance in, 21
Euglena, herbicide resistance genes in, 88
Explant material, use in cocultivation, 9
Extensin, genes encoding, 54

F

Fatty acids, engineered removal from seed crops, 219–220
Field trials, on transgenic plants, 159–161, 223–224
Fluorescein isothiocyanate, use in detection of coat protein antibodies, 126
FM gene, in transgenic plants containing insect control proteins, 67, 69, 70, 75
Foods, from genetically engineered plants, 219–220
Foreign proteins, production by genetic engineering, 220

G

Geminiviruses, DNA of, 162
Gene(s)
 amplification and overexpression of, for herbicide resistance, 98–100
 engineering for plant protection, 117–124
 herbicide-detoxifying, 100–104
Genetic engineering
 for crop improvement, 1–47
 for disease-resistant plants, 117–169, 223
 for herbicide resistance, 87–100, 222, 227
 for improved plant nutritional quality, 171–217, 219–220, 221
 for insect-resistant plants, 49–84, 222–223
 of seed storage proteins, 172–206
 for virus-resistant plants, 117–169, 223
Gene transfer
 for crop improvement, 1–47
 by electric discharge particle acceleration method, 34
 by electroporation, 14–23
 by microinjection, 27–28
 by microprojectile bombabrdment method, 28–34
Gentamycin resistance gene, in plant virus vectors, 5
Gigiratia sanguinalis, herbicide resistance in, 87
Gliadin
 amino acid sequences in, 194
 properties and isolation of, 173, 193
α-Gliadins, conformation of, 195
β-Gliadins, conformation of, 195
γ-Gliadins, conformation of, 195
ω-Gliadins, conformation of, 195
γ-Globulin, human, *Pdlec*1 gene comparison to pseudogene for, 187–188
Globulin(s)
 DNA sequences regulating, 208
 of oat, properties of, 173
 of rice, properties of, 173
 as seed storage proteins, 171, 207
7S Globulin
 assembly and processing of, 208
 of soybean, genes for, 175–177
8.2S Globulin, of soybean, 175
11S Globulin, of soybean, 175
 genes for, 175–176
12S Globulin, of oat, genes for, 176
N-Glucosyl transferase, herbicide detoxification by, 101
β-Glucuronidase gene, see GUS gene
Glufosinate ammonium, commercial product of, 102
Glutamine, in cereal prolamines, 192
Glutamine synthase, as target for herbicide-resistant plant production, 89, 100
Glutathione-*S*-transferases
 herbicide resistance based on, 87, 91, 101
 isoenzymes of, 87
Glutelin-2, γ-zein as, 200
Glutelins
 genes for, 196–197
 as rice storage proteins, 196–197
 as seed proteins, 171
Glutenin
 amino acid sequences in, 194
 HMW-type, 195, 196
 properties and isolation of, 173, 193
Glycine max, see Soybean

Glycinin
  genes for, 174, 176
  isolation of, 174
  properties of, 173, 174–175
  subunit molecules of, 174
Glyphosate
  bacterial degradation genes for, 102
  engineered plant resistance to, 86, 89, 91–92, 97–98, 99, 102
  herbicidal properties of, 85
  nonselectivity of, 87
Gould model, for pest adaptation to resistant plants, 72, 73
Grasses, herbicide resistance in, 87
Growth hormones, engineered production of, 220
GUS gene
  advantages of, 8
  in chimeric genes
    containing legumin gene, 179
    for proteinase inhibitor studies, 54
  cotransformation with, 14
  disadvantages of, 8
  expression in transformed cells, 8, 26, 33, 191, 203, 204, 205, 223
  introduction by electric discharge particle acceleration method, 34
  introduction by microprojectile bombardment, 29, 30, 31, 32, 34
  as reporter gene, 7, 8, 24
Gypsy moth, Bt endotoxin activity against, 64

H

Helianthinin, sulfur deficiency effects on level of, 178
Heliothis sp., trypsin inhibitor toxicity to, 55
Heliothis virescens
  pI protoxin effects on, 57
  transgenic tobacco plant toxicity to, 55
Heliothis zea, see Tomato fruitworm
Helium gas, use in microprojectile distribution, 36
Herbicide(s)
  amino acid biosynthetic pathways as targets for, 92
  gene detoxification of, 100–104
  ideal properties of, 85
  nonselective, 87–88
  selective toxicity of, 85

transgenic plants resistant to, see Herbicide-resistant plants
Herbicide-binding protein, 88
Herbicide-resistant plants
  commercial development of, 222, 227
  strategies for development of, 87–98
    alteration of level and sensitivity of target, 88–98
    alternate, 98–104
    by gene amplification and overexpression, 98–100
Herring sperm, GUS gene expression by, 8
Hilmet phaseolin, expression of, 191
Hordein
  in mutant barley, 172
  properties of, 173, 196
Hordeum vulgare, see Barley

I

Ice minus (*Pseudomonas syringae* deletion mutant), field tests on, 224
*ilv*2 gene, as ALS gene, 93, 94
  spontaneous mutations, 94
*ilv*G gene, as ALS gene, 93, 94, 95
*ilv*M gene, as ALS gene, 94
Imazapyr, plant resistance to, 86
Imidazolinones
  ALS as target of, 93
  glucose conjugates of, 87
  metabolism of, 87
  plant resistance to, 89, 91, 97, 99
  as selective herbicides, 85
Inclusion matrix protein, functions of, 148
Indoleacetic acid, oncogene coding for, 2
Inhibitor 1, as serine proteinase inhibitor, 50
Insecticidal crystal protein (ICP), of *B. thuringiensis* varieties, 58, 64–70, 74
Insect-resistant plants, genetically engineered, 49–84, 222–223
  ecological aspects of, 71–73
  pest adaptation to, 72–73
Irradiation, effects on electroporation, 23

J

Jack pine, electroporation studies on, 22, 23, 24

K

Kanamycin resistance gene, in plant virus vectors, 5, 8, 13, 14, 25, 26, 32, 33, 64, 66
Keiferia lycopersicella, see Tomato pinworm
Klebsiella ozaenae, herbicide degradation by, 102

L

lacZ' gene, incorporation in T-DNA, 4
LAT25 gene, expression in transgenic plants, 30
Lectin
 genes for, 174, 189
 sulfur-containing amino acids in, 182
*Leg*A gene, 179
 expression in transgenic plants, 179, 180
Legumes, storage proteins of, 174–192
Legumin
 genes for, 176, 178–179
 properties of, 173, 177
 sulfur-containing amino acids in, 182, 184
 sulfur level effects on, 178
Legumin box, in legumin genes, 177, 179
Lepidopteran pests, Bt endotoxin in control of, 55, 56, 57, 61, 65, 222, 227
Lettuce
 engineered sulfonylurea resistance in, 97
 transgenic plants from, by electroporation, 25
Lolium multiflorum
 chemical-mediated DNA uptake by, 12
 transgenic, production by electroporation, 25
luc gene
 expression after biolistics bombardment, 33
 in plasmid constructs, 10–11
 as reporter gene, 8
Luciferase
 assay for, in transgenic plants, 11, 24
 gene for, see *luc* gene
Lupins
 legumin gene regulation in, 178
 sulfur deificiency effects on 11S globulin of, 180
Lycopersicon esculentum, see Tomato
Lycopersicon peruvianum, transgenic plants from, by electroporation, 25

Lysine, low level of in cereal prolamines, 192

M

Maize
 anthocyanin pigment, molecular studies of gene for, 29
 electroporation studies on, 24
 gene transfer to, 36
 GST genes of, 91
 herbicide resistance in, 87, 99
 seed storage proteins of
  aggregation, 208
  properties, 173, 192
Maize streak virus, DNA of, 162
Manduca sexta
 Bt endotoxin activity against, in transgenic plants, 65, 75
 control by proteinase inhibitors, 50
Marker genes, use in genetic engineering, 221
Melon, engineered sulfonylurea resistance in, 97
Metallo-proteinase inhibitors, 49
Metribuzin
 detoxification by *N*-glucosyl transferase, 101
 glucose conjugate of, 87
Microinjection, DNA delivery by, 27–28, 36
Microprojectile bombardment method
 development of, 29
 DNA delivery by, 28–34, 36
 high-velocity, 34
Millet, see *Pennisetum*
Mixed function oxidases, herbicide detoxification by, 101
Monocots, transformation by *Agrobacterium* cocultivation method, 9
Moth bean, transgenic plants from, 14
 using electroporation, 23, 25

N

Napier grass, electroporation studies on, 24
Napin, sulfur deficiency effects on level of, 178
Napoline synthase, *nos* gene for, 14, 126
Narcissus mosaic virus, RNA encoding viral coat protein of, 135
Nematodes, Bt endotoxin use in control of, 222

Index

neo gene
  cotransformation with, 14
  as plant selection marker, 5, 65
Neomycin phosphotransferase gene, see NPTII gene
Neomycin resistance gene, in plant virus vectors, 5
Neuropeptides, production by genetic engineering, 220
Nicotiana benthamiana, CP gene introduction into, 125
Nicotiana debneyii, transgenic, PVS coat protein gene expression in, 135
Nicotiana plumbaginifolia, *Leg*A gene transfer to, 180
Nicotiana sylvestris, tobacco mosaic virus infection of, 130
Nicotiana tabacum, see also Tobacco
  ALS gene introduction into, 96, 97
  chemical-mediated DNA uptake by, 12, 13
  Pat gene introduction into, 103
  sulfonylurea-resistant mutants of, 93
  transgenic, 35
    PR1b gene expression in, 151
Nicotiana tabacum var. *xanthi*
  CAT gene expression in, 7
  CP gene expression in, 140, 143, 144, 145, 146–147
Nitrilase gene
  effect on herbicide tolerance, 87–88
  role in bacterial herbicide metabolism, 102
nos gene
  in chimeric gene for PVC CP, 132
  in plasmids, 14, 25
NPTII gene
  delivery by microprojectile bombardment, 29–30, 32, 34
  disadvantages of, 8
  introduction by electric discharge particle acceleration method, 34
  legumin gene fused to, 179
  plasmids carrying, 13, 14
  as selection marker in plants, 5, 7, 65
  transformation by
    using electroporation, 24, 26
    using microinjection, 27
Nutritional quality, of plants, improvement by genetic engineering, 171–217

O

Oat, seed storage proteins of, properties, 173
Odontoglossum ringspot virus, TMV CP-conferred resistance to, 130, 145
Off-flavors, in foods, removal by antisense nucleic acid technology, 220
Oils, engineered improved yields and quality of, 219–220
Oncogenes, in T-DNA, 2
Opines, T-DNA encoding for, 2–3, 5
Orchard grass, transgenic plants from, by electroporation, 25
orf1 and *orf2*, in *cryB1* gene, 58
Organelles, transformation of, using microprojectile methods, 30
Oryza indica, transgenic, 34
Oryza japonica, transgenic, 34
Oryza sativa, see also Rice
Oryza sativa cv. Taipei 309, transgenic, production by electroporation, 26

P

pABD1, use in genetic engineering of plants, 2, 13, 14, 25, 26
Panicum dichotomifolium, herbicide resistance in, 87
Panicum maximum, transgenic plants from, by electroporation, 24, 25, 26
Papaya spp., cross-protection role in diseases of, 162
Particle bombardment method, for transgenic plant production, 30
Patatin, potato tuber storage of, 50
Patents, on transgenic plants, 220–221, 226
pat gene
  bar gene comparison to, 103, 107
  construction of plant vector containing, 106
  introduction into tobacco, 103
Pathogenesis-related genes, role in viral resistance, 151
Pathogenesis-related proteins, synthesis of, 154
pBL1103–4, use in plant transformation, 25
pBR322, *Bacillus thuringiensis* var. *thuringiensis* plasmids cloned into, 58
pCaMVCAT, delivery by electroporation, 24
pCaMVneo, delivery by electroporation, 26
pCT1T3, use in transformation studies, 20

*Pdlec*1 and *Pdlec*2 genes
  polypeptides encoded by, 187–188
  transfer of, 190
Pea
  2,4-D tolerance in, 101
  globulin genes in, 176
  seed storage protein properties of, 173, 180–181
Pea albumin 1
  gene for, 183
  sulfur-containing amino acids in, 182
  sulfur stress effects on, 184
Pea early browning virus, resistance to, in tobacco rattle virus CP(+) plants, 130, 143
pEB2, in plasmid pES1.1, 104
Pectinophora gossypiella, Bt endotoxin effects on, 56
PEG, see Polyethylene glycol
Pennisetum spp.
  electroporation studies on, 26
  seed storage proteins of, 192
  trispecific hybrid of, electroporation studies on, 24
Pennisetum purpureum, see Napier grass
Pepper mild mottle virus, TMV CP-conferred resistance to, 130, 144–146
pES6.1 and pES6.1 plasmids, construction of, 105
pES1.1 plasmid, restriction map of, 104
Petunia
  glyphosate resistance in
    by gene amplification, 99
    in transgenic plants, 98, 100
  seed protein gene transfer to, 172, 175, 176, 177, 179, 203
Petunia hybrida
  seed protein gene transfer to, 175
  transformation studies on, 14, 24, 35
Petunia monococcum, electroporation studies on, 24
pG11 plasmid, of *Bacillus thuringiensis* var. *thuringiensis*, 58
pG12 plasmid
  of *Bacillus thuringiensis* var. *thuringiensis*, 58
  nucleotide sequence of, 61
pG13 plasmid, of *Bacillus thuringiensis* var. *thuringiensis*, 58
pG1161, pG1163, pGS1151, and pGS1152 plasmids, construction of, 65

pG2260 plasmid, use in production of Ti plasmids for Bt endotoxin activity, 65
pGL2 plasmid, transformation studies using, 26
pGV0422, *nos* gene in, 14
Phaseolin
  genes for, 174, 176–177, 185, 187–192
  structure, 189, 191
  transfer, 189, 191
  intracellular transport and processing of, 185–186
  synthesis and glycosylation of, 184–185
β-Phaseolin, gene for, 191
Phaseolus vulgaris, see Bean
Phaseolus vulgaris cv. Greensleeves, PHA-L gene of, 183, 188
Phaseolus vulgaris var. *aboriginecus*, arcelin of, 192
Phosphinothricin
  engineered plant resistance to, 89, 100, 103
  as inhibitor of glutamate synthase, 5
Phosphinothricin acetyltransferase, *bar* gene coding for, 5, 102, 103
Phosphinothricin tripeptide, *Pt* resistance gene for, 103
Phytohemagglutinin
  biological activities of, 186–187
  genes for, 174, 187–192
  intracellular transport and processing of, 185–186
  synthesis and glycosylation of, 184–185
pIB16.1, construction of, 106
Picea glauca, see White spruce
Picea mariana, see Black spruce
Picloram, plant resistance to, 86
Pieris brassicae, BT endotoxin toxicity to, 75
Pinus banksiana, see Jack pine
Pisum sativum, see Pea
pJL81
  plasmid map of, 51
  transfer of proteinase inhibitor I by, 52
pKR612, construction of, 14
pKR612B1, NPTII gene contained in, 2
pKR612N1, gene insertion into, 14
Plant cells, bacteria cocultivation with, 9
Plants
  disease-resistant, 117–169, 223
  herbicide-resistant, 87–100, 222, 227

Index

improvement of nutritional quality by genetic engineering, 171–217, 219–220, 221
Plant viruses
as biological vectors, 1–2
selection markers in, 5–8
engineered resistance against, 117–169, 223
selection of transformed plants, 126
using coat proteins, 131–147
genome replication and dissemination of, 153–159
Plasmids
of *B. thuringiensis*, 58–59
DNA in, effect on electroporation, 21–22
size effects on electroporation, 19–21
pLGVneo11, use in electroporation studies, 23
pLGVneo2103
protoplast uptake of, 14
use in transformation studies, 14
PM gene, in transgenic plants containing insect control proteins, 67, 69, 70, 75
*pm*2 gene, activation of, 53
pMOG131 binary vector, structure of, 4
pMON563, delivery by electroporation, 24
pMON8059, construction of, 157
pMON8126, pMON8127, pMON8128, and pMON8129A plasmids, components of, 119, 120
pMON9749, delivery by microprojectile bombardment, 33
pMON9909, delivery by microprojectile bombardment, 33
pMS1, zein gene in, 14
pNOSCAT, delivery by electroporation, 24
pOCA18, use in transformation studies, 20
Pollen, expression of clonsed genes in, 29
Pollen tube pathway, for physical gene transfer, 36
Polyethylene glycol, use in delivering plasmids to plant protoplasts, 7, 8, 12, 13, 21, 24, 25
Poly-L-ornithine, use in delivery of plasmids, 12, 13
Polymerase chain reaction, use in cloning of reconstructed *Bz2* gene, 29
Polymeric materials, from plants, genetic engineering of, 220
Polyvinyl alcohol, use in DNA uptake mediation, 12, 13

Polyvirus, CP-mediated resistance to, 161
Potato
Bt endotoxin gene intoduction into, 64–65, 66
CP-mediated protection to viruses in, 161
electroporation studies on, 24
patatin in tubers of, 50
proteinase inhibitor II of, 50
cloning and expression, 51
gene induction for, 54
response to insect attack in, 53
transgenic
bar gene expression in, 102, 103
coat protein expression and resistance in, 120
cross-pollination in, 227–228
Potato virus M, coat protein-mediated resistance to, 135, 138
Potato virus S
coat protein-mediated resistance to, 134–136, 137, 138
genome organization of, 134, 136
Potato virus X
in CP(+) plants
field trials on, 159, 161
resistance in, 124, 127, 128, 130, 131, 132–134, 135, 139, 146, 147, 156
cross-protection against, 149
protection against in transgenic plants, 124, 127, 156
viral replicase protein of, 135
Potato virus Y
in CP(+) plants, field trials on, 159, 161
genome structure of, 133
resistance to, in CP(+) plants, 127, 128, 130, 132–134, 146
symptoms of, 132
Potexvirus, CP-mediated resistance to, 161
P1 protoxins, of *B. thuringiensis*, 56–57
PPT, engineered plant resistance to, 102
pRiHR1 origin of replication, in plant virus vectors, 5
pRK290, smaller derivatives of, 5
Prolamines
amino acids of, 192
of maize, 197–206, see also Zeins
regulation of genes for, 208
as seed storage proteins, 171, 207

of wheat
  classification, 193–194
  comparison to hordeins, 196
Proline, in cereal prolamines, 192, 194
PROSAMO, studies on genetically altered plants by, 225
Proteinase inhibitor I gene, activation of, 53
Proteinase inhibitor II gene
  extensin gene comparison to, 53
  fusion to CAT reporter gene, 54
  promoter for, 53, 54
  regulation of, 52
Proteinase inhibitor II K, wound-inducible promoter from gene for, 55
Proteinase inhibitor inducing factor, function of, 50, 53
Proteinase inhibitors
  advantages and limitations of, 49–50
  development by genetic engineering, 49–55
  effectivenes for reducing insect damage, 54
  induction of, 50
  as insect-control agents, 49–50
Protein bodies, of seed storage proteins, 208
Proteins, engineered improved amino acid composition in, 220
Protoplasts
  immobilization of, 28
  manipulations of, 86
  physiology effects on electroporation, 23
  regenerable, use in cocultivation, 9
Protoxins
  of $B.\ thuringiensis$, 55, 74
  DNA sequence of gene for, 74
P19S, as CP gene promoter, 125
P35S, as CP gene promoter, 125
$psab$A gene
  location in plastid genome, 90
  mutant, in herbicide-resistance engineering, 89
  Q_B protein coded by, 90
  transfer into nuclear genome, 90–91
psbA gene
  coding of herbicide-binding protein precursor by, 88
  mutants of, 88–91
$Pseudomonas$, glyphosate degradation by, 102
$Pseudomonas\ fluorescens$, Bt endotoxin insertion into, 64, 74–75
$Pseudomonas\ syringae$, field tests on deletion mutant of, 224
PSII herbicides
  cross-resistance among, 89
  dislodgement of Q_B quinone from binding niche, 89–90
  Q_B protein insertion into, 89
pSVB20, Ptt^R cloned into, 105
pTiACH5, chemical-aided delivery of, 13
pTiB6S3SE, potato virus S coat protein gene introduction into, 136
pUC19
  $lacZ'$ gene in, incorporation in T-DNA, 4
  luc gene constructs of, 10–11

Q

Q_B protein, see also Herbicide-binding protein
  herbicide-resistant, amino acid changes in, 90
  niche for PSII herbicide binding in, 89
  as target for herbicide-resistant plant production, 89
  cross-resistance with other PSII herbicide classes, 89

R

Rapeseed, see also $Brassica\ napus$
  engineered sulfonylurea resistance in, 97
  legumin gene regulation in, 178, 179
  transgenic, 36
    ecological data on, 225, 229
    pharmaceutical peptides from, 220
    production by microinjection, 27
    2S albumin expression in, 206
$Rhodopseudomonas\ viridis$, photosynthetic reaction center of, 89
Ribgrass mosaic virus, TMV CP-conferred resistance to, 130, 145, 146
Ribozyme molecules, creation of, 124
Ribulose bisphosphate small subunit, TMV CP gene promoter from, 149, 156
Rice
  seed storage proteins of
    aggregation, 208
    properties, 173, 193, 196–197
  transgenic plants from
    by electric discharge particle acceleration, 34
    by electroporation, 24, 25

Rice dwarf virus, genome of, 162
Rice tingro virus, genome of, 162
RNA dot blots, use in viroid genome
  detection, 127
RNA viruses
  satellite RNAs of, 121–122
  use in genetic engineering, 2
Root cutworm, biotoxin insertion in
  Pseudomonas fluorescens, 64
Rubber, engineered production of, 220
Rugose mosaic, from dual infections of
  potato viruses X and Y, 132
Rye, seed storage proteins of, 192

S

Saccharomyces cerevisiae
  sulfonylurea-resistant mutants of, 93
  yeast artificial chromosome maintenance
    in, 20–21
Saccharum spp., transgenic, production by
  electroporation, 25
Saccharum officinarum, see Sugarcane
Salmonella typhimurium
  ALS isozymes in, 93, 94, 95
  mutant *Aro*A gene of, 98
  sulfometuron methyl studies on, 93
  sulfonylurea-resistant mutants of, 93
Satellite nucleic acids
  description and properties of, 121–122
  use in engineered plant virus protection,
    118, 121–123
Secale spp., see Rye
Seed storage proteins
  engineered nutritional quality
    improvement of, 171–217
    problems of, 209–210
  properties of, 173
  synthesis and deposition of, 208, 210
  types of, 171
Selection markers, in plant virus vectors,
  5–8
Serine proteinase inhibitors, effects on
  insects, 49
Setaria italica, herbicide resistance in, 87
sh-gus activity, expression in plants, 8
sh promoter, use in control of GUS gene
  expression, 8
Shuttle cloning vector, stability in
  Agrobacterium tumefaciens, 5
Solanum brevidans, see Wild potato
Solanum nigrum

  proteinase inhibitor I gene transfer to,
    50–51, 52, 53
  zein gene transfer studies on, 206
Solanum tuberosum, see Potato
Somaclonal variation
  in cell cultures, 86
  in CP gene product levels, 149
Sorghum
  electroporation studies on, 24
  seed storage proteins of, 192, 193
    aggregation, 192, 193, 208
Soybean
  herbicide resistance in, 93
  legumin gene regulation in, 178
  seed storage proteins of, properties, 173,
    174–177
  transgenic plants from, 36
    by electroporation, 24, 25
Soybean mosaic virus
  coat protein-mediated resistance to,
    150–142
  CP gene of, 124, 125
    chimeric gene construct, 140–142
    in transgenic tobacco plants, 129,
      130, 131, 140
Spodoptera sp., trypsin inhibitor toxicity
  to, 55
Spodoptera exigna, sensitivity to Bt insect
  control protein, 66, 75
*spo*0 loci, in *B. subtilis*, 63
Starch, improved yields of, from
  genetically engineered plants, 219,
  220
Storage proteins, in seeds, see Seed storage
  proteins
Streptococcus faecalis, *npt*III gene from, in
  pIB16.1, 106
Streptomyces hygroscopicus, bialophos
  production by, 102, 103
Streptomyces viridochromogenes,
  phosphinothricin tripeptide
  resistance gene from, 103, 104
Sugar beet
  electroporation studies on, 24
  engineered sulfonylurea resistance in, 97
  plasmid uptake by electroporation in,
    22–23, 25
  protoplast electroporation of, 17, 18
  transgenic, wind pollination by, 229
Sugar-beet yellow virus, symptoms of, 154
Sugarcane
  electroporation studies on, 24

herbicide resistance in, 87
Sulfometuron methyl
  engineering plant resistance to, 86, 95, 97
  low toxicity to animals, 93
Sulfonanilides, ALS as target of, 93
Sulfonylureas
  metabolism of, 87
  plant resistance to, 89, 91, 93, 99
    molecular basis, 94
  as selective herbicides, 85
Sulfur levels
  effects on legumin synthesis, 178, 180
  effects on sulfur amino acids in seed proteins, 183–184
Sunflower
  Bt endotoxin gene intoduction into, 65
  globulin genes in, 176
  legumin gene regulation in, 178, 179
  seed protein genes of, 189
  zein gene transfer to, 202–203
Sunn hemp mosaic virus
  CP encapsidation of TMV RNA by, 153
  CP homology to TMV CP, 130, 145, 146
SURA and SURB loci, on mutant ALS genes, 96, 97
SURB-HRa gene, use to produce sulfonylurea-resistant plants, 97
Suspension culture, use in cocultivation techniques, 9

T

T-DNA
  binary plasmid construction from, 3
  napoline synthase gene from, 126
  recombinant, introduction and analysis into *Agrobacterium*, 4
  role in tumor induction by *Agrobacterium tumefaciens*, 2
  separation from *vir* region, 3, 13
  sequence removal from, 3, 13
  structure of, 6, 12
  transfer into cereal cells, 9
*tfd*A gene, in herbicide-resistance engineering, 89
Thiol proteinase inhibitors, as plant-defense mechanism, 49
Thiol proteinases, as insect digestive enzymes, 49
3' end sequence, importance in CP genes, 125–126
Ti plasmid(s)
  disarmed, 3, 5, 175
  role in tumor induction by *Agrobacterium tumefaciens*, 2, 3, 12
  use in cocultivation techniques, 9
  use in transformation studies, 13, 25, 52, 172, 175, 206
TL-DNA, use in transformation studies, 13
Tn5, APHII gene from, 13
Tobacco, see also *Nicotiana tabacum*
  ALS genes of, 96
  Bt endotoxin gene insertion into, 64
  CAT protein accumulation in wounded tissue of, 55
  cross-protection role in diseases of, 162
  glyphosate resistance in, 86
  GUS gene expression in, 8, 54, 205, 223
  pOCA18 delivery of DNA to, 20
  proteinase inhibitor II gene transfer to, 51–52
  protoplast electroporation of, 17, 19
  seed protein gene transfer to, 172, 175, 179, 181–182, 184, 188–189, 190, 203, 204, 205, 206
  seeds, zein protein synthesis in, 203
  transgenic
    atrazine-metabolizing GST gene in, 91
    bar gene expression in, 102, 103
    bxn gene expression in, 102
    coat protein expression and resistance in, 120, 121, 128, 129, 133, 136, 137
    cross-protection in, 148
    insect control protein genes in, 68–70, 74, 75
    pat gene expression in, 103
    potato virus X protection in, 124
    PR1b gene expression in, 151
    from satellite RNA, 122
    toxicity to *Manduca sexta*, 65
    using electroporation method, 23, 24
    using particle bombardment method, 30, 31, 32, 33, 34
  Xanthi "nc" cultivar, TMV resistance in, 128
Tobacco budworm, see *Heliothos virescens*
Tobacco etch virus, resistance to, in CP(+) plants, 128, 130, 131, 140–141
Tobacco hornworm, see *Manduca sexta*

Index

Tobacco mild green mosaic virus, TMV
    CP-conferred resistance to, 130, 131, 145, 146
Tobacco mosaic virus
  CP gene of
    introduction into tobacco, 121, 125, 143, 148
    PR1 gene expresion and, 151
    resistance mediated by, 131, 138, 143–147, 155–156, 159
    role in cross-protection, 147, 148–149
  in CP(+) plants
    dissemination of, 154–159
    field trials on, 159–161
  ELISA assays of, 145
  genome of, 143
  infection mechanism of, 117, 143
  resistance to
    in CP(+) plants, 127, 128, 129, 130
    in transgenic plants, 118, 119, 121, 152–153
  strain LS-1, dissemination in plants, 154, 157
  symptoms of, 154
Tobacco rattle virus
  genome structure of, 142
  nonstructural protein role in resistance to, 151–152
  resistance to pea early browning virus by CP(+) plants of, 130, 143
Tobacco ringspot virus, use in tobacco plant transformation, 122–123, 124
Tobacco streak virus, coat protein-mediated resistance to, 131, 139–140, 149
Tobacco vein mottling virus, transgenic plants containing RNA of, 126
Tobamoviruses, CP-mediated resistance to, 130–131, 143–144, 146, 161
Tobravirus, CP-mediated resistance to, 161
Tomato
  bar gene expression in, 103
  coat protein-mediated resistance in, 143
  CP gene expression in, 126, 129, 136, 144
  electroporation studies on, 24
  engineered sulfonylurea resistance in, 97
  EPSPS protein of, 99
  metribuzin tolerance in, 101
  proteinase inhibitors I and II of, 50
    accumulation in wounded tissue, 55
    transfer of, 50, 51, 53
  response to insect attack in, 53

transgenic
  coat protein expression and resistance in, 120, 128, 129
  insect-resistant, 74, 75
Tomato aspermy virus, induction of CMV satellite RNA in transformed tobacco, 122, 123
Tomato fruitworm, Bt endotoxin use in control of, 65, 66
Tomato golden mosaic virus, coat protein-mediated resistance to, 125, 157
Tomato mosaic virus, in CP(+) plants, 143, 146
  field trials on, 159, 160
Tomato pinworm, Bt endotoxin use in control of, 65
Tomato spotted wilt virus, DNA of, 162
Tomato yellow leaf curl virus, DNA of, 162
Transformation, by electroporation procedures, 14–26
Transgenic plants
  commercial value of, 220–224
  ecological risk of, 219–230
  expressing Bt endotoxin gene, 64–70
  field testing and performance of, 223, 226
  herbicide-resistant, 85–115, 222
  insect-resistant, 49–84, 222–223
  patents on, 220–221, 226
  production by electroporation, 24–26
  production by microprojectile bombardment, 30
  viral resistance in, 223
    coat protein expression, 118–127
    factors affecting, 128–131
    growth conditions, 129
    host-dependent factors, 128–129
    viral inoculum concentrations, 128
TR-DNA, use in transfoirmation studies, 13
Triazines
  engineered resistance to, 91
  as selective herbicides, 85
  thylakoid membrane binding of, 88
Tribolium sp., trypsin inhibitor toxicity to, 55
Trichoplasia ni
  P1 protoxin effect on, 57
  sensitivity to Bt insect control protein, 66
Tripsacum spp., prolamines of, 193

DATE DUE

DEC 09 1994			
DEC 13 1995			
NOV 25 2001			
DEC 10 2001			